Catalysis –
Science and Technology

Volume 3

CATALYSIS - Science and Technology

Volume 3

With 70 Figures and 49 Tables

Akademie-Verlag · Berlin 1983

Die Originalausgabe erscheint
im Springer-Verlag Berlin · Heidelberg
New York

Vertrieb
ausschließlich für alle Staaten mit Ausnahme
der sozialistischen Länder:
Springer-Verlag Berlin · Heidelberg
New York

Vertrieb
für die sozialistischen Länder:
Akademie-Verlag Berlin

Erschienen im
Akademie-Verlag, 1086 Berlin, Leipziger Straße 3–4

Lizenznummer: 202 · 100/546/83
Gesamtherstellung: VEB Druckerei „Thomas Müntzer", 5820 Bad Langensalza
Umschlaggestaltung: Eckhard Steiner
Bestellnummer: 763 156 3 (6727) · LSV 1215
Printed in GDR
DDR 128,— M

General Preface to Series

In one form or another catalytic science reaches across almost the entire field of reaction chemistry, while catalytic technology is a cornerstone of much of modern chemical industry. The field of catalysis is now so wide and detailed, and its ramifications are so numerous, that the production of a thorough treatment of the entire subject is well beyond the capability of any single author. Nevertheless, the need is obvious for a comprehensive reference work on catalysis which is thoroughly up-to-date, and which covers the subject in depth at both a scientific and at a technological level. In these circumstances, a multi-author approach, despite its well-known drawbacks, seems to be the only one available.

In general terms, the scope of *Catalysis: Science and Technology* is limited to topics which are, to some extent at least, relevant to industrial processes. The whole of heterogeneous catalysis falls within its scope, but only biocatalytic processes which have significance outside of biology are included. Ancillary subjects such as surface science, materials properties, and other fields of catalysis are given adequate treatment, but not to the extent of obscuring the central theme.

Catalysis: Science and Technology thus has a rather different emphasis from normal review publications in the field of catalysis: here we concentrate more on important established material, although at the same time providing a systematic presentation of relevant data. The opportunity is also taken, where possible, to relate specific details of a particular topic in catalysis to established principles in chemistry, physics, and engineering, and to place some of the more important features into a historical perspective.

Because the field of catalysis is one where current activity is enormous and because various topics in catalysis reach a degree of maturity at different points in time, it is not expedient to impose a preconceived ordered structure upon *Catalysis: Science and Technology* with each volume devoted to a particular subject area. Instead, each topic is dealt with when it is most appropriate to do so. It will be sufficient if the entire subject has been properly covered by the time the last volume in the series appears. Nevertheless, the Editors will try to organize the subject matter so as to minimize unnecessary duplication between chapters, and to impose a reasonable uniformity of style and approach. Ultimately, these aspects of the presentation of this work must remain the responsibility of the Editors, rather than of individual authors.

The Editors would like to take this opportunity to give their sincere thanks to all the authors whose labors make this reference work possible. However, we all stand in debt to the numerous scientists and engineers whose efforts have built the discipline of catalysts into what it is today: we can do no more than dedicate these volumes to them.

Preface

A cursory examination of the current scientific and technological literature is sufficient to show the enormous interest in the possibility of producing liquid fuels from coal. There are, of course, a number of ways in which coal liquefaction may be effected. Many of the important steps are catalytic. The direct liquefaction route, that is, coal hydrogenation, has a long history with origins in the early years of this century. It also has the distinction of being a process which was once operated on a very large scale and which, having died, now shows every prospect of resurrection. The technology which finally emerges will doubtless differ significantly from the original practice, but it is sensible for those currently working in the field to be aware of the achievements of the past. Dr. E. Donath, who was personally involved during the heroic years of coal hydrogenation, has provided an historical account of the subject up to the time immediately following World War II, when the large scale process began its rapid decline to oblivion.

Processes involving catalytic oxidation form a very large and important part of chemical industry. The reactions involved are very varied, ranging from the classical oxidation processes of heavy industry, such as the oxidation of sulfur dioxide or of ammonia, to selective oxidations designed to produce specific organic products from a range of possibilities. The chapter by Professor G. K. Boreskov deals with the subject of catalytic oxidation from a fundamental viewpoint, and will also serve as a background against which more specific process-oriented chapters may be viewed when they appear in later volumes of this series.

The third chapter of the present volume, by Professor

M. A. Vannice, deals with the fundamentals of catalytic chemistry of carbon monoxide. Processes based upon the catalytic hydrogenation of carbon monoxide are, of course, of great importance in present-day chemical industry. However, in addition to this, there are various processes based upon carbon monoxide hydrogenation which are likely to become of increased importance in the future as the need increases to generate liquid fuels and petrochemical feedstocks from non-petroleum based raw materials. One of these processes, the Fischer-Tropsch reaction, has already been dealt with in Volume 1 of the present series, and further chapters dealing with other specific processes will appear in later volumes. The present chapter provides valuable fundamental insights relevant to more specific process-oriented chapters.

Adsorption processes are, of course, fundamental to heterogeneous catalysis. Catalysts may be either metallic or non-metallic, and to understand the phenomenon of catalysis, it is necessary to understand the contributing adsorption processes. The fourth chapter in this volume, by Dr. S. Roy Morrison, summarizes the fundamentals of chemisorption on non-metallic surfaces, while the final chapter, by Dr. Z. Knor, surveys the fundamentals of the chemisorption of hydrogen on metals. From the point of view of quantum chemistry and the electronic theory of chemical binding, a good deal more is known about the chemisorption of hydrogen on metals than about any other adsorbing species. Thus, in discussing the chemisorption of hydrogen, Dr. Knor also illuminates a number of the important concepts which are germane to any discussion of chemisorption on metals: we expect to return to this subject at a more general level in a later chapter of this series.

Contents

List of Contributors

Dr. E. E. Donath
P. O. Box 1068
Christiansted, St. Croix 00820, USA

Professor Dr. G. K. Boreskov
Institute of Catalysis
Novosibirsk 630090, USSR

Professor M. Albert Vannice
College of Engineering
Dept. of Chemical Engineering
The Pennsylvania State University
133 Fenske Laboratory
University Park, PA 16802, USA

Dr. S. Roy Morrison
Energy Research Institute
Simon Fraser University
Burnaby, B.C. V5A 1S6, Canada

Dr. Z. Knor
Czechoslovak Academy of Science
The J. Heyrovský Institute of
Physical Chemistry and Electrochemistry
121 38 Prague 2, Măchova 7, ČSSR

Chapter 1

History of Catalysis in Coal Liquefaction

E. E. Donath

P.O. Box 1068
Christiansted, St. Croix 00820, USA

This chapter is dedicated to the memory of Dr. Matthias Pier. The work covered herein was conducted to a great part in his department at Ludwigshafen under his inspiring guidance. Coal liquefaction is now recognized as a process of immense future significance for the production of alternative liquid fuels and chemical feedstocks. It is a process with a long technical history, and it is important for those currently working in the field to nave an opportunity of seeing at least the catalytic component from an historical viewpoint. This chapter summarizes the industrial experience in direct coal hydrogenation up to the end of the second World War, with most attention being paid to the German experience since it was in this country where the process was most fully developed.

Contents

1. Introduction[1]

The term "Coal Liquefraction" or "Coal Hydrogenation" in this chapter describes the addition of hydrogen to coal or lignite using catalysts at elevated pressure, and temperature and includes the treatment of coal tars and of

---

1 Conversions and definitions:
Barrel: bbl = 159 liters = 0.159 m^3;
Gallon: gal = 3.785 liters;
Nm^3: 1 m^3 Gas at 1 bar and 0 °C;
Ton: 1 t = 1000 kg;
Bar: 1 bar = 10^5 Pascal = 0.1 MPa (about 1 atm).
Asphaltene: Toluene soluble, n-heptane insoluble product fraction.
Aniline Point: Indicator of aromaticity, increases with increasing hydrogen content of product fraction.
Stall: Concrete enclosure of high pressure reactors, hot separator and heat exchangers with adjoining feed stock heater.
Throughput or product rate in kg of feed or product per liter catalyst or reactor volume and hour: $kg\,l^{-1}\,h^{-1}$.

heavy oils. It will further discuss the secondary hydrogen treatment of the primary coal liquefaction products using catalysts. This secondary hydrogen treatment was later adapted to the refining of shale oil or petroleum derived oils. Catalysts for this last process and the many variations revolutionized the petroleum refining industry.

Examination of the ultimate analyses of coals and tars and comparison with commercial petroleum fuels will give a first view of the reactions that must occur in coal liquefaction. The following Table 1 gives an abridged view of the composition of typical coals and tars. By comparison gasoline, the major motor fuel, has more than twice the hydrogen content of coals; it is practically free of heteroelements (O, N, S) and has a molecular weight of about 100. Furthermore all coals contain mineral matter (ash) which must be removed in the liquefaction process.

Thus coal liquefaction catalysis must be concerned with the following types of reactions:

1. Addition of hydrogen (hydrogenation)
2. Removal of heteroelements (hydrorefining) and saturation of unstable hydrocarbons
3. Molecular weight reduction (hydrocracking)

The above reactions have to occur in the proper sequence and rate depending on the feed stock and the desired final product. In coal liquefaction for instance the hydrogen addition must proceed at a sufficient rate to avoid cracking reactions which form coke and decrease the oil yield. In converting oils to gasoline the hydrocracking reactions and the removal of heteroelements should proceed with a minimum of hydrogen addition and consumption, thus avoiding saturation of aromatic rings and formation of hydrogen-rich methane.

In addition to these three reaction types, several catalyst groups accelerate the isomerization of carbon chains and rings

Before describing the development of the industrial process a brief look at earlier experiments may be given. The first catalytic coal liquefaction was reported by M. Berthelot [1] in 1869. He heated coal with hydroiodic acid in a sealed glass tube to 543 K. By the action of "nascent" hydrogen

Table 1. Ultimate Analyses

	Composition/g per 100 g carbon		Molecular weight
	H	O, N, S	
Bituminous Coal	6.7	16	5000
Bituminous Coal low temperature tar	9.0	11.0	400
Brown Coal	7.0	30.0	5000
Brown Coal low temperature tar	12.0	9.0	250
Residual oils	12.5–14.5	2–5	400
Gasoline	15.0	—	100

and the catalytic influence of the acid the bituminous coal was converted into a semiliquid material with an oil content of 60%. Similar results were obtained by F. W. Dafert and R. Niklausz in Vienna in 1911 using hydroiodic acid and phosphorus. In 1917 F. Fischer and H. Tropsch [2] repeated the above experiments at 473 K with a variety of coals ranking from anthracite to highly volatile bituminous coal. While the former was quite inactive, the highly volatile coal gave a 70% yield of chloroform soluble semisolids which were nitrogen free and much richer in hydrogen than the coal. The same authors treated coal with sodium formate which at 673 splits off nascent hydrogen. Even at this higher temperature the coal conversion to soluble products was smaller than with hydroiodic acid. At the time it was assumed that the nascent hydrogen evolving from formate is not as active as the hydrogen from hydroiodic acid. Today the difference would be ascribed to acid catalysis by the hydroiodic acid.

Beginning his extensive work in 1905, V. N. Ipatieff showed that the pyrolysis of organic compounds is profoundly changed by working with hydrogen under pressure in the presence of nickel as catalyst. He found further that the activity of the nickel catalysts is rapidly destroyed by the sulfur of sulfur-containing feed materials. The use of high pressure hydrogen in this work did not become immediately suitable for large scale application, the main reason being hydrogen embrittlement of available steels.

Since 1910 at the Badische Anilin & Sodafabrik (BASF) in Ludwigshafen, Germany (later one of the founding companies of the IG Farbenindustrie AG), the ammonia synthesis invented by F. Haber was developed and industrialized under the direction of C. Bosch. It was found that the iron catalyst for this process was poisoned by even traces of sulfur compounds. Thus it became generally believed that sulfur poisons hydrogenation catalysts.

Historic Outline of the Bergius-Pier Process

Germany is rich in coal but has only very small petroleum resources. This made conversion of coal into petroleum products of interest. The first systematic studies of coal liquefaction at high hydrogen pressure were conducted by F. Bergius (1884–1949), Figure 1. Studying the constitution of coal, he attempted to make "synthetic" coal by heating cellulose with water under pressure. Then he compared the hydrogenation of his product with that of coal. For the hydrogenation he used an initial hydrogen pressure in his autoclave of 100 bar at elevated temperatures. Together with his coworker J. Billwiller, he showed in 1911 that under similar conditions of temperature and pressure heavy oils are converted to low boiling oils without significant formation of coke. Later in 1913 they applied for patents [3] for the conversion of coal and other materials into oil or soluble organic products by reaction with hydrogen at elevated temperature and high pressures.

Figure 1. Friedrich Bergius (1884–1949), Nobel Laureate in chemistry 1931

At 150 bar hydrogen pressure and 673 to 723 K about 80% of the coal was converted into gases, oils and benzene soluble products. At lower pressure, below about 50 bar, significant conversion to coke occured and made a continous process inoperable. The autoclave experiments showed that the reaction is exothermic and that care is needed to avoid uncontrolled temperature excursions. Oil derived from coal was added to the coal to facilitate stirring and improve temperature control. This step was of great importance for continuous operation and later commercialization: it made it possible to pump the coal-oil slurry continuously into reactors. It also became important for the heat economy of the process by permitting the use of heat exchangers between feed and product streams. Thus the liquefaction of coal at high hydrogen pressure and temperature and the use of a slurrying oil were the *two basic inventions of Bergius*. He was a solitary inventor and engineer of genius, and as will be seen below an outstanding entrepreneur.

Bergius made his initial experiments in his private laboratory at the technical university in Hannover. Beginning in 1914 the experimental work was conducted at the Th. Goldschmidt plant in Essen and at the same time a larger pilot plant was built near Mannheim. Because of World War 1, operation in this plant began only in 1921 with financial assistance from other companies, mainly the Shell group. Until 1925 many coals were treated there usually with the addition of a few percent of "Luxmasse", the iron oxide residue from alumina production, for the absorption of sulfur compounds evolved from the coal. Luxmasse otherwise was used for the removal of H_2S from coke oven gas.

Table 2. Coal liquefaction yields, Bergius data

15%	Naphtha 503 K
20%	Middle oil
6%	Heavy oil
21%	Gases
7.5%	Water
0.5%	Ammonia
35%	Coke and oil residue (pyrolysis gave 8% fuel oil in addition)

It was found that brown coals and bituminous coals with a high content of Volatile Matter (VM) were suitable feedstocks, whereas coals with more than 85% carbon on a moisture and ashfree basis (maf) were difficult to convert. As an example, the hydrogenation of a coal with 25% VM gave the yields in weight percent given in Table 2.

A semicommercial plant was built in 1924 near Duisburg using horizontal, stirred reactors [4]. Operating results of this plant were disappointing because of the low throughput and the unsatisfactory yield and quality of the oils produced from coal without a catalyst. Operation there ceased in 1930 after the BASF in 1924 had successfully liquefied coal with catalysts and then in 1926 acquired the Bergius patent rights which were important for a commercial coal hydrogenation process.

At the BASF plant in Ludwigshafen, ammonia synthesis and methanol synthesis had been developed to commercial scale by the hydrogenation of N_2 or CO with H_2. The experience from these catalytic high pressure hydrogenation processes was available there, as well as high pressure equipment and machines and large amounts of hydrogen. The rapidly growing motorization of the highway traffic following the example of the USA made the motor fuel production from coal in Germany increasingly important. At the same time, such a process would help to decrease the oppressive level of unemployment after World War 1 and alleviate the shortage of foreign exchange. Furthermore, knowledgeable American geologists predicted rapid exhaustion of the US crude oil reserves. This made production of liquid fuels from abundant coal an international problem. Therefore the top executives of the BASF, C. Bosch and C. Krauch, decided after the rapid success of the methanol synthesis in 1923 to study the hydrogenation of coal [5].

It was M. Pier (1882–1965), Figure 2, who had developed a mixed zinc-chrome oxide as a selective catalyst for methanol synthesis. He found that the methanol catalyst was less sensitive to sulfur poisoning than the ammonia catalyst. This unexpected result encouraged him in the revolutionary idea to search for sulfur resistant catalysts for the hydrogenation of coal and to try sulfides as hydrogenation catalysts.

Experiments for the discovery of catalysts for the liquefaction of coal began in Ludwigshafen in 1924 using cresylic acid from coal tar as a model substance. It was a convenient feed liquid since it contains oxygen, sulfur and nitrogen compounds, and the reaction products can be speedily analyzed

Figure 2. Matthias Pier (1882–1965)

by determining their boiling range and alkali solubility. Already in 1924 Pier and coworkers found that sulfides and oxides of molybdenum, tungsten and cobalt, and sulfides of iron, were active hydrogenation catalysts at 200 bar hydrogen pressure, and were not poisoned by sulfur or other heteroelements in the cresylic acid.

The experiments were then extended to brown coal tar made locally in gas producers. With a molybdenum oxide catalyst this tar was converted without coke formation into essentially pure gasoline at 200 bar pressure and 723 K. This experiment was the *first milestone* in the development of the direct catalytic liquefaction of coal. At the same time experiments with coal itself were started and they will be described later.

The equipment used for the hydrogenation of cresylic acid, tar and oils is shown in Figure 3. It will appear primitive to the research worker of today but is was a rather accurate tool when operated by skilled workmen. The feed flows a pressurized, heated tank through a sightglass as measuring device for the counting of drops and is joined by hydrogen. The mixture flows from to the electrically heated high pressure tube-reactor and after preheating reverts through the catalyst bed containing 50 cc of catalyst granules. The products leave through a cooler into a sight-glass separator from which liquid products and gases are withdrawn separately and measured. By 1926 forty such units, see Figure 4, were in operation for the systematic testing of catalysts using middle oil as feedstock.

Nearly all elements of the periodic system were tested alone or in combinations. Later the testing of supports and supported catalysts became important. Using conversion to gasoline as a measure of activity, molybdenum and tungsten retained their first place followed by elements of the fifth and

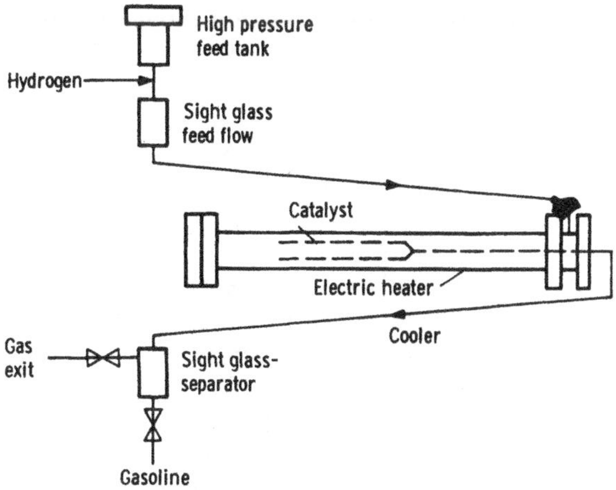

Figure 3. Catalyst testing reactor, 50 cm^3

Figure 4. Battery of catalyst testing reactors

sixth groups of the periodic table. Iron, nickel and cobalt were mainly tested in combination with molybdenum or tungsten. They increased the hydrorefining activity of these elements [6]. Noble metals were only cursorily tested because of cost and limited availability in Germany. The first outstanding combinations in activity and physical properties were molybdic acid, zinc oxide and magnesia tested in April 1926, and shortly thereafter molybdic and chromic acid. As a curiosity it may be mentioned that the newly separated rhenium was tested and found active.

Larger units were built and operated to obtain engineering data, and to obtain larger amounts of products using recycle of the unconverted feed.

In the experiment with brown coal tar called the first milestone, a small throughtput and large excess of hydrogen were used. These conditions turned out to be not practical nor economical for large scale use. Higher feed rates and lower hydrogen excess however led to an excessive loss of catalyst activity. This was attributed to the high boiling fractions and especially the asphaltenes in the tar. These are benzene soluble and heptane insoluble compounds of high molecular weight. After extensive experimentation with tar and brown coal, the solution to this problem was found: it was to divide the coal liquefaction process into two stages [7].

1. The "Liquid Phase Hydrogenation" for coal and oils boiling above 598 K using impregnated or finely divided catalyst in suspension, and
2. The "Vapor Phase Hydrogenation" for lighter, aphaltene-free oils using lump catalyst in a fixed bed.

This division into two stages was the *second milestone* in the development of the direct catalytic coal liquefaction.

It should be noted that in the beginning with catalysts then available, the rigid 598 K boiling range limit was very useful for the conversion of coal, tar and heavy oil into gasoline, while catalysts developed later can be used with long catalyst life for the hydrotreating of higher boiling fractions as well.

These two milestone inventions induced the BASF management in 1925 to embark on a large program for the commercialization of "catalytic coal hydrogenation", a process that is called today the "Bergius-Pier" process. As had occurred in the earlier methanol synthesis, experiments began in the historic building "Lu 35" in Ludwigshafen in which previously Bosch had brought the ammonia synthesis to maturity. Lu 35 contained laboratories, concrete stalls for high pressure reactors, compressors, pumps and other equipment. Hydrogen and nitrogen were supplied from the nearby Oppau ammonia plant by high pressure pipeline. It was well suited for the assembly and use of new high pressure units.

Pier reported regularly and frequently to C. Krauch who always supported the coal liquefaction program, urging rapid progress. Pier's earliest coworkers were K. Winkler on the chemical side and W. Rumpf on the engineering part of the work. The "High Pressure Experiments" group consisted by the end of 1924 of four chemists, two mechanical engineers and forty fore- and workmen. The largest size was reached in 1927, when it consisted of sixty graduate chemists, physicists, mechanical engineers and metallurgists, and about two thousand workmen. At this time several large scale development units were operated with the additional purpose of training personnel for the Leuna plant. As time progressed experienced men were sent to manage and staff new plants. On the average, the staff at Ludwigshafen consisted of about sixty graduate people and one thousand workmen.

The development of the coal liquefaction process with its multiphase reactions and ever changing raw materials proved to be a much more difficult task than the single phase methanol synthesis. It challenged the ingenuity and perseverance of Pier and his coworkers and the support of BASF specialists in many fields of mechanical engineering, heat transfer and materials sciences as well as the confidence and patience of the management. Of great importance was the close cooperation with the Leuna plant.

The decision to build the first commercial coal and tar hydrogenation plant at Leuna (at the site of the ammonia plant of the affiliated Ammoniakwerk Merseburg GmbH) with its rich brown coal deposits was made in 1926. Operation of this plant designed for the production of 100,000 t y^{-1} of gasoline started in 1927 with brown coal and tar as raw materials. This was the prototype plant for the later coal hydrogenation plants. In it the required large scale equipment was perfected and operating procedures established. The catalysts used there were produced in Ludwigshafen and supplied to Leuna and later plants. These plants belonged to other companies and were licensees of the IG.

Already in 1925 the management of the IG realized that marketing of gasoline from coal in Germany without the cooperation of the oil companies would be difficult. On the other hand, the catalysts developed for the hydrotreating of coal oils would be of interest to all petroleum refiners. Moreover the US Oil Conservation Board in a 1926 memorandum had estimated that the USA oil supply may become exhausted in about six years. These were the main reasons that led in 1927 to a first cooperative agreement between the IG and the Standard Company Oil of New Jersey (later Exxon). The Standard Oil Company built a 100 bbl d^{-1} experimental oil hydrotreating plant in Baton Rouge and later one 5000 bbl d^{-1} commercial plant there and a second plant in Bayway [8]. These plants were used mainly for the quality improvement of kerosenes and lubricating oils beginning in 1929. The then available IG catalysts produced a lubricating oil with an increased viscosity index (flatter temperature-viscosity curve) that was marketed as the premium product (Essolube). Later this was replaced by selective solvent refined oils available at lower cost. During World War 2 the US hydrogenation units were used to produce isooctane from diisobutylene.

In the years 1929 to 1931, the cooperation between the IG and the Standard Oil Company of New Jersey was extended to include other oil companies and also the British ICI (Imperial Chemical Industries Ltd.). To exploit the hydrogenation patents, two companies were formed in the Hague. One, was a patent licensing company which could license the pooled patents outside Germany and the USA. The second, the engineering company

IHEC, was to build hydrogenation plants. Its data base was an important and useful innovation: all participants exchanged their plant and development experience. This arrangement made any new development speedily available to all licensees. For its implementation technical representatives of the Standard Oil, the ICI and the IHEC were stationed at Dr. Piers laboratory in Ludwigshafen and in steady contact with personnel there. This technical information exchange ended with the beginning of World War 2.

Already in 1929 it was found in Leuna that the hydrogenation of brown coal tar was more economical than the direct liquefaction of brown coal and became a reliable process for the production of motor fuels. The direct liquefaction of brown coal itself only reached a comparable degree of perfection in 1932. In 1933 the production of motor fuel at Leuna was 300,000 tons per year, mainly from brown coal: it increased in the following years to 650,000 tons per year.

Beginning in 1933 the production of motor fuels from other domestic raw materials was increased with government guaranties. In addition to bituminous coals from the Ruhr, brown coals especially from central Germany were suitable raw materials. The German brown coal industry founded the Braunkohle Benzin AG, (BRABAG), which decided to built two plants for the hydrogenation of brown coal tar using the experience gained in Leuna. The coproduct of low temperature carbonization of the central German brown coal, the char (Grude), was used advantageously in newly built power plants. Already in 1936 two tar hydrogenation plants in Boehlen and Magdeburg with an annual production of 200,000—250,000 tons of motor fuels began operation.

A third brown coal tar hydrogenation plant was completed in Zeitz for the production of 280,000 tons per year of petroleum products. Experiments in Ludwigshafen and Leuna had shown that the TTH process (see later) converts brown coal tar into high quality lubricating oils and paraffin wax in addition to motor fuel by hydrogenating the tar with a new, more active fixed bed catalyst rather than in the normal liquid phase system.

The largest tar hydrogenation plant was built in Bruex (Most), Czechoslovakia, using brown coal from local mines, for the production of gasoline and Diesel fuel.

Already at the beginning of the international cooperation and exchange of development data, the ICI [9] described lengthy laboratory experience with the catalytic liquefaction of bituminous coal. Especially tin in combination with hydrochloric acid were found to be active liquefaction catalysts. After an experiment with English coal in the 300 bar demonstration plant at Ludwigshafen and the success of the brown coal hydrogenation at Leuna, the ICI began design and construction of the first commercial bituminous coal hydrogenation plant at Billingham in England. The plant began operation in 1935 and produced later 150,000 tons gasoline per year from coal and tar distillates.

The first German bituminous coal hydrogenation plant was built in the Ruhr district at Scholven for the Hibernia AG Mining Company. The operating pressure was 300 bar. The Leuna experience was the basis for the design of the liquefaction stage. The tin-chlorine combination was used as the catalyst as in the Billingham plant. Production started in 1936 and was so successful that the capacity was later increased to 280,000 tons of motor fuels per year.

All the hydrogenation plants mentioned so far were built for an operating pressure in the range of 200 to 300 bar. Beginning in 1935 experiments for the catalytic liquefaction of bituminous coal at 600 bar hydrogen partial pressure (corresponding to 700 bar total pressure using hydrogen recycle) were initiated at Ludwigshafen. The object was to determine whether the somewhat cumbersome hydrochloric acid cocatalyst used in Billingham and Scholven could be eliminated. It was expected from autoclave experiments and then confirmed on larger scale that at 600 bar the use of an iron catalyst leads to satisfactory liquefaction, asphaltene conversion and a lower off-gas formation. This reduces hydrogen consumption and thus mitigates the higher equipment cost.

At this increased pressure in 1937 the comparatively small plant of Stinnes AG at Welheim began operation using coal tar pitch as raw material. The middle oil obtained from pitch was converted into a gasoline with a high aromatics content at the same pressure using a newly developed vapor phase catalyst.

Another small plant of the Wintershall AG at Luetzkendorf produced motor fuels from various oils and tars at 500 bar pressure.

The first bituminous coal liquefaction plant for 700 bar pressure was built by the Vereinigte Stahlwerke AG at Gelsenberg in 1939. It reached a production of 400,000 tons gasoline per year.

An even larger plant was built by the German oil companies Deutsch-Amerikanische Petroleum Gesellschaft and Rhenania-Ossag Mineraloelwerke AG and the IG at Poelitz near Stettin (now Szczecin, Poland). It was designed to hydrogenate imported distillation and cracking residues as well as Silesian bituminous coal into gasoline. Since it began operation in 1940 after the start of World War 2 coal became its main raw material.

For silesian coal and low temperature tar the plant at Blechhammer was built in 1943 but did not reach full production. Its main product was to be navy fuel oil in addition to gasoline. In the meantime in 1941 the Rheinische Braunkohle Kraftstoff AG built the second plant for the direct catalytic liquefaction of brown coal. It was located in Wesseling near Bonn and had a capacity of 250,000 tons per year of gasoline and Diesel fuel. For the Rhine district brown coal with its low hydrogen and bitumen content, the pressure of 700 bar proven satisfactory for the liquefaction of bituminous coal was selected.

It will be helpful to summarize in Table 3 the main data for the 12 hydrogenation plants built by Germany. The total annual capacity of these plants in 1943 was about 4 million metric tons of motor fuels.

Outside of Germany, the United States and Britain, two high pressure hydrogenation plants were built in Italy. They were located in Bari and Livorno. Each had a capacity of producing 100,000 tons per year of gasoline from heavy oil residues as feed material. They started operation in 1936.

Table 3. German Hydrogenation Plants 1943/44

Startup date	Location	Raw material	Pressure bar Liquid phase	Pressure bar Vapor phase	Product capacity/t per year 1943/44
1927	Leuna	Brown coal Brown coal tar	200	200	650,000
1936	Boehlen	Brown coal tar	300	300	250,000
1936	Magdeburg	Brown coal tar	300	300	220,000
1936	Scholven	Bituminous coal	300	300	280,000
1937	Welheim	Coal tar pitch	700	700	130,000
1939	Gelsenberg	Bituminous coal	700	300	400,000
1939	Zeitz	Brown coal tar	300	300	280,000
1940	Luetzkendorf	Tar oils	500	500	50,000
1940	Poelitz	Bituminous coal oils	700	300	700,000
1941	Wesseling	Brown coal	700	300	250,000
1942	Bruex	Brown coal tar	300	300	600,000
1943	Blechhammer	Bituminous coal and tar	700	300	420,000
	12 Plants				about 4 million

After this description of large scale developments of the Bergius-Pier process we shall turn now to catalyst development in primary coal liquefaction and coal oil refining.

Primary Coal Liquefaction

The interest of the IG initially was centered around the liquefaction of brown coal from the rich deposits near Leuna as was mentioned above. There, in addition large quantities (measured on the gasoline consumption of that time) of brown coal low temperature tar were available. Its conversion into gasoline became a major task for the experiments at Ludwigshafen by M. Pier and his coworkers. As occasion arose other raw materials, bituminous coals, their tars and pitches as well as petroleum and shale oils and their fractions and residues were studied. The conversion of all the high molecular weight materials was done by "liquid phase hydrogenation". The common denominator of these processes was the similarity of the equipment and reaction conditions and the use of catalysts that were impregnated on the coal or that were pulverized and used in suspension in the feed material. This distinguished this reaction step from "vapor phase hydrogenation" for lower boiling oils using catalyst in a fixed bed.

The first equipment used for catalyst testing in liquid phase reactions were rotating autoclaves containing stainless steel balls for the stirring of the charge. It was observed that the same elements found active in fixed bed catalysts used for the hydrogenation of cresylic acid and tar oils, were also active in coal liquefaction. They increased coal conversion to oils thus reducing the amount of solid residue that had to be separated from the oil product. They reduced the formation of hydrocarbon gases and thereby the hydrogen consumption, and they produced lower boiling oils with smaller asphaltene content and better stability. The main drawback of autoclave experiments is that they do not readily lend themselves to the recycle of the slurrying oil used in preparing the coal slurry, and that the residence time of the products that evaporate is the same as that of those that remain liquid. Both characteristics differ from the conditions prevailing in continuous flow reactors.

When using a molybdenum catalyst in concentration of 20–25% of the coal slurry, practically refined products were obtained from coal with fresh catalysts. However it was obvious that molybdenum was too expensive to be used in such large concentrations unless efficient catalyst recycle was possible. Since this was difficult small amounts of molybdenum, that is 0.02–0.06% MoO_3 of the coal, were used and impregnated on the coal from ammonium molybdate solution. With Leuna brown coal having a high content of calcium humates the same observation was made as with vapor phase catalysts: the presence of alkali and alkali earth compounds decreased molybdenum activity. It was observed that neutralization of the calcium with sulfuric acid was effective in enhancing the activity of the molybdenum catalyst. Furthermore it was found that other elements such as titanium, tin and lead were active. In larger amounts iron oxide, several iron ores and iron sulfate when impregnated on the coal gave satisfactory results.

It may be of historic interest to show in Figure 5 a tilting reactor which was used for some initial brown coal hydrogenation experiments and catalyst tests before it was concluded that the use of coal slurries was superior to the use of dry coal. Double screened brown coal in a screen cartridge was placed in the cold end of the reactor and pressurized. After the hot end reached

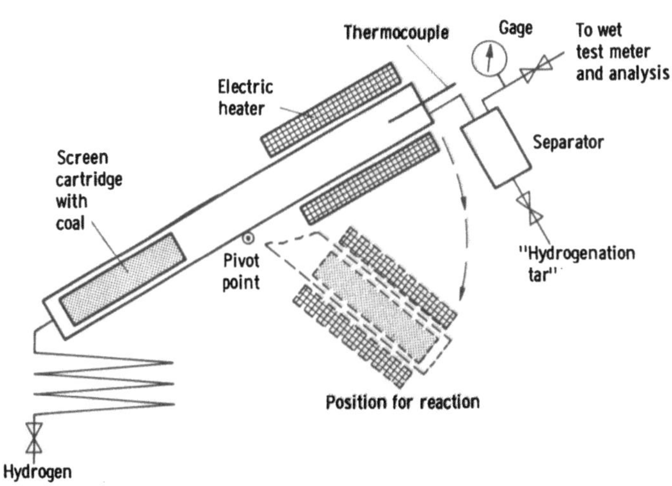

Figure 5. Tilting coal hydrogenator

reaction temperature the hydrogen flow was started and the reactor tilted to move the cartridge into the heated zone. The reactor gave rapid coal heating and cooling and well defined reaction times. The small amount of coal products obtained was not diluted by oils derived from the slurrying oil used in other experiments. However the system was limited to small charges to maintain the reaction temperature inspite of the exotherm, and was suited only for non swelling or melting coals such as brown coal. The product distribution was different from that in coal slurry experiments since fractions obtained from slurry oil hydrogenation were absent.

For the testing of catalysts under conditions approaching those of commercial liquid phase operation continuous flow reactors of 1 and 10 liter volume were used routinely. From time to time, to obtain more accurate results a similar reactor of 100 liter volume was taken into operation. This was beside the large scale experiments in pressure vessels of 300, 500 and 800 mm ID and up to 12 m length which initially were used for the development of equipment and in later years for the testing of new coals or tars before the design of a new commercial plant was finalized. In such tests usually personnel of the new plant took part.

In the 10 liter reactor a stirrer was used in order to avoid excessive settling because of the smaller flow velocity than that prevailing in the much longer commercial converters. In the 1 liter units the reactants were recycled from the hot catch pot to similarly avoid excessive settling of solids. Since the backmixing rate in commercial reactors could initially not be foreseen these two units gave indications of the differences between plug flow and fully backmixed reactors. Coal slurries or heavy oils or tars and catalyst were fed together with hydrogen through the preheater coil into the reactor. The reaction products were separated in a vessel at a temperature about 20 °C below that of the reactor. Solids containing residue was withdrawn

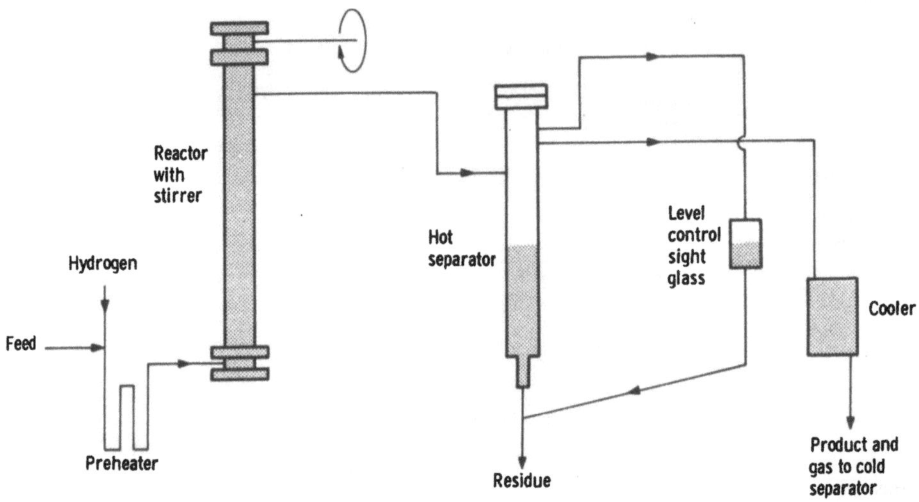

Figure 6. Catalytic coal liquefaction pilot plant

Table 4. Liquid phase tar oils with Mo and Fe catalysts

Component	0.5% Catalyst (in tar residue) containing	
	2% MoO_3	10% $FeSO_4$
Phenols in naphtha to 483 K	3.3%	8.7%
Phenols in oil to 598 K	9.8%	12.0%

at the bottom while oil vapors and gases exited at the top to a cooler and a cold separator from which gases were withdrawn at the top and product oil at the bottom. A schematic diagram is shown in Figure 6.

For the conversion of the brown coal tar residues, at first pulverized MoO_3–ZnO–MgO catalyst was used in a concentration of 25–30%. The hot separator bottoms containing the catalyst were recycled and only the distillate from the cold separator with drawn as product. It soon became apparent that the catalyst activity decreased and that solid condensation products were deposited on the catalysts. The catalyst consumption was excessive. Then operation with very small amounts of catalyst was tried and the catalysts discarded after one use. It was found that satisfactory catalyst activity with low molybdenum consumption was achieved if the molybdenum oxide was distributed on a support. The lowest cost, readily available, and most suitable support was brown coal char obtained as fly-dust from the fluidized bed Winkler gas producers. At first char impregnated with 2% molybdic acid was used after adding sulfuric acid to neutralize the calcium oxide in the char ash. Later the molybdenum was replaced by 10% ferrous sulfate. The hydrocracking and asphaltene reducing activity of these catalysts is about equal. However there is a difference in their hydro-refining activity. This can be seen in the phenols content of the products obtained from brown coal tar residue as shown in Table 4 [10].

It is evident that the molybdenum catalyst has higher refining activity but the difference is not significantly detrimental for the iron catalyst for the conversion to gasoline. The vapor phase hydrorefining catalysts have sufficient activity to overcome the poorer efficiency of the iron catalyst.

A number of problems had to be solved when experiments on a larger scale were begun. Late in 1924 the first large scale reactor was installed to provide greater amounts of oils. It was a pressure vessel of 300 mm inside diameter with inside insulation and a complicated and troublesome electric heater. In the first test in February 1925 a molybdena-alumina catalyst was used for the conversion of brown coal tar. Large heat losses and settling of solids because of low velocity were the main difficulties.

The next larger test reactor of 500 mm ID was equipped with a central electric heater of the type used in the ammonia synthesis. It preheated the hydrogen to a temperature above that required for the reaction to supply additional heat to the partially preheated feed. The feed, crude oil, was preheated in a coil immersed in a molten salt bath. It became apparent that in liquid phase hydrogenation reactors any inserts serve only as places for the deposition of solids and catalyst and that a sufficient velocity must be maintained either by a stirrer or, preferably, by a high throughput of the feed. Further tests with reactors of 800 mm ID led to similar conclusions. The reactors used in the commercial plant at Leuna had the same diameter and were 18 m long. The volume of the forging was 9 m^3 and with inside thermal insulation this was reduced to 7 m^3. Two, three and usually four reactors in series were used in one high pressure stall.

For efficient heating heat exchangers were developed. The heat exchanger became a bundle of parallel tubes installed in a pressure vessel of 600 mm ID and 18 m length. It was very efficient for the heating of vaporzable or non-vaporzable oils with hydrogen and for coal-oil slurries, as long as their viscosity was not excessive or became so during heating. The feed materials flowed upwards outside of the tube bundle while the hot product vapors and the hydrogen descended inside the tubes.

Finally, for heating from the heat exchanger exit temperature to the temperature at which the reaction started, usually 703 K for liquid phase reactions, two types of heaters were developed. Where continuous heat supply during operation was needed as in liquid phase stalls, gas fueled heaters were used that contained finned tubes heated by convection with recycled flue gas [11]. The fins increased the outside heat transfer surface by a factor of 20 and permitted use of a flue gas temperature that did not exceed the temperature at which the yield strength of the tube metal became insufficient. This assured safe metal temperatures even if inside deposits reduced heat transfer which increased the metal temperature. For vapor phase stalls, especially the hydrorefining units with a high heat of reaction, electric heaters were used. The heater tubes themselves provided the resistance element for the low voltage current. These units were used in stalls that ran autothermally after the initial heat-up. In the design of all the heating equipment care was taken to avoid dead spaces in which settling of solids or condensations reactions could occur. These were the cause not only of plugged equipment but also of catalyst deactivation.

The use of diluted (supported) catalysts on active carbon of large surface area was a major advance in the hydrogenation of brown coal tar and other residues. Analytical data for a brown coal tar from central Germany are given in Table 5.

Table 5. Analytical Data for Brown Coal Tar from central Germany

Specific Gravity at 323 K	0.929
Asphaltene/%	1.3
Solids/%	0.5
Boiling range, to 598 K/%	45
Carbon weight/%	83.4
Net Hydrogen/g per 100 g C	12.9
Specific Gravity at 373 K of resid. 598 K	0.913
Vacuum Distillate to 598 K/%	69.8

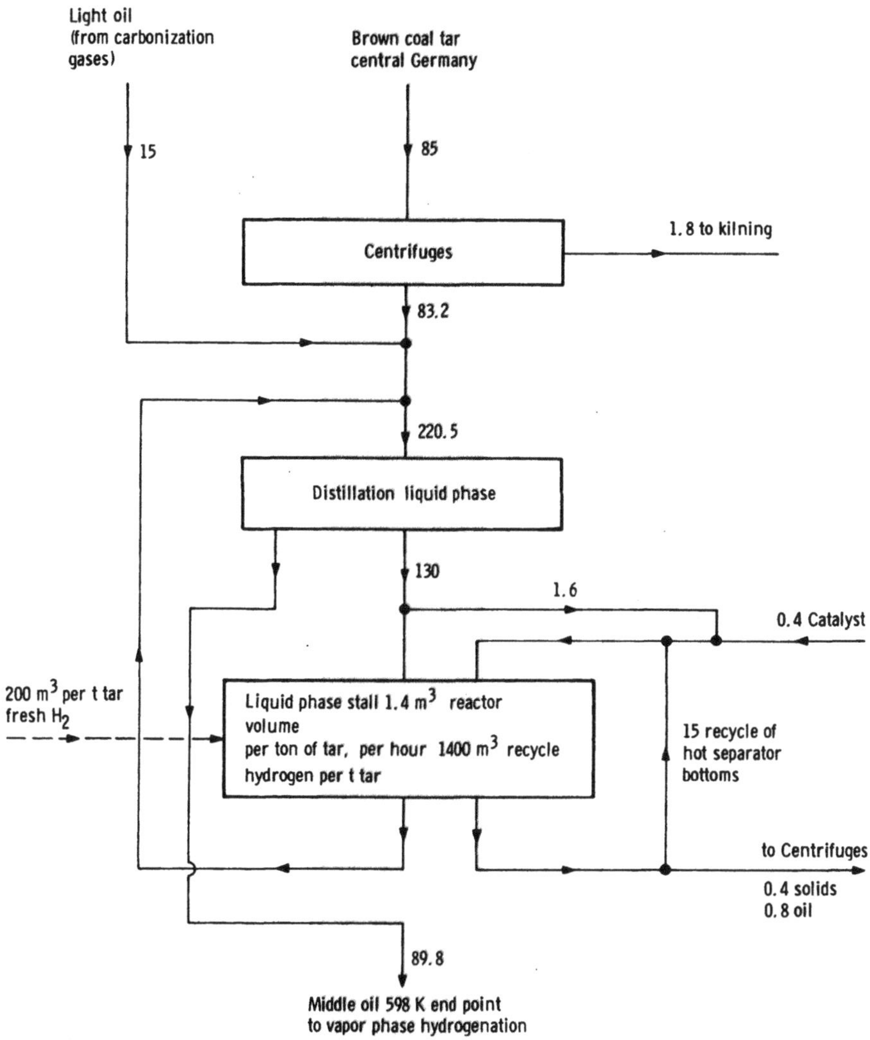

Figure 7. Schematic flow sheet, brown coal tar catalytic hydrogenation in liquid phase: (figures in tons, exept when otherwise stated)

A schematic flow sheet of the hydrogenation of such tar at 230 bar pressure and 743 K is shown in Figure 7. The light oil recovered from the low temperature carbonization gases is treated together with the tar. With a yield of almort 90% the tar is convented to a middle oil which is a suitable vapor phase feed material for the conversion to gasoline or diesel fuel.

The liquefaction of brown coal itself posed major problems in catalyst selection and the separation of the residual solids from the oil. The flow sheet in Figure 8 will be helpful in seeing the problems and constraints on finding and testing catalysts for coal liquefaction to middle oil with high yield in the framework of the equipment that had been developed.

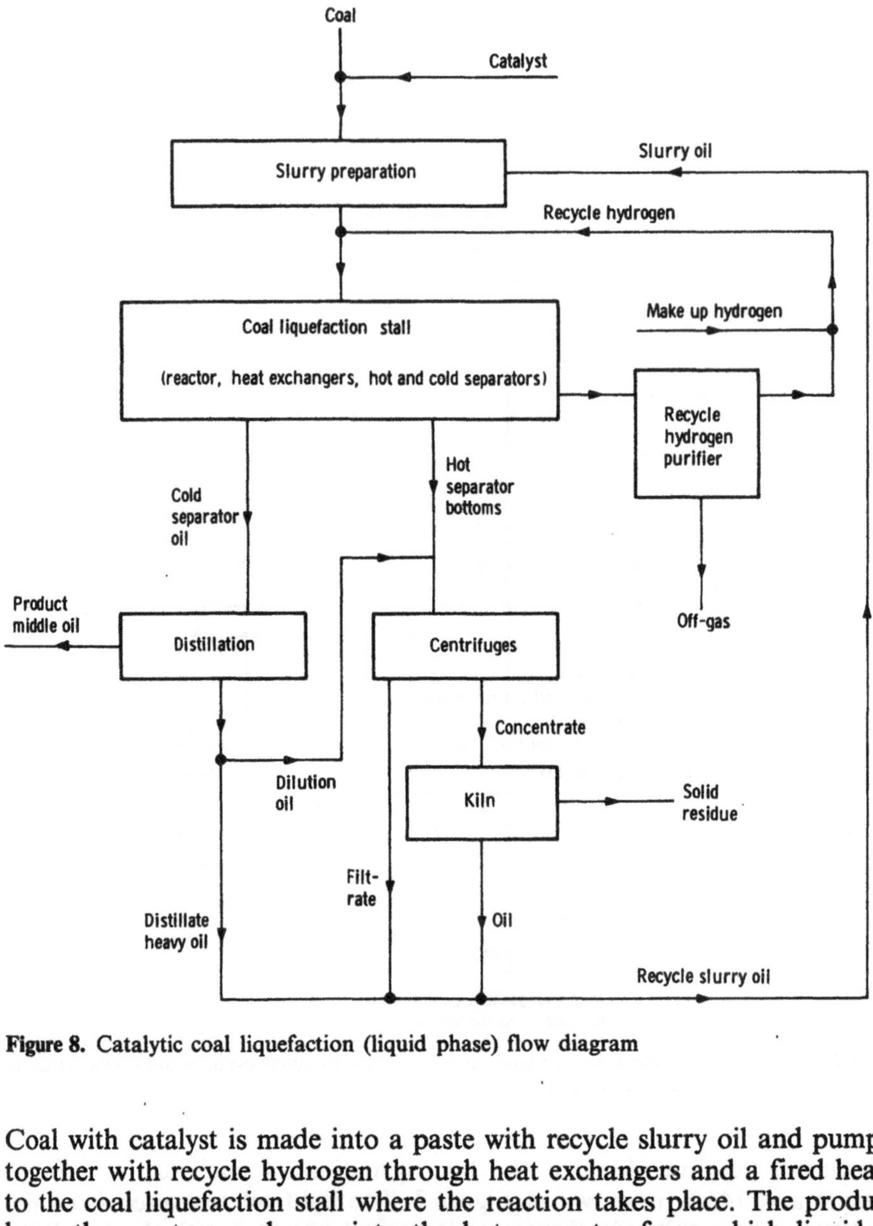

Figure 8. Catalytic coal liquefaction (liquid phase) flow diagram

Coal with catalyst is made into a paste with recycle slurry oil and pumped together with recycle hydrogen through heat exchangers and a fired heater to the coal liquefaction stall where the reaction takes place. The products leave the reactors and pass into the hot separator from which liquid oil and unreacted solids are withdrawn as bottoms, while oil vapors and hydrogen are cooled and enter the cold separator. Here the lighter oils are separated from recycle hydrogen. The cold separator oil is distilled into middle oil (including naphtha) as overhead and feed for the conversion to gasoline and bottoms. These are partly recycled as slurry oil component and part is used to dilute the hot separator bottoms with a solids content of about 35% to a solids content of 15–20% as centrifuge feed. The deep hydrogenation of coal to middle oil leads to a hot separator product from which solids are difficult to separate by filtration. Either the filtration rate was low or the consumption of filter aid too large or the amount of diluting oil and the oil loss excessive. Satisfactory solids separation was possible with continuous centrifuges operating at 413–433 K. A filtrate with a solids content of 2–12% was obtained as a slurry oil component and a residue with 38–40% solids. The latter went to kilns that pyrolized the residue at about 823 K. Special kilns were developed for this purpose. They had either an inside conveyor screw to scrape coke from the walls or were rotary kilns charged with heavy steel balls to avoid or dislodge coke deposits. They were sensitive to the feed composition and required for satisfactory operation a feed that contained at least about 5 parts of solid per part of asphaltenes. The function of these cumbersome units would today be taken over by residue coking processes developed by the petroleum industry or by vacuum distillation whereby the overhead would return to the slurry oil while the residue would find other uses.

The necessity of obtaining maximum distillate fuel production from existing plants was fulfilled by the above flowsheet Today freedom in liquefaction and hydrogen production plant design would eliminate or at least alleviate the restraint imposed by the high asphaltene conversion required at that time: Vacuum distillation of the solids-containing residue would provide an asphaltene-free slurry oil component and a pumpable residue that could be used for the production of hydrogen. At the same time the asphaltenes of limited stability (possibly because of a content of free radicals carried over from the coal) would not return to the hydrogenation units where they are the one component which not only is most difficult to convert into oil but also may cause precipitations or deposits of insoluble material.

A main object of coal liquefaction experiments was the finding of catalysts, their combinations and the amounts required that within the asphaltene limits established in commercial operation, would result in the highest throughput, best conversion of the solid coal into liquids, and the lowest formation of gaseous hydrocarbons. These limits vary widely for different raw materials. The following figures for the asphaltene content of the total oil boiling above 598 K were obtained from commercial operation and used as guidelines:

Brown Coal	5– 8% Asphaltene in Heavy Oil
Bituminous Coal, Ruhr District	9–11% Asphaltene in Heavy Oil.

Close coordination of small and large scale experiments in Ludwigshafen with the commercial operation in Leuna lead to great improvements which for the years 1927 to 1932 are summarized in Table 6.

The most important factors in achieving this improvement were the development of a low cost disposable iron catalyst, the development of new steels and equipment, and the optimization of the interplay of the processing steps. These included careful control of the temperature distribution, and control in the heating and reaction zones using the newly developed

Table 6. Brown coal liquefaction Leuna 1927 to 1932

Result	Initial	1932
Liquefaction/% maf coal	65	98
Middle oil in product oil/%	70	100
Centrifuge feed/t/t product oil	2.5	0.7
Asphaltene in product oil/%	10	0

equipment for heat exchange and for the final heating of the coal slurry with hydrogen. These improvements, together with newer more active vapor phase catalysts (see later) made it possible to produce in Leuna in 1943 650,000 tons of motor fuel by adding reactors and preheaters to the existing stalls but without changing the original lay-out.

It may be noted that these improvements were achieved without increasing the operating pressure which was given by the original ammonia synthesis plant. The pressure at the stall inlet was 230 bar corresponding to a hydrogen partial pressure in the reactors of about 170 to 180 bar. Later experiments showed that a pressure increase to 325 bar at the stall inlet which corresponds to about 250 bar hydrogen partial pressure would have brought considerably improved liquefaction results. In Leuna the disadvantage of the low pressure was partially compensated by an increase of the amount of iron catalyst added to the coal. The increased pressure became a standard for the later 300 bar liquefaction plants.

In addition to the brown coal deposits in central Germany there were also large brown coal seams in the Rhine district near Cologne. As mentioned above, coal from this district was later used in the liquefaction plant at Wesseling. A comparison of coals from these two areas in Table 7 shows that the Rhine brown coal is lower in hydrogen, sulfur and tar than that from central Germany. It was found that with the available iron catalyst a pressure of 650 bar was more satisfactory than 300 bar for a sufficient reduction of the asphaltene content of the heavy oil to assure satisfactory oil recovery from the solids of the hot separator bottoms. Comparison of the operating results in commercial plants with these two brown coals are

Table 7. Analyses of two German brown coals

Origin	Central Germany	Rhine District
Ash, dry basis/%	12.8	5.9
Volatile matter/maf basis %	57.9	53.5
Carbon/maf basis %	71.9	68.7
Net hydrogen/g per 100 g C	4.1	2.4
Sulfur/dry basis %	5.0	0.4
Low temp. tar yield/dry basis %	17.1	10.0

Note: maf = moisture and ash free. Net hydrogen = total hydrogen minus hydrogen equivalents of oxygen, nitrogen and sulfur.

Table 8. Catalytic liquefaction of two brown coals

Coal origin	Central Germany	Rhine District
Operating pressure/bar	230	650
Catalyst/% of maf coal	up to 9% Bayer Masse	6% Iron ore + 1.2% sulfur
Temperature/K	753–763	748–753
Throughput/maf coal kg h^{-1} l^{-1}	0.46	0.52
Recycle gas/Nm^3 kg^{-1} maf coal	2.5	2.8
Coal to slurry oil, weight ratio	50/50	40/60

shown in Table 8. Because of the lower hydrogen content, the hydrogen consumption for the Rhine brown coals was 2000 Nm^3 per ton of middle oil product, considerably higher than the 1450 Nm^3 for the Leuna coal.

The activity of the catalyst used for the low sulfur content Rhine brown coal was increased by sulfiding. Thus smooth operation without formation of deposits or crusts was assured.

A vexing and at first puzzling problem in brown coal liquefaction was the formation of the so called "caviar". These were small spheres of a few millimeters diameter of high ash content which settled in the first converter until it was plugged. A simple remedy was the withdrawal of 0.2 to 1% of the feed from the first reactor, the so-called "desanding". The cause of "caviar" formation seemed to be the conversion of calcium humates into calcium carbonate which was deposited often on sand grains contained in the Leuna coal. Later "caviar" was observed in plants using Rhine brown coal and bituminous coals. Although the composition and probably the mechanism of formation there was different, the remedy was the same. Attempts to avoid "caviar" formation from Leuna brown coal by catalysis seemed successful in small scale experiments. Impregnation of the coal with $FeSO_4$ solution decomposed the calcium humate into stable gypsum which did not form aggregates. However verification on a large scale was not possible before operations ended in 1945.

In order to discuss the liquefaction of brown coal without interruption we have progressed far ahead in time since the Wesseling plant started operation in 1941. After this digression we turn to the liquefaction of bituminous coal.

Early autoclave experiments showed that the catalysts found active in the liquefaction of the Leuna brown coal increased the oil yields and lowered the asphaltene concentration when using bituminous coals. The asphaltene content was higher than that observed with Leuna coal (see page 19) but this did not lead to asphaltene precipitation since high boiling paraffins obtained from the tar rich Leuna coal were absent. Continous small scale tests with molybdenum and iron catalysts were successful although the solids separation led to larger oil losses because of the higher asphaltene content of the heavy oil. The experiments with bituminous coals progressed slowly at Ludwigshafen until the interest of the ICI and the intention to bouild a commercial plant caused urgency. Experiments at the ICI had shown that one British coal gave much higher conversion than expected from its analysis (rank). It was found that this coal contained volatile chlorine compounds [9] which formed HCl during the reaction. In combination with 0.06% tin oxalate as cocatalyst better results were achieved than with other cocatalysts *e.g.* Mo or Fe compounds. These results were confirmed with other British coals and repeated in Ludwigshafen. To obtain firm data for the design of a commercial plant (to be erected at Billingham) a large scale test with a throughput of 20 tons per day of British coal was made at Ludwigshafen in 1931 for use by the ICI.

Hydrochloric acid in vapor form at coal liquefaction reaction conditions and temperatures of 723 K or higher is not excessively corrosive in the presence of hydrogen sulfide. However during cooling, when fractions containing amines and water condense, the corrosion becomes extremely rapid in steel equipment. It was therefore necessary to neutralize the acid by injecting a suspension of sodium carbonate in oil into the hot separator. Thus the hydrochloric acid was neutralized before it could enter the cooling tract.

For the first German bituminous coal liquefaction plant at Scholven [10] the gained experience was used and a pressure of 300 bar and the tin-chlorine catalyst combination selected. The operating results fulfilled the expectations from the experimental work. The most important data are summarized in Table 9. The results were satisfactory and the plant expanded to a motor fuel production of 280,000 tons per year. During the war tin was partly replaced with fair success by a lead ore. Withdrawal of some hot separator bottoms with its high asphaltene content for use as briquetting medium instead of coal tar pitch, improved the results because of the increased asphaltene purge. Beginning in 1935 at Ludwigshafen experiments at 600 atm hydrogen partial pressure were begun especially for the liquefaction of bituminous coals. The use of the high pressure for the liquefaction of the Rhine brown coal was discussed above. For most bituminous coals and their tars as well as aromatic, highly asphaltic petroleum residues, the high pressure gave satisfactory results with low cost readily available catalysts

Table 9. Catalytic liquefaction of Ruhr coal to middle oil

Pressure/bar	300	700
Carbon/% in maf coal	82.8	83.9
Catalyst/% of maf coal	0.06 SnC_2O_4 + 1.15 NH_4Cl	1.2 $FeSO_4 \cdot 7\,H_2O$ + 2 Bayer Masse + 0.3 Na_2S
Reaction temperature/K	749	758
Coal throughput (maf)/kg l^{-1} h^{-1}	0.3	0.42
Recycle gas per maf coal/m^3 kg^{-1}	4	4.5
Coal liquefraction/% maf	92–94	97
Middle oil production rate/kg l^{-1} h^{-1}	0.19	0.26
Off-gas/% C of liquefied C	26	25
Asphaltene in total heavy oil/%	11	9

such as iron compounds added to coal or to char as support. Comparison of the results for the liquefaction of Ruhr coals at these two pressure levels is shown in Table 9.

The high pressure gives higher coal conversion and a lower asphaltene content in the total heavy oil with following advantages in the liquefaction and solids separation steps.

1. The lower viscosity oil leads to a higher solids content of the centrifuge residue and thus to a smaller amount of feed for the kilning operation.
2. The higher solids to asphaltene ratio improves the operation and maintenance requirements of the kilns.
3. The lower viscosity and asphaltene content increase the heat transfer rate in the high pressure heating units and reduce the danger of deposit formation.
4. The lower off-gas yield reduces the hydrogen consumption.
5. The absence of chlorine obviates the need for the soda ash neutralization step.

Since the 700 bar reactors had a smaller volume than the 300 bar reactors the stalls had about the same annual production of middle oil regardless of the pressure. A coal stall with four converters in series produced sufficient middle oil for the conversion to about 70,000 tons per year of gasoline. This corresponds to about 2000 barrels per day of gasoline and LPG (liquefied petroleum gases, mainly propane and butane). The latter was also used as motor fuel.

The great activity of the tin-chlorine catalysts combination is also apparent at 700 bar. Table 10 [12] gives results from laboratory scale experiments in reactors of 10 liter volume. The increased HCl corrosion at 700 bar pressure prevented use in commercial plants.

Thus the catalysts used for bituminous coal liquefaction were standardized: tin and hydrochloric acid were used at 300 bar pressure and iron catalysts at 700 bar pressure. The main iron catalyst was Bayer Masse, an iron oxide containing usually titanium, a residue of the alumina production from bauxite. In addition ferrous sulfate was impregnated on the coal and for some coals containing small amounts of volatile chlorine compounds some sodium sulfide was added to avoid corrosion. It was observed that the sodium sulfide in this combination seemed to have catalytic activity of its own.

Table 10. Pressure and catalyst for Ruhr coal liquefaction

Pressure/bar	300	700	700
Catalyst/% maf coal	0.06 SnC_2O_4 1.15 NH_4Cl	$FeSO_4$ Fe-Oxide	0.06 SnC_2O_4 1.15 NH_4Cl
Peak reaction temp/K	738	750	748
Coal Liquefaction/% maf	93	96	97
Middle oil product/kg l^{-1} h^{-1}	0.18	0.27	0.32
Off-gas/% C of liquefied C	25	21	18

The operation of the liquefaction plants required very accurate control of the reaction temperature to assure uniformity of the products and smooth operation of the solids separation units, centrifuges and klins. Special attention was paid to the temperature at which the coal slurry entered the first converter. The temperatures were maintained within 2 °C and often debates of the operating supervisors occurred to decide whether a 2 °C temperature increase would be advisable.

The liquid phase processes described so far had as their object the production of middle oil that was suitable for conversion to gasoline or Diesel fuel. When demand appeared for fuel oil for the navy the bituminous coal liquefaction process was directed to the production of a heavy oil with a low viscosity and low pour point having a specific gravity above that of sea water and a high volumetric heating value. The plant at Blechhammer was built for this purpose. It used local coal of Upper Silesia. Final basic data for this plant which began operation in 1943 were obtained in a large pilot plant at Ludwigshafen in 1941. It had a reactor volume of 1.6 m^3 and a coal throughput as high as 25 t d^{-1}. Table 11 shows results obtained when operating in the middle oil mode and when increased throughput gave additional heavy oil. A throughput increase of about 50% and a small temperature increase gave about the same middle oil production rate and in addition distillate heavy oil. In order to replace withdrawn heavy distillate oil some middle oil had to be added to the slurrying oil.

In these tests the asphaltene purge was kept as low as possible to obtain maximum oil production with limited centrifuging and kilning capacity.

Table 11. Middle and heavy oil from bituminous coal: operation at 700 bar with iron catalyst

Main Products: Middle Oil/%	100	65	50
Heavy Oil/%	0	35	50
Temperature/K	749	751	753
Coal Throughput/kg l^{-1} h^{-1}	0.43	0.62	0.63
Slurry Oil: Haevy Oil/%	100	85	73
Middle Oil/%	0	15	27
Coal Liquefaction/% maf coal	96	96	96
Product rate/kg l^{-1} M. O.	0.26	0.26	0.22
H. O.	0	0.14	0.19
Total	0.26	0.40	0.41
Off-Gas/%C of liquefied C	23	20	18.5
Asphaltenes/% in hot separator bottoms H. O.	14.7	18.0	19.3

Analytical data:	Middle Oil	Dist. Heavy Oil
Specific Gravity (K)	0.964 (298)	1.05 (323)
Phenols/%	15	4.4
IBP/K	464	603
50%/K	543	513 Vac.
FBP/K (%)	598 (98)	573 (87 · 4) Vac.
H/C ratio	1.33	1.1
N/%	0.59	0.47
S/%	0.01	0.05

Table 12. Bituminous coal extract and feed coal

	Extract	Coal
Ash/%	0.05.	5.5
Volatile matter/% dry basis	41	30
Carbon/% dry basis	88.5	85.2
Hydrogen/% dry basis	5.4	5.2
Sulfur/% dry basis	0.6	0.9
Softening point/K	473–513	
Solubility/% in benzene	35	
tetralin	69	
cresol	100	

Increasing the removal of asphaltene-rich heavy oil permits operation with iron catalysts at 300 bar pressure. The large asphaltene purge is made attractive by progress in distillation and gasification processes. Vacuum distillation removes oil from the hot separator bottoms and leaves a pumpable vacuum residue. Gasification especially at elevated pressure makes it a desirable source of the hydrogen used in the liquefaction. This variation of the Bergius-Pier process is at present being studied in Germany for large scale use [13].

At this point mention should be made of processes using solvents to convert coal into soluble or meltable solids or heavy residuals with a minimum of molecular weight reduction and hydrogen consumption. In this process, the coal ash was usually the only catalyst, however free radicals in the coal initiated hydrogen transfer from the solvent to the coal. A. Pott and H. Broche [14] found that bituminous coal can be made soluble and freed of ash at about 100 bar and 673 K in hydrogen donor solvents such as an 80:20 mixture of tetrahydronaphthalene with cresol or with middle oils containing phenols from liquid phase coal or tar hydrogenation. These solvents were stable under the above reaction conditions when coal was absent, but they converted coal into an "extract" which remained in solution in the solvent and could be separated by filtration from ash and unconverted coal. The used solvent requires catalytic partial rehydrogenation of aromatics for reuse. Some properties of coal and its extracts are given in Table 12.

Similar results were obtained by F. Uhde and Th. W. Pfirrmann [15] by hydrogenating a bituminous coal slurry in a phenols-containing oil at 300 bar pressure with a small amount (1 Nm^3 per kg coal) of coke oven gas or hydrogen slightly above 673 K. Filtration of the product at 423–473 K gave an extract in a yield of 75 to 90% of the maf coal with a lower softening point than the Pott-Broche extract. Liquid phase hydrogenation of these extracts did not indicate better oil yields than obtained from the original coal.

Both processes have gained recently attention in the USA for the manufacture of low ash and sulfur "Solvent Refined Coal" (SRC). Outside of the organizations mentioned above which industrialized the coal liquefaction process, research went on in many countries. These studies are mainly connected with the names of H. H. Storch and L. L. Hirst of the US Bureau of Mines; H. H. Lowry of the Coal Research Laboratory of the Carnegie Institute of Technology, US; F. S. Sinnatt and J. G. King, The Fuel Research Station, Great Britain; F. Vallette, France; T. E. Warren, Canada and M. Kurokawa, Japan.

Refining of Coal Oils

While the selection of catalysts for the primary coal liquefaction was severely limited by questions of availability and by the complex testing equipment, the development of coal oil refining catalysts showed much greater variety.

With the decision in 1926 to build the Leuna plant it became urgent to optimize and finalize the composition of a vapor phase catalyst of sufficient activity, selectivity and mechanical strength for the large scale conversion of brown coal middle oil to gasoline and diesel fuel.

The first such catalyst adopted in 1927 for all-round use consisted of equimolecular parts of the oxides of molybdenum, zinc and magnesium, it was catalyst 3510. The oxides of high purity were mixed as an aqueous paste, put on sheets in a thin layer, cut into irregular cubes of about 1 cm side length and dried. For use in small scale experiments these cubes were crushed and screened to the desired size. At 200 bar pressure and 723 K a gasoline production rate of 0.1–0.2 kg per liter of catalyst per hour was obtained in Leuna. The gasoline concentration in the product was about 30%, thus recycle operation was necessary for complete conversion. The catalyst activity declined slowly with time and this was compensated by a gradual increase of the operating temperature although the off-gas formation which initially was 20% of the converted feed increased thereby.

Commercial operation in Leuna indicated somewhat smaller catalyst activity than observed in Ludwigshafen in small scale experiments. The explanation was found later: in the commercial plants more ammonia was recycled with the recycle hydrogen. Addition of organic nitrogen compounds to the feed which then are converted to ammonia confirmed that the activity of catalyst 3510 is reduced by basic nitrogen compounds.

Attempts to increase the activity of the 3510 catalyst failed for some time. A somewhat more active molybdenum catalyst containing 20% kaolin, chromic acid and 0,5% aluminium powder appeared satisfactory in small scale experiments but failed dramatically by disintegration in a large scale unit because of its slower heat-up rate.

Used catalysts always contained molybdenum in a sulfided form. Sulfidation however was not the reason for the decrease of the activity of 3510. Many catalyst compositions were treated with hydrogen sulfide before use with mixed results. By such treatment of a tungsten-zinc oxide catalyst at 200 bar hydrogen pressure and a high hydrogen sulfide content, a catalyst of higher activity than 3510 was obtained. However the use of this successful activation method was impractical for large scale catalyst production. A major breakthrough in catalyst preparation came in 1930 when the thermal decomposition of oxygen-free ammonium sulfotungstate gave a tungsten disulfide of outstanding catalytic activity.

This was catalyst 5058 [16] which gave about three times the conversion rate of middle oil to gasoline in comparison with 3510 at an operating temperature around 673 K, that is fifty degrees lower than the older catalyst. In addition the conversion into off-gas was cut in half and, surprisingly, the ration of iso- to normal-butane was about 1:1 and thus above that given for the equilibrium for the isomerization of butane. This indicates that the isobutane is formed by the splitting of suitable ring or branched structures. Furthermore the aromatics content of the gasoline was reduced to about 10% even when starting with highly aromatic middle oils. The catalyst has many properties that were later attributed to dual function catalysts.

More than 1500 catalyst samples were tested to come from 3510 to 5058. All tests were made in high pressure flow reactors (Figure 3), and some life tests in larger units with continous recycle of the unconverted oil. They served also for the determination of the optimum temperature, throughput and conversion rate. Samples that were not satisfactory were replaced after 2 or 3 days. In this way some 1000 catalyst preparations could be tested per year. About five times more preparations were made but not tested in high pressure units either because other similar samples in a series had indicated them to be not promising or the mechanical properties were unsatisfactory.

Catalyst 5058 came at an opportune time for Piers department and the Leuna plant. It was the depth of the economic depression in Germany, and cost reductions became a necessity. Therefore this major process improvement made continuation of the hydrogenation work more secure. Furthermore, at that time liquefaction of bituminous coal gained general interest and special importance because of the participation of the ICI in the international cooperative effort. Catalyst 5058 converted bituminous coal middle oils with ease into gasoline while 3510 did so only with low throughput, small conversion and high off-gas formation. In addition the life span of cata-

Table 13. Catalyst 5058 reactions

Feed	Product	Temperature/K	Reactions conditions		
			Pressure/ bar	Rate/ kg l^{-1} h^{-1}	Conversion/ %
Diisobutene	Isooctane	489	250	2.0	99
Naphthalene	Decalin	609	200	0.9	90
Brown Coal Middle Oil	Prehydrogenated M. O.	653	200	1.0	
Phenols 18%					99.5
N-Bases					99.5
Sulfur 2.5%					99.0
n-Butane	Isobutane	673	200	0.5	35
Paraffins 533–593 K	Gasoline 453 K EP	681	220	1.0	90

lyst 5058 was greater and a catalyst life of 1–2 years and more became the rule.

Some of the reactions that show the versatility of catalyst 5058 are shown in Table 13. Saturation of olefins occurs already near 470 K, that of aromatic rings in naphthalene requires higher temperatures. Temperatures around 670 K are needed for the hydrorefining of brown coal middle oil and the almost complete removal of heteroelements including nitrogen. Similar temperatures are used for the conversion of normal to isobutane, and for the hydrocracking of a paraffinic oil to gasoline. Full scale use of 5058 since 1931 in Leuna fulfilled all expectations. Its use virtually trebled the capacity of the vapor phase hydrogenation units. As was the case with 3510 the higher content of ammonia in the Leuna recycle hydrogen reduced the activity somewhat but did not cause any further or progressive deterioration and was compensated by a small temperature increase.

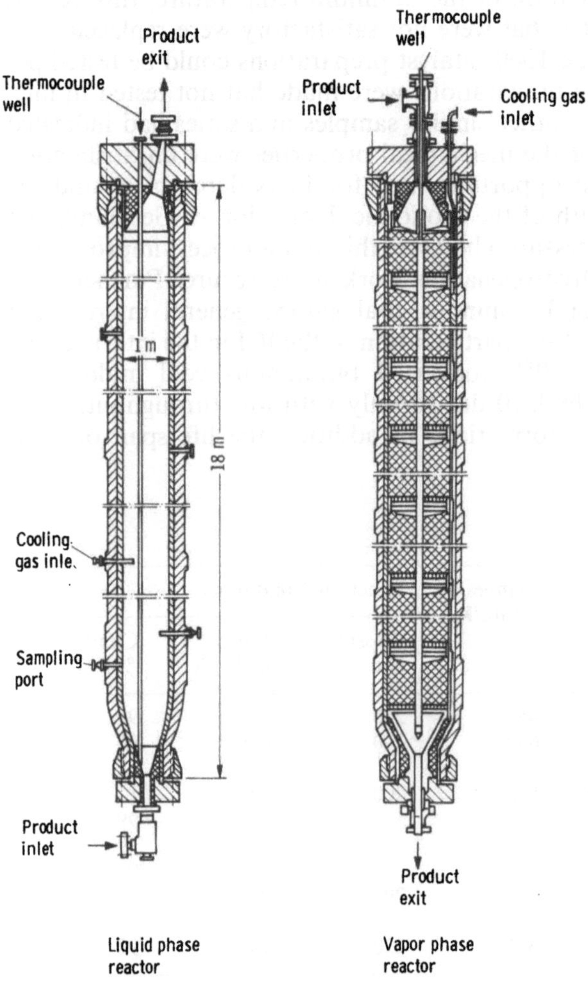

Figure 9. Cross section of a commercial fixed bed catalyst (Vapor Phase) reactor in comparison with a liquid phase reactor

The physical properties of WS_2 as those of other important catalysts were characterized by various methods [17]. Specialists in various BASF laboratories provided their skill for such studies. X-ray analysis showed a hexagonal latice and a primary crystallite size of 3×10^{-7} cm in height and twice that in width. Microscopic examination showed that the crystals are pseudomorphous with the monoclinic ammonium sulfotungstate crystals from which they were obtained. This was asumed to be the cause for the large surface area of the catalyst. Argon adsorption isotherms indicated a surface area of $50\ m^2\ cm^{-3}$ (using the method of E. Hueckel [18]), very similar to that of silica gel. Other experiments showed that at room temperature adsorption is selective for hydrogen and unsaturated hydrocarbons. Adsorption of hydrogen depends on the pretreatment with hydrogen and the presence of other gases.

The exotherm of hydrogenation reactions with 5058 and the necessity of accurate temperature control to optimize yields required subdivision of the catalyst bed in commercial reactors into many parts. Between these sections the reactants were mixed and cooled with cold recycle gas. This was necessary since direct impingement of cold recycle gas caused disintegration of the catalyst pellets. The cross section of such a reactor is shown in Figure 9 together with that of a liquid phase reactor.

Catalyst 5058 was satisfactory for the production of motor fuel from coal middle oils as far as life, yield, throughput, conversion, product purity and stability are concerned. The high hydrogenating activity was useful for the production of Diesel fuels but it gave gasolines with a low aromatic content and unsatisfactory octane number. Saturation of aromatic rings caused higher hydrogen consumption than necessary. The isomerizing activity and the content of branched chain paraffins did not compensate for the loss of aromatics in comparison with 3510. Catalyst development thus was concentrated on alleviating these drawbacks and specifically on finding supports for the WS_2 which would lead to catalysts with the same hydrocracking activity but reduced ability to saturate aromatics. In 1934 it was found that "Terrana", a montmorillonite earth, was an active support for WS_2. In attempting to improve the support by "etching" to increase its surface area it was found that it retained the hydrofluoric acid used for this purpose. This gave a support of increased activity.

Thus "Terrana" after treatment with 10% hydrofluoric acid and impregnation with ammonium sulfotungstate to produce 10% WS_2 was a catalyst of about the same hydrocracking activity as 5058 in converting middle oil into gasoline if that middle oil was free of nitrogen bases (limiting value 5 ppm NH_3 equivalent).

The new catalyst 6434 was found in the tenth year of continuous catalyst development. Gasolines obtained with it had about a 5 points higher octane number than those obtained with 5058. The isobutane content of the C_4-fraction was a high as 75%. The feed oils to this catalyst were practically sulfur free (coming from prehydrogenation) and addition of sulfur was necessary to maintain the catalyst at peak activity.

The requirement of nitrogen-free oils from coal liquefaction for catalyst

6434 led to a subdivision of the vapor phase hydrogenation into the hydrorefining step, which at the time was called "prehydrogenation", and the hydrocracking step called "splitting hydrogenation" or "benzination".

After the introduction of 6434, catalyst 5058 was used for the hydrorefining of coal middle oils. On account of its hydrocracking activity it produced in that step some gasoline of low octane number. This is due to the high temperature of 653 K, and that at 200–250 bar hydrogen pressure a middle oil feed rate of 0.8 kg l^{-1} h^{-1} is needed to remove organic nitrogen to the required low level while hydrocracking begins already at 623 K.

It may be mentioned that in the brown coal tar hydrogenation plants in Boehlen and Magdeburg, use of the 5058/6434 catalyst combination gave a gasoline that was marginal in octane rating and specific gravity. In addition the use of two different catalysts caused a shortage of ready spare equipment. It was therefore attempted to use 6434 not only for the hydrocracking but also for the hydrorefining step, and indeed at a temperature about 40 °C above that for hydrocracking the hydrorefining activity of 6434 was sufficient to reduce the nitrogen content of the brown coal tar middle oil below 5 ppm. In this connection it was found that 6434 at 600 bar and even higher temperatures of 723–773 K had high and sustained hydrocracking activity for nitrogen-containing oils.

In the search for more selective hydrorefining catalysts it was found that combinations of tungsten disulfide with sulfides of nickel and, in a decreasing degree, with sulfides of cobalt and iron, had improved refining and decreased splitting activity. They were suitable — and used commercially — for special reactions, such as the diisobutene saturation and sulfur removal from gasolines or coke oven light oil. To reduce catalyst costs various supports were studied. Supports were tested for the various basic reactions, and activated alumina was found most suitable. The first hydrorefining catalyst on activated alumina as support, catalyst 7846, contained 10% MoO_3 and 3% NiO. There was similar activity with an alumina catalyst containing 27% WS_2 and 3% NiS found in 1940 and numbered 8376. Its hydrofining activity was sufficient at the same temperature at which previously 5058 was used, but it had a much lower hydrocracking activity. Thus the amount of low octane "prehydrogenation" gasoline was reduced. Catalyst 8376 was universally used for prehydrogenation in the German coal

Table 14. Hydroefining of coal middle oil (453–598 K): nickel in catalyst 8376 replaced by cobald and iron

	Gasoline to 453 K/ wt%	Hydrorefined Middle Oil		
		Aniline Point/K	Phenols/%	Nitrogen compounds/ ppm NH_3 equivalent
Nickel	23	320	0.01	3
Cobalt	23	284 · 5	0.03	40
Iron	20	267 · 5	0.05	100

liquefaction plants since tungsten was more readily obtainable than molybdenum. It was also used for the direct hydrorefining of brown coal tars and shale oils using the TTH process which will be discussed later.

The activation of this catalyst by nickel sulfide is very specific and replacing nickel by equal amounts of cobalt or iron gives catalysts of lower hydrogenating activity and less complete reduction of nitrogen compounds. This is shown in Table 14 for the hydrorefining of bituminous coal middle oil. The nickel containing catalysts increases the hydrogen content of the feed as indicated by the Aniline Point. The nitrogen content is reduced to an acceptable level for subsequent hydrocracking with 6434 even with a safety margin. Replacing nickel by cobalt leads to a catalysts that increases the hydrogen content much less, reduces phenols and sulfur sufficiently but does not decrease the nitrogen content to an acceptable level. The iron-containing catalyst has only a slight hydrogenating activity, comparable to that of the catalyst without iron. The three catalysts reduce the nitrogen content in rough numbers to 0.1%, 1% and 10% of that of the feed oil.

The cobalt-molybdenum catalysts have excellent refining activity and, although not sufficient at the time for coal oils, found later wide use in the petroleum refining industry.

Already at the beginning of the cooperation with the Standard Oil Company, interest was shown in the improvement of lubricating oils. Experiments indicated soon that high boiling petroleum vacuum distillates can be hydrogenated at 300 bar and 623–673 K with long catalyst life and a throughput of 1 kg l^{-1} h^{-1} or more with molybdenum or tungsten catalysts such as 5058, or with hydrorefining (prehydrogenation) catalysts. The volumetric yield is 103–108% and the product contained 60–80% lubricating oil. Table 15 gives the properties of a Columbia crude oil vacuum distillate and of the lubricating oil obtained from it. The main object is an increase of the viscosity index, that is, obtaining an oil whose viscosity decreases less with tem-

Table 15. Lubricating oil improvement by refining hydrogenation (yield 60—80%)

		Columbia Crude Distillate	
		Feed	Product
Lubricating Oil			
Specific Gravity (298 K)		0.922	0.892
Viscosity/Cp	at 311 K	304	129
	at 323 K	148	67
	at 372 K	15.2	11.0
Viscosity index		36	81
Flash point/K		496	522
Pour point		258	272
Conradson carbon/%		0.94	0.15
Sulfur/%		0.91	0.10
g H/100 g C		13.8	14.7

Table 16. TTH-Process for brown coal tar

	Brown coal tar (Central Germany)	
	Feed	TTH-product
Specific gravity (323 K)	0.918	0.828
Boiling range		
% to 473 K	0	11.2
% 473–623 K	51.2	53.1
above 623 K	47.8	35.6
Phenols/%	7	0.2
Middle oil fraction		
Phenols/%	14	Trace
Cetane number	20	above 50

perature. Use of a somewhat higher temperature or lower throughput increases the viscosity index further at a decreased lubricating oil yield.

A similar application is the hydrogenation of heavy catalytic cracking recycle oils. By increasing their hydrogen content they can again be cracked catalytically with high conversion and low coke yield and thus the feed material converted completely to motor fuel.

Another process that treats feed materials boiling above middle oil is the TTH process (*T*ief-*T*emperatur-*H*ydrierung). It uses raw materials with a low asphaltene content such as brown coal tars or shale oils. These are treated after centrifuging to remove solids over a fixed bed catalyst in mixed liquid-vapor phase. The small amounts of asphaltenes are reduced to oils and O, N, and S-containing compounds to hydrocarbons. Temperature, pressure, throughput, catalyst and hydrogen rate must be carefully selected to obtain optimum yield structure and product quality.

For brown coal tar at 300 bar pressure the temperature of the catalyst was kept at 553 K at the catalysts inlet to reduce first the most unstable compounds such as diolefins before thermal condensation reactions can take place. The heat of reaction is then permitted to increase the temperature to 623–648 K. A comparison of the properties of brown coal tar with those of the TTH product is given in Table 16.

It shows that the change of the boiling range is mainly caused by the reduction of the phenols. The product yield is 92–94% by weight of the tar with 3% off-gas formation. The remaining difference is due to the removal of heteroelements. The final products obtained at the Zeitz TTH plant

Table 17. TTH products from brown coal tar

25%	Gasoline	Octane number 58 (MM)
50%	Diesel	Cetane number 50–55
7%	Spindle oil	25.5 Cp at 293 K
4.5%	Lubricating oil	29.2 Cp at 323 K
13%	Paraffin wax	325–327 K Melting point

Table 18. Gasolines from bituminous coal hydrogenation at 300 bar with various catalysts

Catalyst			
Hydrorefining	5058	5058	8376
Hydrocracking	5058	6434	6434
Gasoline			
Specific gravity	0.735	0.745	0.770
°API	61	58.4	52.3
ON-RM	67	75	78
ON-MM	66.5	74	75

from brown coal tar and ist light oil at 300 bar pressure with catalyst 5058 are shown in Table 17.

By increasing the final catalyst bed temperature by 30–50 °C and slightly decreasing the throughput a product containing about 35% gasoline, 60% Diesel fuel and 5% paraffinic residue is obtained. The off-gas formation in this so called MTH process is about 6%.

To return to gasoline production. The properties of motor gasolines obtained from bituminous coal middle oil with the catalysts 5058, 5058/6434, and 8376/6434 are shown in Table 18. The improvement in gasoline octane number above that with 5058 alone was significant especially since no additional catalyst volume was required and the hydrogen consumption decreased. The results were obtained with the tungsten containing hydrorefining catalyst 8376. However it should be emphasized that equal results were obtained with molybdenum containing catalysts.

The catalyst combination 8376/6434 remained the workhorse of the German hydrogenation plants. Therefore yield figures for various raw materials are presented in Table 19. The motor gasoline met all specifiactions existing at the time. The aviation gasoline was only satisfactory until the widespread use of isooctane and alkylate demanded gasolines with a leaded octane number of 100, which improved the performance of aircraft engines

Table 19. Yields of motor and aviation gasoline

	Yield/weight percent gasoline			
	Referred to middle oil product		Referred to raw. material	
	Motor	Aviation type ON 87	Motor	Aviation type ON 87
Raw material				
Bituminous coal	92	83	62	55
Brown coal	92	81	50	44
Coke oven tar	93	84	74–78	67–71
L.T. tar, bit. coal	90	81	79	71
L.T. tar, brown coal	90	80	82	73
Crude oil	94	84	88	79

materially over that attainable with the ON 87 fuels. The isooctane and alkylate production in Germany was very small and more difficult than in the U.S.A. because of the absence of raw materials from a large petroleum refining industry which were obtained them as by products.

On the other hand one ring aromatics boiling above benzene have a high octane number and could be made available in the coal liquefaction plants. The flexibility of the coal liquefaction process with different catalysts readily permitted conversion into gasolines with low content of aromatics to the manufacture of highly aromatic aircraft fuels. In view of the high aromatics content of the gasolines obtained at 723 K or above with the old catalyst 3510, new catalysts with the specific purpose of producing highly aromatic gasolines were readily developed. Two catalysts reached commercial use.

In Ludwigshafen an active carbon specially prepared at high temperatures was impregnated with 15% chromium oxide and 5% vanadium pentoxide and used at 300 bar pressure. It converted bituminous coal middle oils into an aviation fuel with about 50% aromatics. The yield of such gasoline obtained at about 773 K was about 10 weight percent lower than that of gasolines obtained around 673 K because of higher formation of gasous hydrocarbons. In mixture with 10% isooctane and tetraethyllead it gave a valuable aviation fuel that, especially in rich mixture, gave high engine performance.

The other development occurred at the coal tar pitch hydrogenation plant of the Ruhroel GmbH at Welheim. For reasons of simplicity, this comparatively small plant used for the vapor phase hydrogenation the same pressure of 700 bar as that required in liquid phase for the conversion of pitch. Thus only one gas recycle circuit was used. This also required new catalysts for the hydrocracking of the middle oil to gasoline. The catalyst developed there consisted of "Terrana" with small amounts of HF, Zn, Cr, Mo and S. At the high pressure and a temperature of 773 K the supported catalyst with fullers earth as base retained its activity inspite of a high nitrogen content of the feed material and gave an aromatic rich gasoline. These aromatic aviation fuels gave satisfactory performance in supercharged engines. Both the catalyst referred to above and an operating pressure of 700 bar were selected as the newest developments for the 200–300 barrels per day United States Bureau of Mines coal-to-oil Demonstration Plant in Louisiana, Missouri, in 1947. The object of the plant was to test the suitability of the coal liquefaction process for American coals, and to use or develop suitable American equipment, machinery and instruments. Four typical American coals and lignites were hydrogenated successfully at 500 to 700 bar pressure with high liquefaction and moderate off-gas production. A good grade of aromatic type gasoline [19] was produced using the Welheim catalyst. The catalyst was reproduced at the Bruceton Station of the BoM using American ingredients and tested in experimental equipment with good result.

Another route to gasolines with a high aromatics content was the DHD process, a variation of the hydroforming process as developed at the same time in the USA. This used 3–10 bar hydrogen pressure to dehydrogenate gasoline with an alumina-molybdena catalyst with short reaction periods.

The DHD process was developed in Ludwigshafen in connection with previous attempts to dehydrogenate gasolines from coal liquefaction with a tungsten-nickel catalyst. The activity of this catalyst was maintained for a rather long time if the treatment severity remained low. But this catalyst could not be regenerated *in situ*. It was then found that the high content of naphthenes and the low content of sulfur of the hydrogenation gasolines from coal permitted reforming with the alumina-molybdena catalyst at high severity with comparatively long operating periods of 40–200 hours at 50 bar pressure.

In the DHD process only the fractions boiling above 353–373 K were treated to give a product with 65–70% aromatics. After mixing with light ends a DHD gasoline with 50% aromatics was obtained. The alumina used as support for 10% molybdenum as catalyst was precipitated under carefully defined conditions from aluminate as well as aluminium nitrate solutions. The latter had a slightly smaller hydrocracking activity but also less thermal stability.

The DHD gasolines had a higher octane number than the "aromatizaton" gasolines of the same aromatics content since their paraffinic components had a higher proportion of branched chain molecules. They had also better stability because of the lower content of olefins. This was achieved by an additional "refining" reactor in the DHD units with the same catalyst which saturated only the olefins at a temperature of 473 K.

The DHD process was preferred over the "aromatization" process especially in plants that did not use bituminous coal as raw material. Firstly, it did not require change over of the vapor phase "benzination" units to aromatization, with the consequent temporary production decrease. Further, the process was superior for raw materials such as brown coal or crude oils that are not as highly aromatic in nature as bituminous coal. Thus the DHD process could be called a forerunner of present day catalytic reforming processes that use supported noble metal catalysts of superior activity.

In addition to these last processes of aromatization, dehydrogenation or reforming of gasoline to improve the knock resistance, attempts were continued to improve hydrorefining catalysts such as 8376 towards reducing the hydrogen addition ability while maintaining the capacity to remove heteroelements, and to improve catalyst 6434 towards increased conversion to single ring aromatics and branched chain paraffins to obtain higher octane number gasolines.

At the ICI a fullers earth catalyst was developed which contained iron in addition to hydrofluoric acid. The gasoline obtained from coal tar oil had a two points higher octane number (77 MM) than that obtained with the 6434 catalyst [9]. Similar catalysts were found under the direction of G. Kaftal [20] by the ANIC in Italy for the hydrocracking of petroleum middle oils obtained by hydrogenation. In Ludwigshafen catalysts containing 20% FeF_3 on "Terrana" gave gasolines with a higher octane number but also a higher content of olefins.

Considerably greater increases of the aromatics content and improvements of the octane number were obtained with catalysts based on synthetic alu-

Figure 10. Reactor and high pressure stalls at Leuna

Figure 11. Line of high pressure stalls at Gelsenberg

minum silicates as support. Their operating temperature, throughput and off-gas formation were in the same range as used for catalyst 6434 or 5058. These catalysts did not come into use in commercial plants before World War 2 ended.

To give the reader an impression of the industrial plant Figure 10 shows a high pressure reactor at Leuna and the row of 18 high pressure stalls which housed these reactors since 1927. Figure 11 gives a similar picture of the Gelsenberg plant which started operation in 1939. The picture shows the crane and the fired heaters on each side of the twin stalls.

The catalyst and process development then branched out from the laboratories of the coal liquefaction plants to those of the petroleum refining industry and its suppliers. The catalysts were prepared to fill specific and even individual requirements. Increasingly new analytical techniques for the study of surfaces, their area, particle size, acidity, pore size, pore distribution, and other factors added to the art of catalyst preparation more incisive scientific tools. Thus hydrorefining, hydrocracking and reforming catalysts and processes under various names became a growing and indispensable part of the petroleum refining industry. It is the author's hope that these catalysts which have their roots in Pier's work on coal liquefaction will in the future help again to make coal an important factor in our supply of liquid fuels.

Acknowledgment

The author thanks Dr. Maria Hoering in Heidelberg for many valuable suggestions and a thorough review of the manuscript.

References

Summary Publications

(A) Catalytic Pressure Hydrogenation of Coal, Tars and Oil. Work of the "High Pressure Experiments Department" 1924 to 1945 at the BASF, Ludwigshafen/Rh. Germany, with foreword by M. Pier. 8 Vols. (English translation: FIAT Final Report) No. 1319, ATI No. 88, 723. March 1951)

(B) Kroenig, W.: Die katalytische Druckhydrierung von Kohlen, Teeren und Mineraloelen. Heidelberg: Springer-Verlag, 1950

(C) Hoering, M.; Donath, E. E.: In: Ullmanns Encyklopädie d. techn. Chem. **10**, 483, München-Berlin: Urban and Schwarzenberg 1958

(D) Donath, E. E.: Chemistry of Coal Utilization (Lowry, H. H., ed.). New York: Wiley 1963, p. 1041

(E) Wu, W. R. K.; Storch, H. H.: Hydrogenation of Coal and Tar. US Bureau of Mines Bulletin No. 633 (1968) (*c.f.* NTIS PB-233, 396, 1968). See also US Bureau of Mines Bulletin No. 485

Individual References

1. Berthelot, P. E. M.: Bull Soc. Chim. **11**, 278 (1869)
2. Fischer, F.; Tropsch, H.: Abhandl. Kohle **2**, 154 (1917)
3. Bergius, F.; Billwiller, J.: German Patent 301, 231 (1919)

4. Bergius, F.: Proc. World Pet. Congr. Vol. 2, p. 282 (1933)
5. Hughes, T. P.: Technological momentum in history: Hydrogenation in Germany 1898 to 1933. Past and Present No. 44, August 1969, p. 106. The Past and Present Society
6. Krauch, C.; Pier, M.: Zeit. Angew. Chem. **44**, 953 (1931): also presented at 3rd Internat. Coal Conf. Pittsburgh, 1931. Also German Patents 608, 466; 609, 538; 619, 739; 633, 185; 643, 141; 651, 203; 664, 563: all assigned to IG Farbenindustrie
7. French Patent 612, 504 (1925); German Patent 608, 467 (1926): both assigened to IG Farbenindustrie
8. Haslam, R. T.; Russel, R. P.; Asbury, W. C.: Proc. World Pet. Congr. Vol. 2, p. 302 (1933)
9. Gordon, K.: T. Inst. Fuel **9**, 68 (1935); **20**, 42 (1946)
10. See (A) above: Vol. 3
11. See (A) above: Vol. 7
12. See (C) above: p. 495
13. Strobel, B.; Koelling, G.: IGT Symp. Adv. Coal Utiliz. Technol. Louisville, 1978
14. Pott, A.; Broche, H.: German Patents 632, 631 (1927); 719, 439 (1936)
15. Uhde, F.: French Patent 800, 920 (1929)
Uhde, F.; Pfirrmann, T. W.: French Patent 819, 660 (1937)
16. Pier, M.: Oel u. Kohle **13**, 916 (1937)
Pier, M.: Z. Elektrochem. **53**, 291 (1949)
Reitz, O.: Chem.-Ing.-Tech. No. **21–22**, 413 (1949)
17. Donath, E. E.: Advances in Catalysis **8**, 239 (1956)
18. Hückel, E.: Adsorption u. Kapillarkondensation, Leipzig: Akademischer-Verlag, 1928
19. Hirst, L. L.; Clarke, E. A.; Chaffee, C. C.: US Bureau of Mines Rept. Invest. No. **4676** (1950); No. **4944** (1953); No. **5043** (1954)
20. Kaftal, G.: Chim. Ind. (Italy) **24**, 53 (1942)

Chapter 2

Catalytic Activation of Dioxygen

G. K. Boreskov

Institute of Catalysis, Novosibirsk 630090, USSR

Contents

1. Introduction

Dioxygen is the commonest oxidative agent and a considerable part of oxidation reactions with its participation proceed *via* heterogeneous catalysis. Many of these catalytic reactions form the basis for important industrial processes such as production of sulphuric or nitric acids as well as numerous oxygen containing organic compounds obtained by selective oxidation of hydrocarbons and other organic substances. The catalytic reactions of complete oxidation by dioxygen are extensively used for detoxication of organic substances and carbon monoxide from industrial and motor-transport exhaust gases. Some attempts have been made recently to utilize the catalytic oxidation of fuels for energy production. The interactions of dioxygen with dihydrogen or carbon monoxide often serve as model reactions for fundamental research in heterogeneous catalysis.

A. Electronic Structure of Dioxygen

In the dioxygen molecule, electrons coming from the two oxygen atoms occupy molecular orbitals:

$$(\sigma_1)^2(\sigma_1^*)^2(\sigma_2)^2(\pi)^4(\pi^*)^2$$

This corresponds to the binding order equal to 2 and to the presence of two unpaired electrons on the antibonding oribital, π^*. This is the main triplet paramagnetic state of dioxygen, $^3\Sigma_g^-$. Spin pairing of electrons on the orbital π^* leads to a singlet excited state of dioxygen, $^1\Delta_g$. The energy of this state is 94 kJ mol^{-1} with respect to the energy of the ground state; the radiation lifetime in vacuum is 2700 s and in the gas phase is 5×10^{-2} s at $p_{O_2} = 1$ atm. The energy of the second singlet state of dioxygen, $^1\Sigma_g^t$, is higher by 157 kJ mol^{-1} than that of its triplet state. The lifetime in vacuum is 7 s. The electron affinity of dioxygen is 92 kJ mol^{-1}. Electron binding producing O_2^- (peroxide ion) or O_2^{2-} (superperoxide ion) weakens the bond and increases the distance between the atoms (Table 1). On the contrary, ionization of dioxygen producing O_2^+ strengthens the bond.

Table 1. Binding energy of dioxygen and its ions

	Binding energy/kJ mol^{-1}	Binding order	Distance between atoms/Å	$\nu(O_2)$/cm^{-1}
O_2^+	640	2.5	1.12	—
O_2	494	2	1.21	1580
O_2^- superperoxide ion	394	1.5	1.26	1097
O_2^{2-} peroxide ion	—	1	1.49	802

B. Reactivity of Dioxygen

The high energy of dissociation inhibits the atom-by-atom addition of oxygen. For the atom-by-atom addition to be energetically favourable, it is necessary that each of the oxygen atoms has a binding energy equal to or higher than 250 kJ mol^{-1}.

Reactions of homogeneous oxidation by dioxygen in the gas phase always proceed *via* a chain mechanism. For example, the interaction of dihydrogen with dioxygen follows the scheme of branched chains involving H and OH radicals. The bond breaking in dioxygen is promoted by its interaction with hydrogen atoms possessing large energies compared to H_2. This energy is accumulated due to the formation of water in the preceding steps. Owing to the free energy evolved during water formation from H_2 and O_2, the concentration of radicals, which provide the reaction, significantly exceeds the equilibrium one.

The effect of the surface of some solids on chain reactions in the gas phase has long been known. It manifests itself in both chain termination due to the binding of the radical by a solid, and the generation of active species resulting from chemical reactions on the surface. Heterogeneous-homogeneous reactions involving chain generation on the surface and chemical transformations proceeding partly in the near surface volume of the gas phase have been studied in detail. Thus there exists a class of heterogeneous catalytic reactions of oxidation by dioxygen in which catalysts induce the chain reaction by generating the primary active particles — free radicals. The number of such reactions is, however, insignificant compared to the number of heterogeneous catalytic reactions of oxidation by dioxygen.

Quite often homogeneous liquid-phase oxidation by dioxygen also proceeds *via* the chain mechanism. Dioxygen participates in the chain process without dissociation, for example, *via* the scheme:

$$O_2 + \dot{R} \rightarrow \dot{R}O_2$$

$$\dot{R}O_2 + RH \rightarrow ROOH + \dot{R}$$

where RH is an oxidizable substance, $\dot{R}$ is a free radical, $\dot{R}O_2$ is a peroxide radical, ROOH is a hydroperoxide. The subsequent decomposition of the hydroperoxide with a weak bond between oxygen atoms yields stable reaction products and additional radicals which branch the reaction chain (so-called "degenerated" branching). Formation of primary radicals is a slow step which can be accelerated by catalysts.

Oxidation reactions catalyzed by enzymes do not proceed *via* the chain mechanism. The difficulty associated with a high energy of dioxygen dissociation seems to be obviated by a simultaneous binding of the both oxygen atoms. In the case of monoxygenases one oxygen atom is incorporated into a substrate (X) to yield an oxidized product, the other atom is reduced to give water. For this reaction to occur an "external" hydrogen donor is indispensable (DH_2):

$$X + O_2 + DH_2 \rightarrow XO + H_2O + D$$

The role of hydrogen donors is provided by specific compounds involved in complex cycles of conjugated reactions. Under the action of dioxygenases both oxygen atoms are incorporated into different places of the same molecule of a substrate.

In most cases the heterogeneous catalysis of oxidation by dioxygen on solid catalysts occurs with an equilibrium energy distribution in the system which excludes the possibility of the chain mechanism.

In the case of a more widespread *stepwise* mechanism the two oxygen atoms are bound simultaneously to a catalyst as a result of dissociative chemisorption of dioxygen. On transition metals and their compounds this process occurs at a high rate with a low activation energy. Since in this reaction a strong bond (500 kJ mol^{-1}) is broken in a dioxygen molecule, the high rate indicates a very high degree of energy compensation during dioxygen chemisorption. In the subsequent steps an oxidized substance reacts separately with chemisorbed oxygen atoms. A more difficult step is that of oxygen abstraction from the catalyst. When a stepwise mechanism provides a decrease of the reaction molecularity (*i.e.* of the number of reactant molecules that react *per* transformation of one active complex *at the most difficult step*), this is an advantage due to a decrease in the total energy of bond rupture in that step. In most cases this advantage results in a degree of compensation for the energy of bond rupture in all the reaction steps. A stepwise mechanism is predominant in oxidation reactions on solid catalysts at elevated temperatures. At temperatures close to ambient, oxidation by dioxygen often follows the *concerted* mechanism with dioxygen and molecules of the oxidized substance being involved simultaneously in the active complex on the catalyst surface. Some intermediate mechanisms are found to be also possible. A more detailed consideration of reaction mechanisms will be given later in relation to particular catalytic reactions.

A great number of solid substances, mainly those including metals with nonfilled *d*-shells, are catalytically active toward reactions of oxidation by dioxygen. Less active are substances containing metals with nonfilled *f*-shells. In the metallic state the most active are Pt, Ir and Ag. Nickel and cobalt are retained in the metallic state only in reaction mixtures with a considerable excess of oxidizable substance. Under these conditions their catalytic activity is very high. Owing to their stability and catalytic activity, transition metal oxides are most commonly used as catalysts for oxidation by dioxygen. Most catalysts used in industry for complete or selective oxidation represent various combinations of oxides, mainly of transition metals. Some other compounds of transition metals, for instance carbides, also show a high catalytic activity toward the above reactions.

2. Interaction of Dioxygen with the Surface of Solid Catalysts

A. Metals

A considerable amount of works has been devoted to understanding the mechanisms of dioxygen interaction with metal surfaces [1]. In many cases, however, the experiments were performed using insufficiently clean surfaces. Indeed, recent Auger and electron spectroscopy studies have shown that the pre-treatment procedures used in most of the earlier studies do not ensure complete removal of impurities from the surface, which should eventually lead to a certain misrepresentation of the events on the surface.

Recent years have been characterized by rapid accumulation of information concerning chemisorption on metals whose surface is purified by evacuation (10^{-8} Pa) and whose purity is examined by electron spectroscopy techniques. From these studies it is clear that oxygen chemisorption proceeds very rapidly on clean surfaces of most metals. At room temperature the sticking probability ranges 0.1 to 1 (Table 2) and is close to unity for many metals. This corresponds to a very small value of the activation energy of chemisorption. There are, however, some exceptions: silver surface [9, 10],

Table 2. Initial sticking probability (S_o) for oxygen chemisorption on metals (300 K, $p_{O_2} = 10^{-6}$–10^{-4} Pa)

Metal	S_o	Reference
W (100)	1.0	[64, 65]
Ni (100), (111), (110)	1.0	[66–69]
Ru (0001)	0.75	[19]
Ir (110)	0.25	[70, 71]
Pd (111)	0.3	[72]
Pt 6 [(111) × (100)][a]	0.12	[73]
Pt [10–20(111) × (100)]	0.1	[4]
Pt (111)	10^{-3}	[5, 4]
Pt (111)	7×10^{-7}	[74]
Ag (111)	3×10^{-5}	[75]
Au (111)	no adsorption	[11]
Au 6[(111) × (100)] and	of oxygen	
polycrystal.		[11, 12]

[a] designation of step surfaces proposed by Somorjai[79], Pt 6[(111) × (100)] describes 6 atomic rows of a terrace with orientation (111) followed by the step of a monoatomic height with orientation (100), etc.

Pt(111) and (100) planes [2–8] characterized by a far slower rate of oxygen chemisorption, and gold surface [11–12] on which the activation energy of chemisorption is high.

Chemisorption of oxygen in the molecular form (α-form) is observed on metals at low temperatures [13, 14, 15, 16]. As the temperature is raised to 200–300 K adsorbed molecular · oxygen either desorbs or transforms to a more stable atomically adsorbed stated (β-form). This latter fact prevents direct observations of oxygen molecular chemisorption. The molecular chemisorption of oxygen on platinum at 100 K has been studied most comprehensively [14, 15, 16]. The retaining of isotopic composition after adsorption and desorption gives evidence for the molecular form of adsorption. Desorption occurs according to the first order law with an activation energy of about 25 kJ mol^{-1}. It is likely that during molecular oxygen chemisorption on transition metals, a donor-acceptor interaction takes place between oxygen molecules and metal atoms. A resulting bond seems to be similar to the coordination bond in complexes containing molecular oxygen [17].

The role of molecular forms of adsorbed oxygen in catalytic reactions of oxidation on metals appears to be of minor importance. It may be proposed that the low temperature oxygen homomolecular isotope exchange observed on a number of metals at 200–300 K includes a step of molecular chemisorption. Clarkson and Cirillo [18] have found that on silver supported over porous quartz, oxygen is adsorbed in the molecular form at 300 K partially producing O_2^- (about 0.02%). The correlation between the concentration of O_2^- ions and the catalytic activity of silver has been established for CO oxidation reaction [18]. At room and higher temperatures dioxygen is chemisorbed on metals usually in a dissociative form. Figure 1 illustrates oxygen desorption curves taken after adsorption at 78 K. A low temperature peak corresponds to the molecularly adsorbed oxygen (α-form), and a high

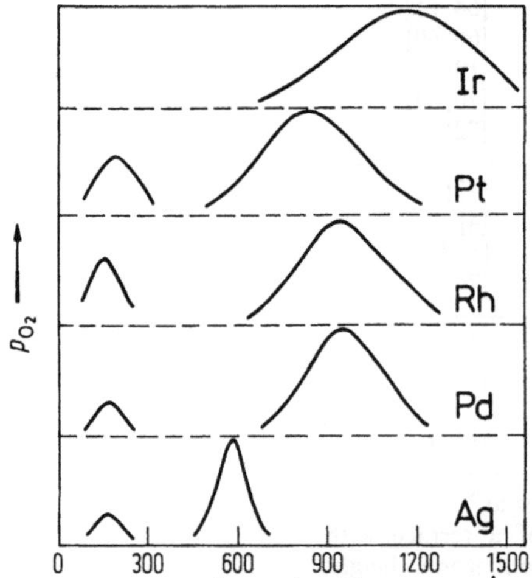

Figure 1. Low-temperature adsorption of oxygen on metals. (Desorption curves after oxygen adsorption at 78 K)

Table 3. Low-temperature adsorption of oxygen on Pt, Rh and Ir

Metal	Adsorption state	T_{des}/ K	E_{des}/ kJ mol^{-1}	$\Delta\varphi$/ eV
Pt	α-molecular	190	40	1.4
	β-dissociative	850	220	
Rh	α-molecular	150	37	0.9
	β-dissociative	950	240	
Ir	α-molecular	not detected		
	β-dissociative	1150	300	1.2

T_{des} — the temperature corresponding to the peak maximum in desorption spectrum,
E_{des} — the desorption activation energy calculated from the thermodesorption spectrum,
$\Delta\varphi$ — the change in the work function during oxygen adsorption at 78 K

temperature peak to dissociatively adsorbed oxygen (β-form). No molecularly adsorbed oxygen was detected on iridium. The main characteristics of adsorbed oxygen are compiled in Table 3. The heat of adsorption for the α-form is much smaller than that for the β-form [80, 81].

Investigation of dioxygen adsorption on the planes of monocrystals have shown that in a number of cases oxygen forms ordered adsorbed layers (Table 4). At small coverages in the region of ambient temperatures the chemisorbed oxygen produces a two-dimensional lattice whose symmetry is determined by that of a metal plane. The surface structure of the metal plane remains almost unchanged and oxygen atoms are covalently bound to the surface layer of metal atoms. These structures are usually called "ideal adsorbed layers". Their formation is not accompanied by the rupture of the interatomic bonds of metals. At ambient temperature ideal ordered adsorbed layers of oxygen are formed on Ni(111), Ru(0001), Pd(111), Ir(111) and Ag(110). At elevated temperatures an adsorbed oxygen layer may undergo reversible disordering which is the case for ruthenium and silver [19, 20]. The ideal layers of adsorbed oxygen on Pd(111) and Ir(111) can dissolve in the metal bulk upon heating [21, 22]. Oxygen chemisorption on Cu(111), Ag(111), (100) and Pt(110) does not produce ordered adsorbed layers even at ambient temperatures. In some cases the higher temperature disordering is not completely reversible.

An ideal adsorbed layer may be transformed to a two-dimensional surface oxide as the amount of adsorbed oxygen increases.

This phenomen is known as a reconstructive chemisorption since it leads to the breaking of the metal bond in the surface layer and to notable changes in the surface structure. For reconstructive chemisorption the valence state of metal atoms bonded to oxygen appears to differ significantly from the initial one, and approaches the valence state of metal atoms in bulk oxides. The changes in the diffraction pattern, heat of chemisorption, and properties of chemisorbed oxygen point to a transformation of the ideal adsorbed layer to the reconstructive form. Two-dimensional oxides are formed rather

Table 4. Surface structures produced by oxygen chemisorption on metals

Metals	surface structures	Reference
W (100)	(2 × 1)O, (5 × 1)O (4 × 1)O (2 × 2)O facets	[64, 65]
Ni (100)	p(2 × 2)O c(2 × 2)O (001)NiO	[66, 67]
Ni (111)	(2 × 2)O ($\sqrt{3} \times \sqrt{3}$)—R 30°—O (111), (100)NiO	[66, 68]
Ni (110)	(3 × 1)O, (2 × 1)O	[66, 69]
Ru (0001)	(2 × 2)O	[19]
Ir (110)	microfacets (111) 2(ID)O structure	[70, 71]
Ir (111)	(2 × 1)O diffraction pattern (2 × 2)	[24]
Pd (111)	(2 × 2)O ($\sqrt{3} \times \sqrt{3}$)R—30°—O	[72]
Pt (111) and step surfaces	(2 × 1)O, diffraction pattern (2 × 2)O	[73, 5, 4]
Ag (111)	not formed at p_{O_2} up to 10^{-3} Pa, $T = 300$ K; (4 × 4)O structure at $T = 450$ K, $p_{O_2} = 10$ Pa	[75, 76]
Au (111)	not formed at $T = 300$ K, structure of the surface oxide at $T = 800$ K	[11]

easily on the surface of nickel ($\theta > 0.25$), silver ($\theta > 0.33$) and ruthenium ($\theta > 0.5$). On Pd, Ir, Pt and Au the surface two-dimensional oxides are formed only at high temperatures. As evidenced by LEED data, oxygen reconstructive chemisorption manifests itself in several forms depending upon the conditions employed. At high temperatures reconstructive adsorption may lead to oxygen dissolving in the near-surface part of the metal bulk and to the further formation of a three-dimensional oxide, provided the reaction conditions correspond to its thermodynamic stability. The formation of this oxide is preceded by an induction period associated with the appearance of nuclei of the three-dimensional phase of the oxide (Figure 2). The induction periods are observed also during the transformation of a two-dimensional structure to another structure. Figure 3 illustrates the change in the sticking probability during oxygen chemisorption on Ni(110) at 300 K and on Ni(100) at 423 K. Formation of every new surface structure

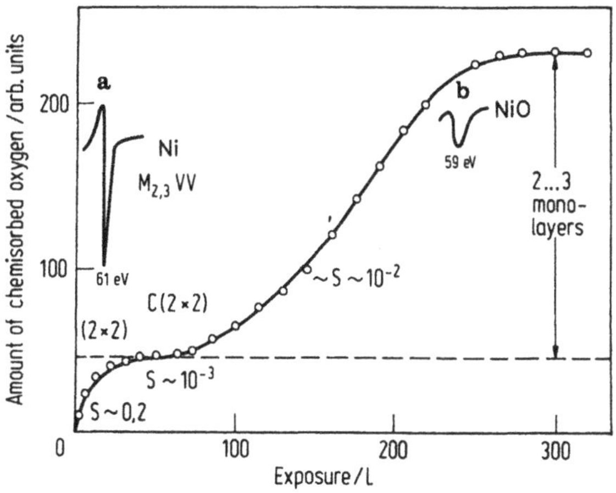

Figure 2. Kinetics of oxygen sorption on Ni (100) at 300 K. Auger-spectra of nickel are given for the region of **a** adsorption and **b** formation of a three-dimensional phase

is preceded by an induction period. Correspondingly, the curve of the sticking probability as a function of coverage shows several minima [23]. For most of the metals the sticking probability initially remains constant as the coverage increases. This may be due to the oxygen binding primarily in the so-called precursor state. Further, this oxygen moves along the surface to become more tightly bound at the boundary of the corresponding two-dimensional phase. The sticking probability drastically decreases as the surface coverage by this phase reaches its maximum.

The surface diffusion of chemisorbed oxygen atoms requires a certain activation energy, which ranges 55 to 80 kJ (g atom)$^{-1}$ [24–26] for metals, that is about 20% of the oxygen binding energy. The activation energy for surface diffusion seems to be related to the change of the oxygen binding energy along the surface. The calculated semiempirical values of the oxygen binding energy for Pd(110) [27] are displayed in Figure 4. The most advan-

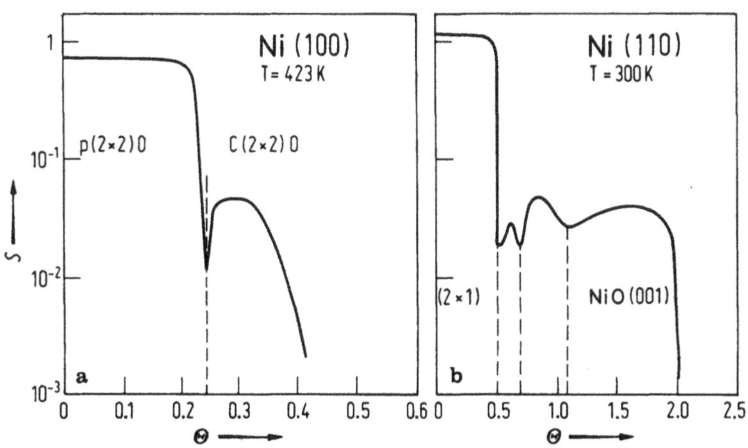

Figure 3. Sticking probability (*S*) *vs.* surface coverage (*θ*) for oxygen chemisorption on Ni (100) at 423 K **(a)** and Ni (110) at 295 K **(b)**

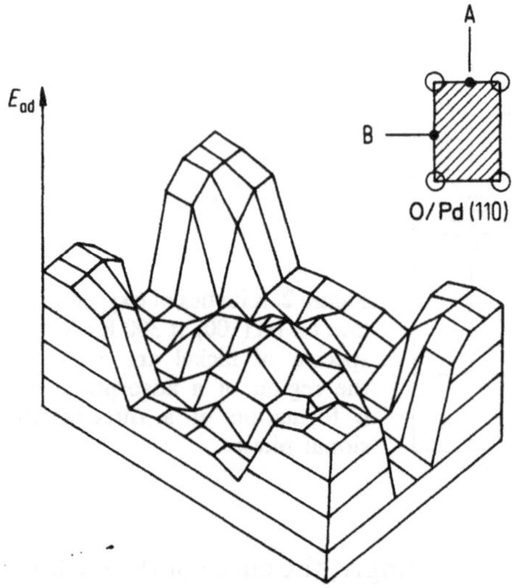

Figure 4. The energy of the oxygen-metal bond as a function of the oxygen atom position on Pd (110) [27]. (After Engle, T., Ertl, G.: Advan. Catal. **28**, 1 (1979) Fig. 26)

tageous are short bridged bonds where an oxygen atom is located between the two metal atoms, as is illustrated in Figure 5 [28]. It should be noted that the dissolved oxygen is far less reactive toward oxidizable substances than its chemisorbed form. Results that prove the formation of ordered adsorbed layers during chemisorption are of fundamental importance for heterogeneous catalysis. Moreover, in some cases they may even require the revision of the existing conceptions on the equilibrium and kinetics of chemisorption. Thus a very important problem is that of the investigation of oxygen chemisorption on clean metal surfaces especially under pressures used for practical performance of catalytic processes.

For catalytic reactions of oxidation by dioxygen of special significance is the value of the energy of oxygen binding to a metal, q_o kJ (g atom)$^{-1}$. This value can be found from the heat of dioxygen adsorption, q_{ads} kJ mol^{-1} and the energy of dioxygen dissociation to atoms equal to 500 kJ mol^{-1}.

$$q_o = 1/2(q_{ads} + 500) . \qquad (1)$$

The heat of adsorption can be determined by direct calorimetric measurements or from the temperature dependence of the equilibrium pressure at constant

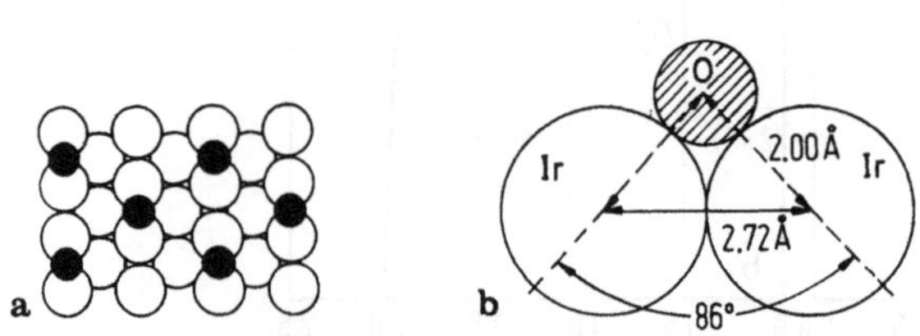

Figure 5. Oxygen adsorption on Ir (110): **a** atomic distribution of oxygen ● and iridium ○, **b** distance between the atoms and the bond angle [28]. (After Engle, T., Ertl, G.: Advan. Catal., **28**, 1 (1979) Fig. 25)

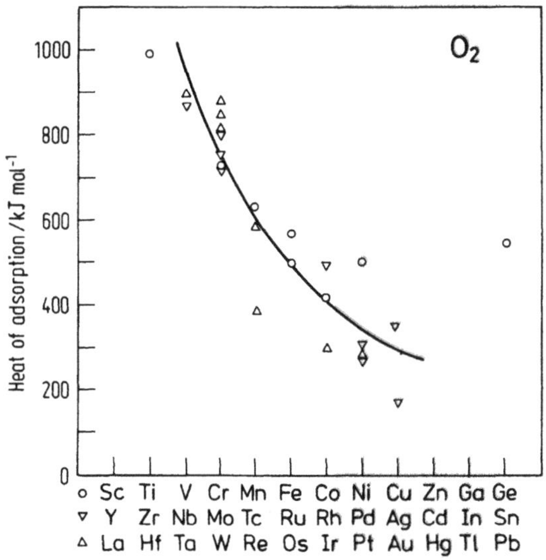

Figure 6. Heats of oxygen adsorption on polycrystalline surfaces of transition metals [29]. (After Toyoshima, J., Somorjai, G. A.: Catal. Reviews, **19**, 105 (1979). For silver new results have been added from [77, 78]

coverage *via* the Clausius-Clapeyron equation or from the thermodesorption data. This latter method allows the determination of the desorption activation energy:

$$E_{des} = E_{ads} + q_{ads} \quad (2)$$

which differs from the heat of adsorption by the value of the adsorption activation energy. Since oxygen is adsorbed on clean metal surfaces with a very low activation energy, *i.e.* $E_{ads} \approx 0$, one may assume that

$$E_{des} \approx q_{ads} \,.$$

The oxygen adsorption heats on polycrystalline samples of transition metals are shown in Figure 6 [29]. A distinct decrease in q_{ads} is observed with increasing the atomic number within individual periods, whereas within the groups, variation in chemisorption heats is negligible. The heat of adsorption on some metals increases as the heat of the formation of higher oxides per one oxygen atom is growing (Figure 7) [29]. It should be noted that for most metals the initial heats of chemisorption coincide with the heats of the formation of higher oxides. For Pt the heat of chemisorption is notably higher. For Ni the initial heat of chemisorption is lower: however, upon the prolonged exposure it approximates the heat of the formation of oxide.

The literature data on the chemisorption heat values on individual planes of monocrystals are rather scarce and equivocal. The difference in the heat values for different planes of the same metal are within the experimental error.

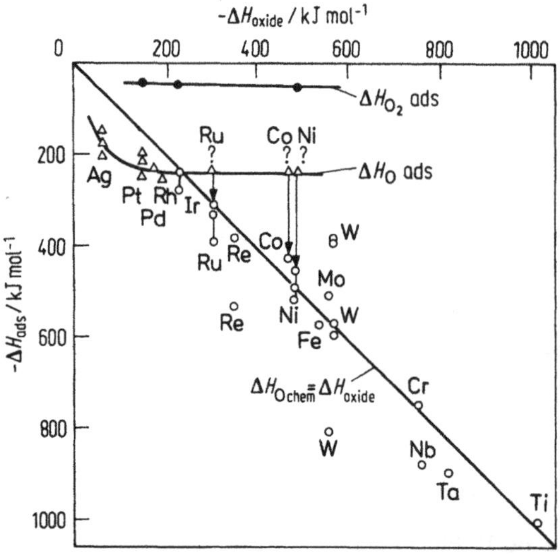

Figure 7. Heats of oxygen adsorption on transition metals as a function of the heat of the formation of higher oxide *per* 2 oxygen atoms. (After Savchenko, V. I.: IIIrd. All-Union Conference on the Mechanism of Catalytic Reactions, 1982)

Oxygen adsorption heats tend to decrease as the coverage increases (Figure 8) [30]. This may be associated either with surface nonuniformity or the interaction of adsorbed atoms or the change in the electronic properperties of the metal surface with the coverage increase.

B. Oxides

In the case of oxygen adsorption on oxides the initial state of the adsorbent remains undetermined since the concentration of oxygen in the bulk and, particularly, in the sub-surface layers of oxides does not correspond in a one-to-one manner to the stoichiometric one, and depends on the temperature and pressure of dioxygen. Treatment *in vacuo* to remove impurities from the surface significantly decreases the oxygen content of the sub-surface

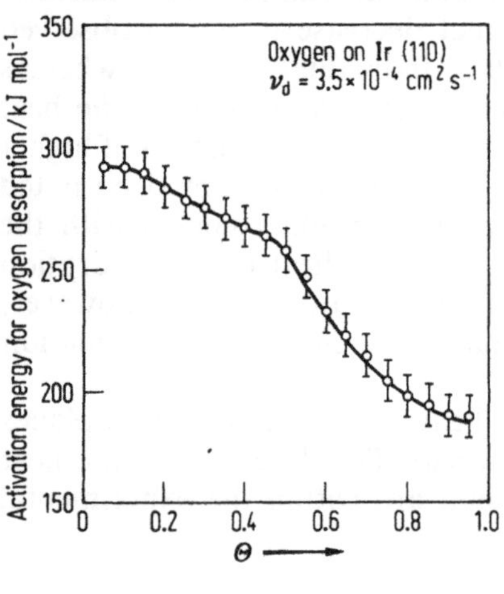

Figure 8. The activation of oxygen desorption from Ir (110) as a function of coverage. (After Taylor, J. L., Ibbotson, D. E., Weinberg, W. H. [30]; and Engle, T., Ertl, G.: Advan. Catal. **28**, 1 (1979)

layers. It is difficult to distinguish between the adsorbed oxygen and the oxygen of the oxide. Therefore the main objective of studies in this field has been to determine the oxygen binding energy on the oxide surface under certain pretreatment conditions.

To determine the oxygen binding energy on the oxide surface, the samples were typically heated at 773 K in vacuum to remove impurities and after this treated in dioxygen (1300 Pa) at the same temperature. The samples were then subjected to cooling down to 323 K and to one-hour evacuation [31–33]. These states were taken as standard. The oxygen binding energy on the surface of oxide was determined from the temperature dependence of the equilibrium oxygen pressure. The amount of oxygen evolved to the gas phase does not exceed 0.1% of monolayer during the measurements. The oxygen binding energy was also determined after certain oxygen amounts had been removed from the sample surface. The results for the standard states of oxides are summarized in Table 5. For iron and lead oxides a partial removal of oxygen does not affect the binding energy. For other oxides an appreciable increase in the binding energy is observed.

The determination of the oxygen binding energy on the surface of oxides was also undertaken by Joly [34] who obtained close values for MnO_2, Fe_2O_3, Co_3O_4, NiO and CuO. At the same time for TiO_2, V_2O_5, Cr_2O_3 and ZnO higher values were obtained which may be associated with different preparation and pretreatment procedures of the samples.

Direct calorimetric measurements of the heats of oxygen removal and adsorption were made for nickel, iron and cobalt oxides. The results obtained were found to be close to isosteric heat values [35]. A dependence of the binding energy of oxygen on its interaction with deeper oxide layers has also been established in this study [35]. For example, upon removal of oxygen from the surface of NiO the oxygen binding energy tends to increase rapidly. However, it notably decreases after a several hour exposure (Figure 9). It is likely that with oxygen removal from the surface, the metal

Table 5. Oxygen binding energy on the surface of oxides as a function of removed oxygen (kJ mol^{-1})

Oxide	Standard State	Amount of removed oxygen, % monolayer		
		1%	3%	5%
Co_3O_4	67	100	151	171
CuO	75	88	113	142
NiO	79	100	125	142
MnO_2	84	125	—	—
Cr_2O_3	109	109	109	155
Fe_2O_3	142	147	151	—
V_2O_5	180	217	—	—
ZnO	226	250 (0.5%)	—	—
TiO_2	247	—	—	—
PbO	167	167	167	—

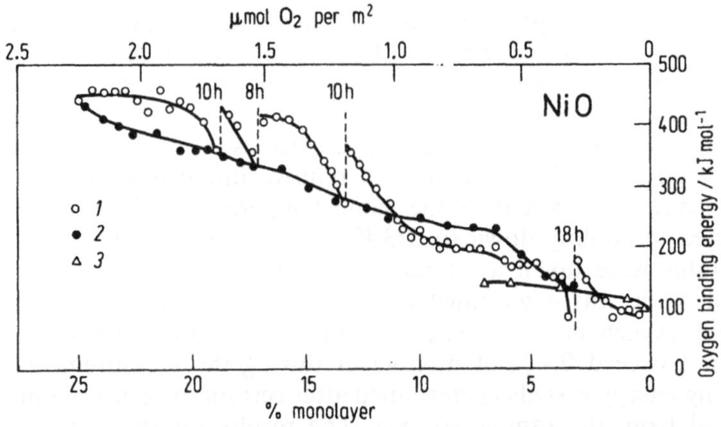

Figure 9. Oxygen binding energy on NiO surface *vs.* the amount of removed oxygen (% monolayer): 1 -- heat of oxygen chemisorption calculated from the heat of its interaction with CO; 2 -- heat of oxygen chemisorption on the reduced surface of the oxide; 3 -- heat of oxygen chemisorption calculated from isosters. (After [32])

cations are displaced into the oxide crystallites thus diminishing the oxygen binding energy on the surface. The thermodesorption method was employed by many scientists to determine the heat values of oxygen binding on oxides. At slow thermodesorption (a rate of temperature increase about 6 K min^{-1}) it is possible to attain equilibrium pressures and thus calculate the oxygen binding energies from isosters. For some oxides the oxygen binding energies vary stepwise depending upon the composition, *i.e.* they remain constant over certain ranges of oxygen concentration variations [36, 37].

In the case of flash desorption the velocity of the sample heating is two orders of magnitude higher (and exceeds 600 K min^{-1}). As a result, adsorption equlibrium is not attained, and the gas evolution is determined by the desorption rate. Proceeding from certain assumptions the desorption activation energy can be calculated from the temperature of gas evolution and the rate of heating. Germain and Halpern [38, 39] have observed several oxygen desorption peaks (from 1 to 3) for most of the oxides studied. This observation confirms the presence of surface oxygen states characterized by different binding energies. According to the increase in the activation energy of desorption, the authors denote these states as 0, 1, and 2. A comparison of oxygen binding energies (the data of ref. [32]) with desorption activation energies is given in Figure 10. The desorption activation energies must be greater than those of oxygen binding by the value of the activation energy of oxygen adsorption. On comparing the curves in Figure 10 it is readily seen that the values under consideration correlate with each other provided the desorption activation energy of state 0 is taken for Co_3O_4, NiO and CuO, and that of state 1 is taken for MnO_2, Fe_2O_3 and ZnO. This comparison shows also that the activation energy of oxygen adsorption is small and equals 17, 25 and 15 kJ mol^{-1} for Co_3O_4, NiO and CuO, respectively (that is, for catalytically most active oxides).

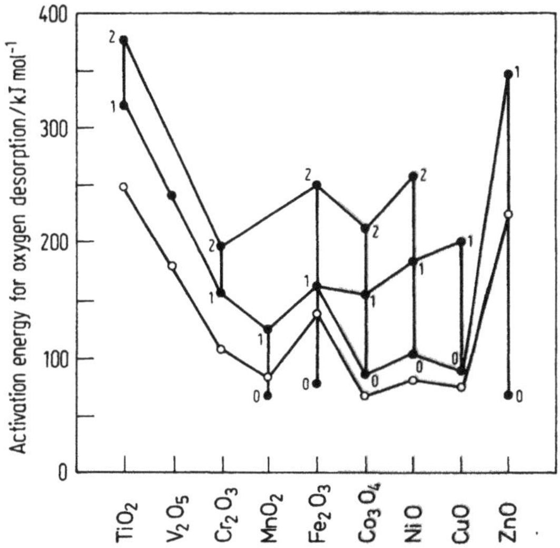

Figure 10. Comparison of the activation energies of oxygen desorption determined by flash-desorption [38, 39] with the heat of oxygen binding [31–33]
–●– 0, 1, 2 — desorption activation energies of oxygen states found by flash-desorption [38, 39],
–○– — oxygen binding energies according to refs. [31–33].
(After Halpern, B., Germain, J. E.: J. Catal. **37**, 44 (1975))

On oxide surfaces, oxygen can be chemisorbed in different forms: molecular or atomic, neutral or charged. Formation of negative ions O_2^-, O_2^{2-}, O^- and O^{2-} is found to be possible due to the high electrophility of oxygen.

The binding of one electron to an oxygen atom producing O^- is accompanied by the energy evolution of about 140.7 kJ (g atom)$^{-1}$. But when the molecular ion O_2^- is produced from O_2, the energy evolution comprises 84 kJ mol^{-1}. The binding of the second electron requires significant energy consumption in both cases. For example, when O^- transforms to O^{2-} the energy consumption is 636 kJ mol^{-1}.

During the course of formation of oxides the energy consumed is compensated by the Coulomb interaction between oxygen ions and metal cations. The larger the charges and the smaller the distances, the stronger is this interaction. The most advantageous in this respect is O^{2-}. For the doubly charged ions the energies of the Coulomb interaction may reach 4000 kJ ion^{-1}, *i.e.* the doubly-charged ions can easily compensate the energy consumed for O^{2-} formation.

In the case of oxygen binding to the surface of oxides the energy of the Coulomb interaction decreases due to a less complete coordination of the oxygen ion but still it remains sufficiently high. One may therefore suppose that in the course of adsorption by oxides the oxygen binding in a charged form also occurs as a result of the Coulomb interaction.

We now compare the energies of formation of O^- and O^{2-} ions on the surface. The energy of formation of one O^- ion from dioxygen is

$$q_1 = J_1 - 1/2D + K_1 - \varphi \quad (4)$$

and that of one O^{2-} ion is

$$q_2 = J_2 + J_1 - 1/2D + K_2 - 2\varphi \quad (5)$$

Here D is the energy of dioxygen dissociation equal to 497 kJ; J_1 is the energy of the oxygen atom ionization equal to 141 kJ; J_2 is the energy of the second electron binding to O^- equal to —626 kJ; K_1 and K_2 are the energies of the Coulomb interaction of O^- and O^{2-} with the catalyst; φ is the electron work function (all energies per mol, g atom or g ion as appropriate). The difference in the energies of the O^{2-} and O^- formation on the catalyst surface is expressed as:

$$\Delta q = J_2 + K_2 - K_1 - \varphi = -626 + K_2 - K_1 - \varphi \quad (6)$$

For approximate estimation let us assume that oxygen coordination on the surface is half that in the bulk and that the Coulomb interaction for singly-charged ions is one quarter of that for doubly-charged ions. Then $K_2 - K_1$ can be taken equal to 1500 kJ. Hence, for $\varphi < 900$ kJ, i.e. 9 eV, the formation of O^{2-} appears to be preferential. For oxides of transition metals of the 4th period the work function is smaller than 9 eV and oxygen interaction with the surface may be described by the equation

$$O_2 + 4\,Me^{n+} = 2\,O^{2-} + 4\,Me^{(n+1)+} \quad (7)$$

It is necessary to note, however, that the above estimates of the energies of the Coulomb interaction on the surface are too approximate to provide any reliable conclusions. Bielanski and Haber [40] have estimated these energies to be 1/6 of the energy of the interaction in the bulk and eventually proposed that singly-charged ions of oxygen, O^-, are formed on the surface. An interesting attempt has been made by Bielanski and Najbar [41] to determine experimentally the charge of adsorbed oxygen for NiO by measuring the ratio between the number of adsorbed oxygen atoms and Ni^{3+} ions formed. They have concluded that at low temperatures and with small residence times it is O^- that is predominantly formed. At 423 K with large residence times a mean charge of oxygen ions tends to increase to 1.5. In oxygen adsorption on MnO [42] and CoO [41] the oxygen charge was found to be 2. The formation of singly-charged forms of chemisorbed oxygen is facile when the concentration of transition metal cations that provide the role of electron donors is small. This is the case for supported oxide catalysts or diluted solid solutions for which the transfer of four electrons according to equation (7) is inhibited. The various forms of oxygen under consideration also may be present on a catalytic surface as intermediates with short lifetimes during oxygen transformation to O^{2-}. During the course of oxygen adsorption the energy of the Coulomb interaction falls whereas the work function increases, which results in a decrease of the adsorption heat.

The first investigations of adsorbed oxygen on the surface of oxide catalysts were based on the measurement of variations in electrical conductivity and work function [43–46]. The late 1950's and early 1960's were marked by many attempts to verify experimentally the ideas of the electronic theory of adsorption and catalysis using the above method [47]. It has been established that negatively charged oxygen species can be formed and the donor centres of oxides contribute to their formation. However this method did not allow an understanding of the nature and the determination of the charge of oxygen anions. This latter problem can be reliably resolved by using the ESR spectra typical for O_2^- and O^-. The O^{2-} and O_2^{2-} ions are nonparamagnetic, and neutral forms O_2 and O, although being paramagnetic, do not induce ESR signals on the surface of solids.

During adsorption of dioxygen over pre-reduced ZnO, TiO_2, SnO_2, ZrO_2 and silica-supported diluted oxides of transition metals, an ESR signal attributed to the molecular ion O_2^- has been observed. The signals are usually anisotropic, their anisotropy decreasing with the increase of the charge of the oxide cation. The hyperfine structure observed in a number of cases makes it possible to determine the binding site of an oxygen ion [49–55]. At elevated temperatures the ESR signal intensity of O_2^- gradually decreases, and above 450 K the signals as a rule disappear due either to desorption of O_2 or its transformation to some other form, most probably, O^{2-} [56].

During adsorption of dioxygen on oxide catalysts, along with the ESR signal from O_2^-, that from O^- is also observed. The species O^- alone is produced by adsorption of N_2O on oxides. Adsorbed O^- is highly reactive towards CO and hydrogen at low temperatures [57, 58, 59]. With the temperature rise, a signal attributed to O^- disappears, apparently, as a result of its transformation to a more stable anion O^{2-}.

Unfortunatelly the ESR technique is inapplicable to paramagnetic oxides, most of which are transition metal oxides important for catalysis. This necessitates the use of IR spectroscopy that allows detection of the surface molecular forms, namely, O_2 and O_2^-. The comparison of the ESR and IR data has suggested that the frequency of vibrations at 1050–1200 cm^{-1} in IR spectra corresponds to the interatomic vibrations in O_2^- ions [60]. These ions have been observed over chromium [61] and nickel [62] oxides. The IR spectroscopy technique has also enabled the observation of neutral molecules of dioxygen at low temperatures [60, 61]. A broad absorption band with several maxima in the 1500–1700 cm^{-1} region ascribed to these molecules is observable in IR spectra due to the polarization effect of the surface of a solid. These bands have been observed at low temperatures on TiO_2 and SiO_2 (oxidized and reduced), NiO, zeolites which contain nickel ions, supported molybdenum oxide catalysts, etc. [62, 63]. The binding of neutral dioxygen to the surface of oxides is weak, and desorption is observed over the temperature range from 300 to 400 K.

3. Isotope Exchange between Dioxygen Molecules

A. General Principles of Isotope Exchange

This simplest reaction of dioxygen corresponding to the reversible symmetric-asymmetric transformation of molecules has been studied in detail using ^{18}O:

$$^{18}O_2 + {}^{16}O_2 \rightleftarrows 2\,{}^{16}O^{18}O \tag{8}$$

The equilibrium constant of reaction (8) is close to 4, which corresponds to the random distribution of isotope atoms.

Since only a single chemical component, namely dioxygen, takes part in the reaction, the activity and kinetics of the reaction can be directly related to the state of oxygen on the catalytic surface.

Homogeneous reaction in the gas phase occurs at a high rate only in the presence of oxygen atoms

$$^{18}O_2 + {}^{16}O \leftrightarrows {}^{18}O^{16}O + {}^{18}O$$

With an equilibrium concentration of oxygen atoms, the rate of exchange is appreciable at temperatures above 1300 K. On the surface of solid catalysts the exchange occurs at far lower temperatures, sometimes even at the temperature of liquid oxygen. The data on the isotope exchange between dioxygen molecules have been reviewed in refs. [82, 83, 86, 96].

In the case of oxide catalysts, and metals, whose surface is covered by oxygen, the isotope exchange in dioxygen (homomolecular exchange) may be accompanied by heteroexchange of atoms between dioxygen and oxygen on the surface of solid catalysts, *e.g.*:

$$^{18}O_2 + {}^{16}O_{surf} = {}^{18}O^{16}O + {}^{18}O_{surf} \tag{9}$$

The rate of homoexchange can be easily determined experimentally when dioxygen and the surface layer of an oxide have equal concentrations of ^{18}O, (α) and (α_s), respectively. For this situation a change in the concentration of asymmetric molecules having mass 34 (C_{34}) is expressed as

$$\mathrm{d}C_{34}/\mathrm{d}t = K\frac{S}{N}(C^*_{34} - C_{34})$$

By integrating one obtains

$$K = N/St \ln (C^*_{34} - C^\circ_{34})/(C^*_{34} - C_{34}), \tag{10}$$

here K is the rate constant of homoexchange equal to the total number of exchanging molecules *per unit surface per unit time*, t is time, S is the catalyst surface area, N is the number of dioxygen molecules in the system, C_{34}, C°_{34} and C^*_{34} are the current, initial and equilibrium concentrations of asymmetric molecules, respectively [82].

Equation (10) is valid always independently of the exchange mechanism. The rate constant in equation (10) depends upon the temperature and pressure of oxygen. The type of these dependences is determined by the exchange mechanism and the catalyst nature.

The above method of determination of K is convenient and is in common use. However sometimes, especially with large amounts of exchangeable oxygen (*e.g.* on oxides) it is impossible to attain equal isotopic compositions of oxygen in the gas phase and on the catalyst surface. Moreover, simultaneous investigation of homo- and hetero-exchange may provide deeper insight into the reaction mechanism. Such types of experiments are carried out with the natural isotopic composition of a catalyst and dioxygen enriched with ^{18}O at a nonequilibrium ratio between symmetric and asymmetric molecules. Simultaneously, the total fraction of the heavy isotope in dioxygen (α) and the difference between the equilibrium and current concentrations of asymmetric molecules, $y = C_{34}^{*} - C_{34}$, can be measured. From the character of variations of these values it has been concluded that the homomolecular catalytic exchange in dioxygen may obey one of three mechanisms depending upon the nature and pretreatment procedure of catalysts [83–86].

I. Without the participation of oxygen of the catalyst surface. Hetero-exchange does not occur, α = const. Variation in y is illustrated in Figure 11.

II. With the participation of one atom of the surface oxygen in each act of displacement. Variations in α and y are illustrated in Figure 11.

III. With the participation of two atoms of catalyst oxygen in each act of exchange. Variations in α and y are illustrated in Figure 11.

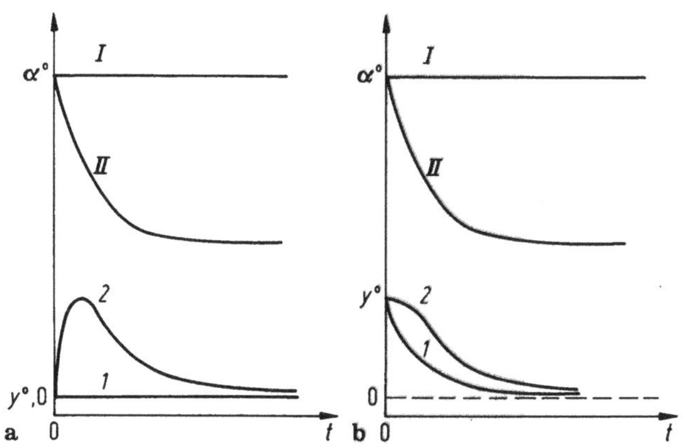

Figure 11. Variation of α and y with time in isotope exchange at eqilibrium (**a**) and nonequilibrium (**b**) distribution of symmetric and asymmetric molecules.
The Type I exchange -- curves I, I;
the Type II exchange -- curves II, I;
the Type III exchange -- curves II, 2

Table 6. Isotope exchange of oxygen on metals

Metal	Homomolecular exchange of oxygen						Exchange between adsorbed and gaseous oxygen			Ref.
	T = 300 K			T = 523 K			T = 523 K			
	p/ Pa	rate/ molec cm^{-2} s^{-1}	E/ kJ mol^{-1}	p/ Pa	rate/ molec cm^{-2} s^{-1}	E/ kJ mol^{-1}	p/ Pa	rate/ molec cm^{-2} s^{-1}	E/ kJ mol^{-1}	
Nickel (film)	13	100×10^{12} unstable activity which is recovered by treatment in vacuum at 723 K								[128–130]
Rhodium (film, several oxygen monolayers were adsorbed)	13	$(1–6) \times 10^{12}$ unstable activity which is recovered by treatment in vacuum at 523 K	12	13	30×10^{12}	12	13	0.11×10^{12}		[119], [130]
Palladium (film, several tens oxygen monolayers were adsorbed)				13	0.04×10^{12}	138	13	0.004×10^{12}	142	[126], [127]
Palladium (powder, several thousands oxygen monolayers were adsorbed, treatment at 823 K, p_{O_2} = 13 Pa	665	10^{12}								[135]

Iridium (film)	13	$(0.1–1) \times 10^{12}$ unstable activity which is recovered by treatment in vacuum at 523 K	25–42	13	10×10^{12}	25–42	13	0.3×10^{12}	100	[130]
Platinum (film)				66	2×10^{12}	67	66	10^{12}	96	[123], [124]
Platinum (powder)	1330	0.1×10^{12} (deactivated when heated to 523 K, activity is recovered by treatment in vacuum at 873 K)		1370	0.08×10^{12}	67				[134]
Silver (film)				13	0.29×10^{12}	134	13	0.26×10^{12}	130	[125]
Silver (powder)				665	2×10^{12}	134				[131–133]
Silver (powder)				800	8×10^{12}		760	3.5×10^{12}	146	[136], [137]
Gold (film)				13	0.0002×10^{12}	71				[129], [130]
Ag—Au alloy 67% at. Ag (film)				13	0.67×10^{12}	130	13	0.7×10^{12}		[129], [130]

The rate of homomolecular exchange on Au was obtained by extrapolating exp. values measured at 823–873 K.

B. Isotope Exchange of Dioxygen on Metals

The possibilitities of investigation are limited by the stability of metals to oxygen. Experimental data for homo- and hetero-exchange at 523 K are shown in Table 6. Within the temperature range from 200 to 900 K at oxygen pressures from 13 to 4000 Pa, no bulk oxide phases are formed on platinum, silver and gold. On other metals, oxygen is adsorbed in quantities corresponding to tens and even thousands of monolayers. Formation of three-dimensional oxide phases is also possible. Monomolecular isotope exchange is observed on Ni, Rh, Pd, Ir and Pt at room temperature and below. However, the catalytic activity is not always stable and may decrease during oxygen adsorption. Initial activity can be restored by treatment in vacuum at 500 to 900 K. During the process the metal surface appears to be blocked by the more strongly bound oxygen that does not desorb under experimental conditions. At high temperatures this oxygen either desorbs or dissolves in deeper metal layers.

The rate of isotope exchange of dioxygen with the oxygen adsorbed on the metal (hetero-exchange) rapidly falls as the extent of exchange increases. This indicates that oxygen is nonuniform on the surface of platinum metals. Less than 25% of the total amount of adsorbed oxygen takes part in isotope exchange with dioxygen from the gas phase at 473 K. The remaining adsorbed oxygen is exchanged at higher temperatures. On platinum films the initial rate of hetero-exchange is close to the rate of homo-exchange. On rhodium, palladium and iridium films the initial rates of hetero-exchange were found to be less than those of homo-exchange. On platinum films the main contribution to homomolecular isotope exchange is made by about 3% of monoatomic layer of oxygen;on rhodium, iridium and palladium this fraction being even smaller. The nonuniformity of adsorbed oxygen on platinum metals may be due to the structural nonuniformity of the metal surface. One may also anticipate the formation on metals of chemisorption complexes of oxygen of various structures.

On silver the hetero-exchange of the total amount of adsorbed oxygen occurs at a constant rate, which evidences for the uniformity of the adsorbed layer. On silver the rate of homo-exchange is equal to that of hetero-exchange. Furthermore, the apparent activation energies and reaction orders with respect to oxygen also coincide. One may therefore conclude that on silver the isotope exchange in dioxygen obeys the type III mechanism (adsorption-desorption).

Alternatively, Kajumov *et al.* [136, 137] have found that on clean and on selenium-promoted silver the rate of homo-exchange is twice as high as that of the exchange with adsorbed oxygen and they proposed that the exchange follows the type II mechanism. Further knowledge is needed to explain the reason for this existing disagreement between the different authors.

On gold the oxygen homomolecular exchange is almost unobservable below 823 K. At higher temperatures a slow exchange occurs, probably, due to the effect of impurities. As shown by the studies on oxygen isotope exchange over films of silver-gold alloys, the rate of exchange as a function of the composition of alloys passes through a maximum at 50–60% gold concentrations. The rate of exchange in the region of the maximum is several times higher than that on clean silver [129]. This suggests that the binding energy of oxygen on the surface of these alloys is lower than on the surface of clean silver.

C. Exchange on Oxides

1. Low Temperature Exchange Without Participation of Oxygen from the Catalyst Surface

The type I homomolecular exchange occurs at a high rate on some oxides pretreated in vacuum. This type of exchange is likely to occur through the formation of a 4-atomic oxygen complex on metal cations. Based on the orbital symmetry conservation rule, Sutula and Zeif [87] have shown that exchange reactions are most probable in complexes having a tetrahedral structure with their tops turned to the catalyst surface. These complexes are formed when an oxygen molecule is added to the ion-radical O_2^-. This latter is linked to the transition metal ion and is perpendicular to the surface. This type of exchange was first observed on zinc oxide treated at high temperatures [88, 89], and later for most of the simple and complex oxides [90, 91]. The only exceptions are the oxides whose oxygen is highly mobile in the bulk (V_2O_5, MoO_3, PbO, Bi_2O_3). The reaction proceeds at low temperatures (down to 78 K) with a low activation energy and low entropy of the active complex, which may result from the complexity of its configuration. Thermal pretreatment in oxygen leads to a decrease in the catalytic activity and finally to the complete deactivation of the catalyst at low temperatures. As an example see in Figure 12 the results obtained for gadolinium oxide [92]. The line BCD corresponds to the activity of the oxide treated in vacuum at 973 K with respect to the Type I homomolecular exchange of oxygen. The activity is high even at very low temperatures while the entropy of the

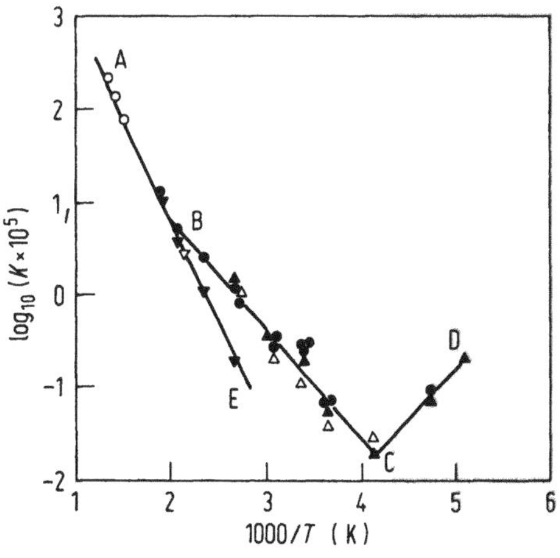

Figure 12. Isotope exchange of oxygen on gadolinum oxide:
Line BCD -- after treatment in vacuum at 973 K;
Line AE -- after heating in oxygen at 473 K

Table 7. Exchange of molecular oxygen on metal oxides

No	Oxide	Surface area/ $m^2 g^{-1}$	Temperature of pre-treatment in oxygen/ K	Temperature range of measurements/ K	$\log_{10} K_{573}$	Activation energy/ $kJ mol^{-1}$	React. order with respect to O_2	Type of mechanism	Ref.
1.	BeO	15.3	993	933–993	3.5	184	0	II	[96]
2.	MgO	9	873	753–803	8.5	105	1	II	[97]
3.	MgO	61.5	723	643–723	8.3	159	0	II	[96]
4.	Al_2O_3 (α)	15	873	773–873	5.0	188	0	II	[96]
5.	Al_2O_3 (γ)	124	873	423–623	10.0	50	1	I	[98]
6.	Al_2O_3 (γ)	124	873	623–748	7.1	155	1	II	[98]
7.	SiO_2	259	973	923–973	4.4	105	0.65	III	
8.	CaO	35	673	573–673	10.2	125	0.7	II	[96]
9.	Sc_2O_3	14.2	793	723–793	7.4	180	0	II	[96]
10.	TiO_2	67	873	803–873	6.0	~209	—	—	[99], [100]
11.	TiO_2 (R)	17	853	773–853	7.0	146	1	II	[96]
12.	V_2O_5	6	773	723–773	5.1	197	0.5	III	[101]
13.	Cr_2O_3	37	873	598–673	10.0	125	0.4	III	[102]
14.	Cr_2O_3	1.4	1473	623–743	9.0	171	0.3	III	[102]
15.	Cr_2O_3	14.2	643	543–642	9.7	142	0	II	[96]
16.	Mn_2O_3 (β)	24.3	823	473–573	12.9	84	—	III	[103]
17.	Mn_2O_3 (α)	4.2	553	473–553	13.6	96	0.3	II, III	[96]
18.	MnO_2	55	673	497–598	12.3	92	0.4	III	[99], [100]
19.	Fe_2O_3	20.0	773	543–643	10.0	117	0.5	III	[104]
20.	Co_3O_4	7.7	673	398–523	13.3	67	0.4	III	[99], [100], [105]
21.	NiO	1	1253	480–548	—	96	0.7	III	[106], [107]
22.	NiO	1	1253	548–663	11.4	130	0.7	III	[106], [107]
23.	CuO	18	673	523–623	11.4	109	0.4	III	[99], [100]
24.	CuO	10.3	603	503–603	15.1	92	0.5	III	[96]
25.	ZnO	1	823	698–798	8.0	167	0.9	II	[99], [100] [106], [108]
26.	ZnO	5.2	683	623–683	7.4	151	1	II	[96]
27.	Ga_2O_3	14.2	893	793–893	6.5	167	0	II	[96]

28.	GeO_2	0.8	973	873–973					[96]
29.	SrO	0.3	593	523–593	11.0	63	0.5	II	[96]
30.	Y_2O_3	3.6	673	523–673	11.1	75	0	II	[96]
31.	ZrO_2	10.4	723	623–723	8.5	125	1	II	[96]
32.	Nb_2O_5	4.2	973	873–973	4.8	171	1	II	[96]
33.	MoO_3	0.25	873	798–873	5.0	213	0.7	III	[105], [109]
34.	CdO	6	623	573–623	10.6	109	—	II	[110]
35.	CdO	5.5	623	523–623	10.8	84	0.9	II	[96]
36.	SnO_2	6.2	793	723–793	8.6	113	1	II	[96]
37.	Sb_2O_4	50	873	813–873	3.9	188	0.65	III	[111]
38.	La_2O_3 (C)	2.4	573	493–573	12.6	46	0	II	[96], [112]
39.	La_2O_3 (C)	28	773	673–723	14.0	105	1.1	II	[94]
40.	HfO_2	7.6	723	623–723	8.3	125	1	II	[96]
41.	Ta_2O_5	3	973	873–973	4.4	167	1	II	[96]
42.	WO_3	22.7	1063	993–1063	1.9	209	—	III	[96]
43.	WO_3	5.6	1073	973–1073	2.9	213	0.4	III	[114]
44.	IrO_2	4.6	473	423–473	15.5	100	0	II	[96]
45.	PbO	1.6	773	693–773	10.0	50	0.35	III	[96]
46.	PbO (α)	1.5	693	623–673	9.1	163	0.5	III	[110]
47.	PbO (β)	0.3	873	673–723	9.1	167	0.6	III	[110]
48.	Bi_2O_3	0.2	873	723–773	7.9	184	0.6	III	[110]
49.	ThO_2	5.9	693	593–693	9.7	92	0.8	II	[96]

K_{573} is the exchange rate constant at 573 K and p_{O_2} = 1330 Pa in molec cm^{-2} s^{-1}

Table 8. Activity of oxides of rare earth elements in isotope exchange of dioxygen

No	Oxide	Surface area/ $m^2 g^{-1}$	Temperature of pre-treatment in oxygen/K	Temperature range of measurements/ K	$\log_{10} K_{573}$	Activation energy/ $kJ mol^{-1}$	Order with respect to O_2	Type of mechanism	Ref.
1.	CeO_2	1.7	973	723–793	9.4	109	0.8	II	[96]
2.	CeO_2	14.8	973	683–823	12.7	121	0.85	III	[113]
3.	CeO_2	28	773	593–693	10.3	130	1.1	—	[94]
4.	Pr_6O_{11}	32	773	473–563	12.7	84	0.85	—	[94]
5.	Nd_2O_3	10.3	583	523–583	11.8	50	0	II	[112]
6.	Nd_2O_3	—	973	573–673	12.6	71	—	—	[115]
7.	Nd_2O_3	26	773	473–643	12.2	58	1.15	—	[94]
8.	Sm_2O_3	1.6	673	573–673	18.0	79	0	II	[112]
9.	Sm_2O_3	10.2	973	423–623	17.7	25	1	—	[94]
10.	Sm_2O_3	42	773	598–693	11.3	58	0.8	—	[94]
11.	Eu_2O_3	6.8	613	643–613	11.3	67	0	II	[112]
12.	Gd_2O_3	11.7	593	513–593	10.8	50	0	II	[112]
13.	Gd_2O_3	4.5	973	473–773	16.5	42	—	—	[92]
14.	Gd_2O_3	31	773	573–643	11.4	63	1.1	—	[94]
15.	Tb_2O_{3+x}	1.8	653	593–653	11.4	100	0.1	III	[112]
16.	Tb_4O_7	49	773	573–643	11.5	100	0.6	—	[94]
17.	Dy_2O_3	1.5	753	673–753	11.4	75	0	II	[112]
18.	Dy_2O_3	—	973	573–673	12.8	54	—	—	[115]
19.	Dy_2O_3	49	773	598–693	10.7	84	0.9	—	[94]
20.	Ho_2O_3	3	753	673–753	8.2	155	0	II	[112]
21.	Ho_2O_3	—	973	573–673	13.5	71	—	—	[115]
22.	Ho_2O_3	46	773	573–693	10.5	105	1	—	[94]
23.	Eu_2O_3	2.2	793	693–793	8.2	151	0	11	[112]
24.	Eu_2O_3	—	973	573–673	12.8	63	—	—	[115]

25.	Eu_2O_3	22	773	473–593	11.6	75	1	—	[94]
26.	Tu_2O_3	6.1	873	753–873	7.4	142	0	II	[112]
27.	Tu_2O_3	22	773	573–693	11.6	71	1	—	[94]
28.	Yb_2O_3	10.7	873	773–873	7.9	125	0	II	[112]
29.	Yb_2O_3	—	973	573–673	13	58	—	—	[115]
30.	Yb_2O_3	34	773	573–693	10.5	84	1.2	—	[94]
31.	Lu_2O_3	15.1	873	573–633	10.6	121	0	II	[112]
32.	Lu_2O_3	—	973	573–673	13	38	—	—	[115]
33.	Lu_2O_3	20	773	573–693	11.5	63	1	—	[94]

active complex formation is very small and the activation energy value is negative. Heating in oxygen at temperatures above 373 K leads to the complete deactivation of the catalyst at low temperatures (the line AE). At high temperatures the rate of the homomolecular exchange is equal to the rate of hetero-exchange with the oxide oxygen. Similar results were reported for zinc [90], nickel [93], lanthanum and samarium [94] oxides. γ-Al_2O_3 exhibits a more stable activity. When a freshly-treated sample is heated in oxygen, its catalytic activity falls by 2–3 orders of magnitude but still remains rather high. The catalyst is completely deactivated only after heating in the presence of water vapor.

This data suggests that heating in vacuum at high temperatures produces certain defects on oxide surfaces (oxygen vacancies, interstitial metal atoms) on which the homo-exchange occurs through the formation of active complex from two dioxygen molecules. Special emphasis should be placed on the possibility of the reaction occurring *via* the rupture of strong bonds in two oxygen molecules at very low temperatures with a small activation energy. On oxide catalysts the degree of compensation in the Type I exchange exceeds 98%. At elevated temperatures the oxygen effect eliminates these defects and thus leads to complete loss of the catalytic activity toward the type I isotope exchange. At high temperatures the isotope exchange on these oxides does occur involving oxygen from the catalyst surface, that is, *via* the Type II or III.

On silica supported vanadium pentoxide the formation of surface ion radicals of oxygen O_2^- and O^- was observed [95]. A proposal has been expressed that the homomolecular exchange occurs with the participation of $^{16}O^-$ in a charged 3-atom oxygen complex, thus:

$$^{16}O^- + {}^{18}O_2 \rightleftarrows ({}^{16}O^{18}O^{18}O)^- \rightleftarrows {}^{18}O^- + {}^{16}O^{18}O \qquad (11)$$

It should be emphasized that in all cases the formation of surface structural defects resulting from heating in vacuum or partial reduction is the necessary condition for the homomolecular exchange to occur at low temperatures.

2. High Temperature Exchange with Participation of Oxygen from the Catalyst Surface

On oxides heated in oxygen until an equilibrium oxygen content is reached in the subsurface layer, the rates of homo- and hetero-exchange differ not more than by a factor of two. Hence on these oxides the homo-exchange involves oxygen atoms of the oxide surface, *i.e.* proceeds *via* the Type II or III mechanisms. The only exception is alumina on which the homo-exchange is partly governed by the Type I mechanism even after heating in oxygen.

The results of a study on the high-temperature oxygen exchange are summarized in Tables 6–8 and in Figure 13. For transition metal oxides the Type III exchange involving two oxygen atoms from the catalyst surface was shown to be predominant. Such an exchange occurs by virtue of dissociative adsorption of dioxygen followed by desorption of molecules with a modified isotopic composition. If the surface atoms mix rapidly enough, the isotopic composition of desorbed molecules corresponds to the ratio of the surface concentrations of isotopes. Leveling of the isotopic composition occurs most often in two or three layers of the near-surface oxygen atoms. However, on some oxides (V_2O_5, MoO_3, PbO, Bi_2O_3), the diffusion inside the bulk is so rapid that the leveling of the isotopic composition occurs in the whole bulk of the oxide during the exchange process.

The rate of exchange of dioxygen atoms with the oxide oxygen (hetero-exchange) is described by the equation

$$-\frac{N}{S}\frac{d\alpha}{dt} = R(\alpha - \alpha_s) \tag{12}$$

The constant number of atoms of the heavy isotope is

$$N\alpha + N_k\alpha_s = \alpha_{eq}(N + N_k)$$

From this we derive

$$\alpha_s = \alpha_{eq}(\lambda + 1) - \lambda \cdot \alpha \tag{13}$$

By substituting (13) into (12) we obtain

$$-\frac{N}{S}\frac{d\alpha}{dt} = (1 + \lambda)R(\alpha - \alpha_{eq})$$

and integration leads to

$$\ln \frac{\alpha^\circ - \alpha_{eq}}{\alpha - \alpha_{eq}} = -(1 + \lambda)\, Rt\, \frac{S}{N}$$

or (14)

$$H = \frac{\alpha^\circ - \alpha_{eq}}{\alpha - \alpha_{eq}} = e^{-(1+\lambda)\frac{S}{N}Rt}$$

Here α°, α and α_{eq} are respectively the initial, current and equilibrium concentrations of the heavy isotope in the gas phase, α_s is the concentration of the heavy isotope on the oxide surface, t is the time, R is the hetero-exchange rate constant (molecules O_2 s^{-1} cm^{-2}), $\lambda = \frac{N}{N_k}$ is the ratio of the total number of oxygen atoms (of the both isotopes) in the gas phase to the number of exchangeable oxygen atoms in the oxide, S is the surface area of the oxide [91].

The main difficulty consists in the determination of N_k. If the surface oxygen exchanges with the oxygen of the bulk oxide much faster than with that of the gas phase, N_k equals N_V, that is, the total number of oxygen atoms in the oxide. In the inverse case N_k equals N_s, that is, the number of oxygen atoms on the surface and in the near-surface layers showing the same rate of exchange. This number can be determined by way of selection so that the experimentally found value $\ln H = \ln\left(\frac{\alpha^\circ - \alpha_{eq}}{\alpha - \alpha_{eq}}\right)$ would linearly depend on t. The value of α_{eq} in equation (14) changes with the variation in N_k. If the rate of diffusion into the bulk is comparable with that of exchange with dioxygen, it is necessary to take into account the rate of oxygen diffusion in the oxide bulk as was done in refs. [120, 121]. When the oxygen of the surface layer is non-uniform and the exchange between individual sites is slow, one obtains

$$-\frac{N}{S}\frac{d\alpha}{dt} = \sum_{i=1}^{n} Z_i R_i (\alpha - \alpha_i) \tag{15}$$

Here α_i is the concentration of the isotope at the sites of the i-th type, Z_i is the fraction of such sites, R_i is the specific rate of hetero-exchange at these sites, n is the number of sites. For the continuous distribution, summation is substituted by integration. With a non-uniform oxide the linear dependence of ln H on t is not achieved. However, deviations from linearity make it possible to estimate qualitatively the non-uniformity of surface oxygen. Quantitative estimation is possible in the case of significant non-uniformity [122].

The reaction activation energy can be calculated from the temperature dependence of the exchange rate constant. For a non-uniform oxide the activation energy of exchange increases with the extent of exchange.

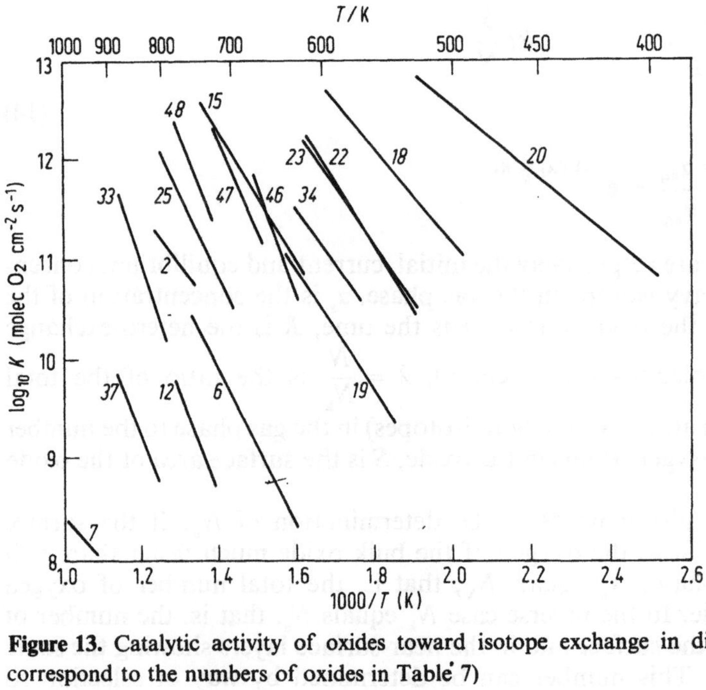

Figure 13. Catalytic activity of oxides toward isotope exchange in dioxygen. (The figures correspond to the numbers of oxides in Table 7)

The rates and activation energies of high temperature homomolecular exchange in dioxygen and of hetero-exchange vary within a wide range depending upon the nature of oxides. Figure 14 illustrates the correlation between the activation energy of isotope exchange and the oxygen binding energy on the catalyst surface. The higher the binding energy the larger is the activation energy of the exchange, and correspondingly, the smaller is the logarithm of the exchange rate. One may therefore conclude that for all the oxides studied the most difficult step of the exchange is that of oxygen displacement from the catalyst surface. Points that correspond to oxides characterized by the Type III mechanism are on the same line and the tangent of the angle of inclination of this line is close to unity. This means that the observed activation energies of the exchange are close to the energies of oxygen bonding to the catalyst surface.

As will be shown below, the dependence of catalytic reaction rates on oxygen binding energies on the catalyst surface is a common feature of a number of oxidation reactions, which in many cases allows prediction of the catalytic action.

In the case of the Type III mechanism, the dependence of the rate of exchange on oxygen pressure corresponds to a reaction order close to 0.5. To explain this dependence it has been postulated [83] that, besides strongly bound atomic oxygen, there is also less strongly bound oxygen adsorbed on the other surface sites, probably, over the layer of the strongly bound oxygen. Both forms of adsorbed oxygen exchange between each other rather rapidly and are in equilibrium. Such processes of oxygen adsorption and desorption can be described by the following equations:

$$O_2 + [\ ] + (\ ) \underset{K_2}{\overset{K_1}{\rightleftarrows}} [O] + (O)$$

$$[O] + (\ ) \underset{K_4}{\overset{K_3}{\rightleftarrows}} [\ ] + (O)$$

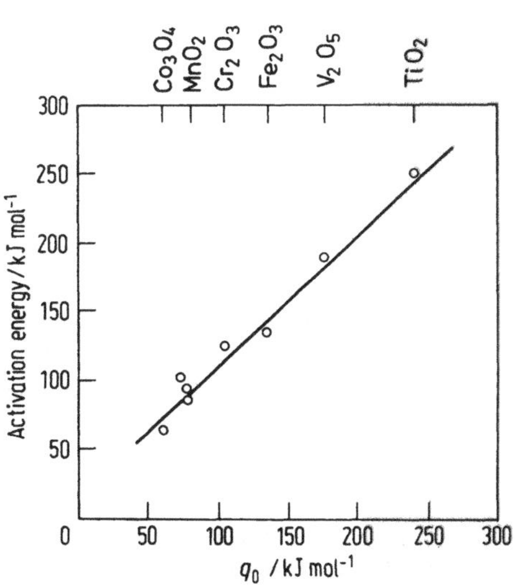

Figure 14. Activation energy of isotope exchange in dioxygen proceeding *via* the Type III mechanism *vs.* oxygen binding energy on the surface of oxide

Here [O] and (O) are the more strongly and less strongly bound oxygen respectively, and [ ] and ( ) are the respective surface sites. K_i stands for the rates of the corresponding elementary processes,

$$\theta = \frac{[O]}{[O] + [\ ]} \quad \text{and} \quad \theta_1 = \frac{(O)}{(O) + (\ )}$$

are the coverages of the sites by the more and less strongly bound oxygen. · Then the balance equation between the gaseous and adsorbed oxygen will be

$$K_1(1 - \theta)(1 - \theta_1)p = K_2\theta\theta_1 \tag{16}$$

and between the more strongly and less strongly bound oxygen will be

$$K_3\theta_1(1 - \theta_1) = K_4\theta_1(1 - \theta) \tag{17}$$

Here p is the oxygen pressure.

From equation (17) it follows that

$$\frac{\theta}{1 - \theta} = \frac{K_4}{K_3}\frac{\theta_1}{1 - \theta_1}$$

By substituting this expression into equation (16) we obtain

$$\frac{\theta}{1 - \theta} = \sqrt{\frac{K_1K_4}{K_2K_3}}\sqrt{p} = a\sqrt{p}$$

where $a = \sqrt{\frac{K_1K_4}{K_2K_3}}$

The rate of exchange equals the rates of adsorption and desorption

$$r = K_1(1 - \theta)(1 - \theta_1)\,p = \frac{K_1}{a}\,\theta(1 - \theta_1)\sqrt{p}$$

For large coverages by more strongly bound oxygen and small coverages by less strongly bound oxygen ($\theta \approx 1$, $\theta_1 \approx 0$)

$$r = \frac{K_1}{a}\sqrt{p}$$

which is in agreement with the experimental data.

In the case of aluminum, zinc, cadmium, alkaline earth metal oxides and some others, the exchange follows the Type II mechanism which involves only one oxygen atom of the oxide surface. The reaction order with respect to oxygen is close to unity. The exchange is likely to occur through the formation of a three-atom complex from an adsorbed molecule and an oxygen atom of the adsorbed layer (Eley-Rideal mechanism)

$$^{18}O_2 + |^{16}O| \rightarrow |^{18}O^{18}O^{16}O| \rightarrow {}^{16}O^{18}O + |^{18}O|$$

On oxides of the rare earth elements the isotope exchange in dioxygen proceeds at a high rate (Table 8) mainly *via* the Type II mechanism (although

Table 9. Catalytic activity of complex oxides in isotope exchange of dioxygen

No	Oxide	Specific area/ $m^2\ g^{-1}$	Temperature of pre-treatment in oxygen/K	Temperature range of measurements/ K	$\log_{10} K_{573}$	Activation energy/ $kJ\ mol^{-1}$	Order with respect to O_2	Type of mechanism	Ref.
1.	$MgFe_2O_4$	18	773	623–673	10.0	79	1	II	[104]
2.	$CoFe_2O_4$	2.8	773	573–623	10.4	109	0.8	II	[104]
3.	$NiFe_2O_4$	20	773	548–673	10.1	96	1	II	[104]
4.	$ZnFe_2O_4$	9.0	773	673–723	9.9	100	1	II	[104]
5.	$ZnMn_2O_4$	2.2	832	523–573	10.5	71	0.7	II	[116]
6.	$ZnCr_2O_4$	32	873	598–648	9.1	146	0	III	[116]
7.	$ZnCo_2O_4$	9.0	523	423–473	12.6	71	0.5	III	[116], [117], [118], [119]
8.	$ZnCoCrO_4$	83	673	523–623	9.5	138	0	III	[118]
9.	$CoMoO_4$	1.6	873	773–823	7.3	176	0.8	III	[105]
10.	$Fe_2(MoO_4)_3$	1.6	873	773–823	5.7	201	0.5	III	[105]

the Type III is also possible). On some oxides subjected to the high-temperature treatment, the low temperature exchange *via* the Type I mechanism cannot be ruled out either.

The isotope exchange has been studied on some complex oxides, *e.g.* ferrites, chromates and molybdates (Table 9) and on vanadium pentoxide — molybdenum trioxide mixtures [109]. The catalytic activity of spinels is mainly determined by the nature of the three-charged cations. The role of the doubly-charged cations was found to be less important. The influence of cations is also insignificant in molybdates.

4. Oxidation of Dihydrogen

The dihydrogen oxidation reaction

$$H_2 + {}^1/_2\, O_2 = H_2O \tag{18}$$

is known to be exothermic (ΔH°_{298} = —242 kJ mol^{-1}) and almost irreversible up to 1300 K (ΔG°_{298} = —248 kJ mol^{-1}, ΔS°_{298} = —10.6 e.u.). In contrast to homogeneous reactions, the catalytic oxidation on solid catalysts occurs with an equilibrium energy distribution. However, at elevated temperatures and with certain reaction mixture compositions, a chain reaction also may occur in the gas phase near the catalyst surface. Transition metals, their oxides and other compounds stable to oxygen are known to be the most active catalysts.

Table 10. Oxidation of dihydrogen on metals in excess hydrogen in the reaction mixture [138–140, 211]

Catalyst	Surface area/cm^2	SCA/ml cm^{-2} hr^{-1} (453 K, conc. O_2 = 1 %)	Activation energy/ kJ mol^{-1}	Reaction order with respect to O_2 at 453 K
Pt	71.1	74	46.0	1
Pd	31.6	63	46.9	1
Ni	305	30.0 (at small conc. O_2)	22.2	1
		2.4 (at large conc. O_2)	58.6	0
Rh	—	16	—	—
Co	252	14.1 (at small conc. O_2)	31.0	1
		4.2 (at large conc. O_2)	—	0.33
Fe	400	1.9 (at small conc. O_2)	18.0	1
		0.15 (at large conc. O_2)	41.8	0.6
Au	—	0.28	—	—
Ag	—	0.22	—	—
Cu	522.5	0.4	50.6	0 (408 K)
		0.2	—	0.6 (527 K)
Mn	312	0.25	53.6	0.4
V	99.5	0.19	40.6	0.6
Cr	50	0.16	39.7	0.4
Ti	2000	4.5×10^{-3}	—	—
Zn	1060	5.5×10^{-3}	—	—

SCA = *S*pecific *C*atalytic *A*ctivity

A. The Catalytic Properties of Metals

Some data [138–140] on the activities of various metals in the oxidation of dihydrogen (453 K, 1% O_2 in the mixture with H_2, atmospheric pressure) are compared in Table 10. Of the metals listed in the Table the most active are Pt and Pd on which the oxidation is fast even at 77 K. The specific catalytic activity (SCA) of Ni, Co and Fe tends to decrease when the oxygen concentration is increased in the reaction mixture. This effect is probably due to the irreversible oxidation of the surfaces of these metals. The data reported for the latter five metals are not completely reliable because the experiments were performed on insufficiently clean metal surfaces. Os, Ir, and Ru have also been found to be highly active.

The specific catalytic activity of metals is determined by their electronic structure, that is, by their position in the Mendeleev's periodic table. The most active are the metals of long periods, their activity increases with the atomic number. The activity sharply decreases when proceeding to the group IB metals, that is, with the complete occupancy of *d*-shells. This dependency is associated with the change of the chemisorption properties and the energy of binding of the dissociatively adsorbed oxygen and hydrogen to the metals. The correlation between SCA and the energy of oxygen binding to the metal surface has been established and reported in (see Figure 15). Moreover a similar change in the SCA of metals with respect to hydrogen isotope exchange [178] indicates that the dissociative adsorption of hydrogen also plays a specific role in hydrogen oxidation.

Metallic catalysts are used to purify hydrogen from oxygen, oxygen from hydrogen and inert gases from oxygen by means of precise dosing of hydrogen. If the reaction occurs in excess hydrogen, nickel catalysts can be used. Platinum and palladium catalysts are used for mixtures enriched with oxygen. For practical purposes platinum and palladium (usually less than 1%) are supported on various carriers such as silica, alumina and others.

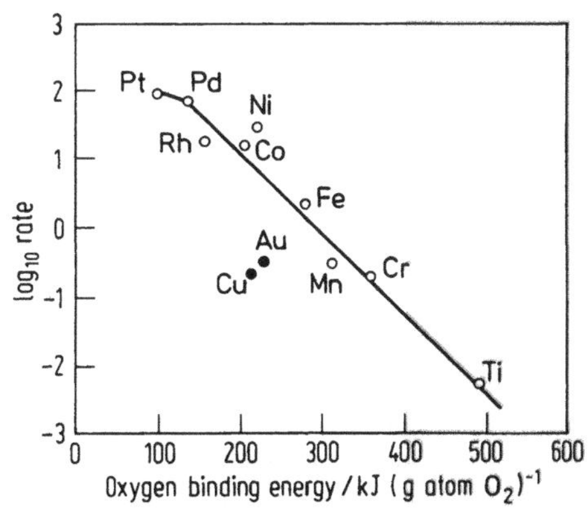

Figure 15. SCA of metals in dihydrogen oxidation *vs.* the energy of oxygen binding to the surface

1. The Catalytic Activity of Platinum

The catalytic activity of platinum in hydrogen oxidation has been studied by many scientists beginning with Davy [141], Döbereiner [142] and Faraday [143]. On platinum dihydrogen is adsorbed dissociatively with an adsorption energy of 65 ± 2 kJ mol^{-1}. At room temperature the sticking probability on Pt(100) is 0.17 and complete coverage corresponds to 4.6×10^{14} atoms cm^{-2} ref. [191]. It does not seem possible to draw any rigorous conclusions concerning the kinetics of hydrogen oxidation on platinum at atmospheric pressure unless one takes into account the effect of the heat and mass transfer processes as well as the reactants' influence on the catalytic properties. A study [144] carried out by using the gradientless method at 7–100 kPa and 323–453 K has shown that in the absence of a distortion effect of macrofactors, in particular conditions a system may have several stable steady states, probably, due to the reactants' effect upon the catalyst. In excess hydrogen (oxygen concentration less than 1 %) the reaction is first order with respect to oxygen. In excess oxygen at temperatures below 373 K first order with respect to hydrogen is observed. However, the system is in an unstable steady state and a temperature rise leads irreversibly to a fall of the reaction rate and to a change of the kinetic dependences. Proceeding from the stoichiometric mixture in nitrogen (2.7 % O_2 and 5.4 % H_2) and gradually substituting nitrogen by oxygen, the reaction rate exhibits a maximum at 20 % O_2. When substituting nitrogen by hydrogen the reaction rate passes through a minimum (Figure 16). For a stoichiometric mixture the activation energy is 28 kJ mol^{-1}. In excess hydrogen the kinetic isotope effect with respect to deuterium is 1.6 and 2 at 453 K and 373 K, respectively. This suggests that hydrogen is involved in the rate determining step.

The two platinum states, oxidized and reduced, differing in their catalytic characteristics have also been observed by Gentry *et al.* [209]. As found by Hanson and Boudart [169] for silica supported platinum, in an excess of one of the components the reaction is first order with respect to the other (deficient) component. In excess hydrogen and oxygen the activation energy is 14.2 kJ mol^{-1} and 7.5 kJ mol^{-1}, respectively.

The same dependences of the rate of hydrogen oxidation on the concentration of both reagents have been also observed by Marshneva [348] for impregnated silica gels over the temperature range 252–293 K. The reaction activation energy was found to be 32.6 kJ mol^{-1} both for excess hydrogen

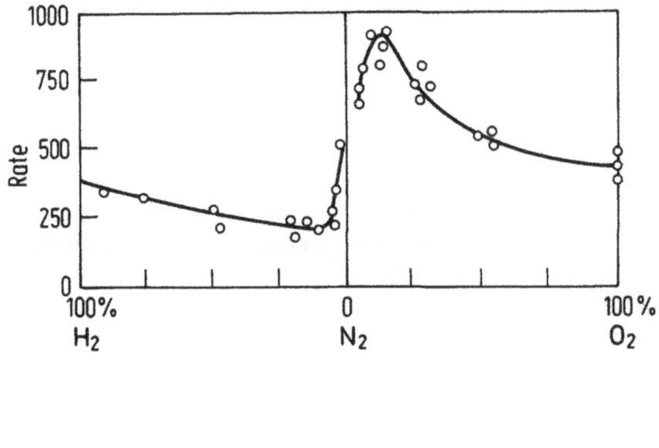

Figure 16. The rate of dihydrogen oxidation on platinum *vs.* the reaction mixture composition

and oxygen. Great interest has been shown, especially in the last few years, in the mechanism of reactions proceeding at low pressures. For example, Kuchaev and Temkin have studied the reaction mechanism by the mass-spectroscopic method at 10^{-3} Pa. Simultaneously they have also measured surface concentrations using the ion-ion emission method [145]. The authors' conclusion is that the reaction occurs *via* two pathways. In excess hydrogen it occurs through the interaction of atomically adsorbed hydrogen with oxygen

$$O_2 + 2[\ ] \rightarrow 2[O]_{ad} \qquad A$$
$$H_2 + 2[\ ] \leftrightarrows 2[H]_{ad} \qquad B$$
$$[O]_{ad} + [H]_{ad} \rightarrow [OH]_{ad} + [\ ] \qquad C$$
$$2[OH]_{ad} + H_2 \rightarrow 2\,H_2O + 2[\ ] \qquad D \qquad (19)$$

where [ ] represents an adsorption site, while in excess oxygen — through the interaction of gaseous dihydrogen with atomically adsorbed oxygen:

$$O_2 + 2[\ ] \rightarrow 2[O]_{ad} \qquad A$$
$$H_2 + [O]_{ad} \rightarrow H_2O + [\ ] \qquad B \qquad (20)$$

Generally, coupling of these mechanisms leads to a rather complicated kinetic equation. At the same time, according to ref. [144], the limiting conditions are described by first order kinetic equations with respect to the pressure of the deficient component.

The conception that only dissociatively chemisorbed oxygen is involved in the reaction is accepted by most of the investigators. However, the authors of ref. [146], who studied the low pressure kinetics (10^{-1}–10 Pa) over a wide temperature range (195–1273 K) on platinum pretreated in vacuum (10^{-8} Pa), have offered a unique interpretation. It implies that the reaction involves the interaction between adsorbed molecules of dioxygen and dihydrogen:

$$[O_2]_{ad} + [H_2]_{ad} \leftrightarrows [H_2O_2]_{ad} + [\ ]$$
$$[H_2O_2]_{ad} + [H_2]_{ad} \rightarrow 2\,H_2O + 2[\ ] \qquad (21)$$

This scheme corresponds to the kinetic equation

$$r = \frac{K b_{O_2} b_{H_2}^2 p_{O_2} p_{H_2}^2}{(b_{O_2} p_{O_2} + b_{H_2} p_{H_2})^3} \qquad (22)$$

which was not, however, confirmed by other researchers. Smith and Palmer [179] have studied the interaction of deuterium with oxygen on Pt(111) using molecular beams. The results obtained in this work show that the reaction involves the dissociative adsorption of hydrogen and a subsequent interaction of four hydrogen atoms with an adsorbed molecule of O_2. Pacia and Dumesic [180] have used molecular beams to study the oxidation of H_2 on a polycrystalline platinum ribbon at 10^{-4} Pa over a wide temperature range from 300 to 1700 K. The authors' conclusion is that the reaction occurs *via* an adsorption (L-H) mechanism, that is, between hydrogen and oxygen atoms on a Pt surface. The interaction *via* the E-R mechanism, *i.e.* between a H_2 molecule from the gas phase and the surface oxygen atoms, does not occur. The true activation energy of the reaction between adsorbed atoms is 84 kJ mol^{-1}. Norton [181] used the XPS method to measure variations in the surface concentrations of oxygen on a platinum foil. Norton has suggested the following kinetic equation to describe the reduction of adsorbed oxygen with dihydrogen:

$$\frac{d\theta_o}{dt} = Kp_{H_2}\theta_o(1 - \chi_o)$$

where θ_o is the fraction of the surface covered with oxygen, and t is the time. The water production probability with H_2 molecule colliding on the surface, γ, is 0.02 ($\theta_o = 0.2$, T = 273 K). When the temperature is lowered to 200 K, γ decreases by a factor of 1.8. According to Norton, the reaction is due to the weak dissociative adsorption of hydrogen, that requires one vacant site on the surface, followed by the interaction between the hydrogen and oxygen atoms. This interpretation seems not to be persuasive. It suffices to note that the dissociative adsorption should be strong enough to ensure the compensation of the bond breaking energy in the molecule by the energy of interaction between atoms and the metal surface. It is more plausible to suggest that the reaction occurs *via* collision of an H_2 molecule with an adsorbed oxygen atom in the vicinity of a vacant site. From the temperature dependence of γ it follows that the activation energy of such an interaction is 3.7 kJ mol^{-1}.

Collins *et al.* [182] have also made an attempt to measure γ using UPS at 295 K. The energy distribution curve shifts as a result of oxygen adsorption. Oxygen removal by hydrogen restores the curve position. For this to occur, it is sufficient to attain 0.2 L (L = 10^{-6} Torr s), which corresponds to $\gamma = 0.7$. Bernasek and Somorjai [183] have used molecular beams to determine the probability of the dissociative adsorption of small molecules (including dihydrogen and dioxygen) on the stepped face of platinum, Pt-(S)[9(111) × (111)]. When a beam of D_2 molecules was directed onto the surface previously exposed to oxygen, the reaction probability was only 2×10^{-4} for a surface temperature of 1000 K and a beam temperature of 3000 K. When an oxygen molecular beam was directed onto the surface previously exposed to hydrogen, $\gamma = 2.2 \times 10^{-5}$. The authors suppose that for a smooth (111) plane γ is even smaller (by three orders of magnitude). The low value of γ observed in these experiments may be caused by small surface concentrations of oxygen resulting from the high temperature of the surface.

A serious question arises whether the above conclusions are valid for the reactions proceeding at higher pressures. Boudart *et al.* [184] have determined the value γ (about 10^{-7}) from the data obtained at high pressures for silica supported dispersed platinum. This value is much less than that found at low pressures. The authors suppose that this phenomenon is not associated with a change in the reaction mechanism. Rather it is due to a more dense surface coverage of the catalyst with chemisorbed species at high pressures. With such a coverage the reaction rate may be significantly slower than that observable at low pressures.

Tompkins *et al.* [207] have studied the interaction of the atomically adsorbed oxygen with dihydrogen and of the atomically adsorbed hydrogen with dioxygen by measuring during the reaction the platinum surface potential. As a result they have proposed the following scheme of elementary steps:

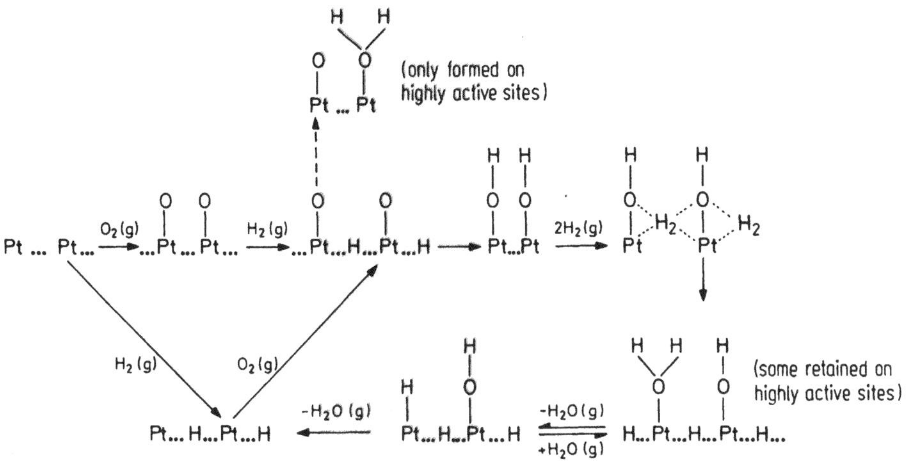

The considerable adsorption of water has been observed at 195 K on both the clean surface of platinum and on the surface covered with oxygen atoms.

2. *Other Metals*

The other metals are characterized by the kinetic laws in oxidation of dihydrogen similar to those described for platinum. On easily oxidized metals the steady state reaction can be studied only in excess hydrogen and in these conditions its rate is proportional to the pressure of oxygen.

For silver which does not adsorb hydrogen the oxidation of the latter occurs only through the interaction of dihydrogen with adsorbed oxygen. The main portion of surface oxygen is equivalent with respect to this interaction. It has been established that the reaction of adsorbed oxygen and dihydrogen falls into two steps:

$$2\,O_{ads} + H_2 \rightarrow 2\,OH_{ads}$$

$$2\,OH_{ads} + H_2 \rightarrow 2\,H_2O\,,$$

the rate constant of the second step being two orders of magnitude less than that of the first step [147].

Benton and Elgin [148] have found that the reaction rate is proportional to hydrogen pressure and independent of oxygen pressure. Resting on the assumption that dihydrogen reacts with the dissociatively adsorbed oxygen, the authors of ref. [149] have derived the equation:

$$r = \frac{K_1 K_2 p_{O_2} p_{H_2}}{K_1 p_{O_2} + 1/2\, K_2 p_{H_2}}$$

where K_1 is the rate constant of oxygen adsorption and K_2 is the rate constant of dihydrogen interaction with adsorbed oxygen. The activation energy of the overall reaction is 33.5 kJ mol^{-1}.

3. Mechanism of Dihydrogen Oxidation on Metal Catalysts

The mechanism of dihydrogen oxidation on metal catalysts involves the formation of chemisorbed oxygen atoms on the catalyst surface. The dissociative chemisorption of dioxygen on clean surfaces is rather rapid.

Dihydrogen may react being both in the molecular state and in the atomic state chemisorbed on the catalyst. The former case corresponds to the following scheme of elementary steps:

$$O_2 + 2[\ ] \rightarrow 2[O]_{ad}$$
$$H_2 + [O]_{ad} \rightarrow H_2O + [\ ] \quad \text{or} \quad H_2 + 2[O]_{ad} \rightarrow 2[OH]_{ad}$$
$$H_2 + 2[OH]_{ad} \rightarrow 2\,H_2O + 2[\ ] \qquad (23)$$

According to theoretical calculations [210], the interaction of dihydrogen with chemisorbed oxygen (the Eley-Rideal mechanism) is characterized by an activation energy close to zero. The reaction route with the intermediate formation of hydroxyls has been proved for silver catalysts.

The situation where hydrogen reacts in the atomically adsorbed state is described as:

$$O_2 + 2[\ ] \rightarrow 2[O]_{ad} \qquad A$$
$$H_2 + 2[\ ] \rightarrow 2[H]_{ad} \qquad B \qquad (24)$$
$$2[H]_{ad} + [O]_{ad} \rightarrow H_2O + 3[\ ] \qquad C$$

or

$$2[H]_{ad} + 2[O]_{ad} \rightarrow 2[OH]_{ad} + 2[\ ] \qquad D$$
$$2[OH]_{ad} + 2[H]_{ad} \rightarrow 2\,H_2O + 4[\ ] \qquad E$$

or

$$2[OH]_{ad} + H_2 \rightarrow 2\,H_2O + 2[\ ] \qquad F$$

On platinum catalysts hydrogen reacts being in both molecular and atomic states. This can be represented schematically as:

$$\left.\begin{array}{ll} O_2 + 2[\ ] \rightarrow 2[O]_{ad} & A \\ H_2 + 2[\ ] \rightarrow 2[H]_{ad} & B \\ [O]_{ad} + 2[H]_{ad} \rightarrow H_2O + 3[\ ] & C \\ H_2 + [O]_{ad} \rightarrow H_2O + [\ ] & D \end{array}\right\} \qquad (25)$$

Based on these schemes it is possible to explain some critical phenomena observed in the course of hydrogen oxidation such as, *e.g.*, multiplicity of steady states in reactions on platinum and nickel [150], and creation of the reaction oscillatory regime. The first evidence for the existence of the oscillatory regime of hydrogen oxidation came to light in 1972 [151, 152, 158]. The relevant review articles have appeared recently [154, 155]. In spite of some discrepancy in the results obtained by different investigators, it is clear that over a certain interval of the reaction mixture compositions and temperatures, the oxidation of hydrogen on platinum and nickel occurs at a periodically changing rate (Figure 17). The new data on oscillatory regimes in the catalytic oxidation of hydrogen are reported in refs. [156–158]. The oscillatory regime can also be observed under certain conditions during CO oxidation on platinum catalysts. On the assumption of ideal adsorption on a homogeneous surface the kinetic equations corresponding to the steps of scheme (25) have a single solution. In order to explain the critical phenomena and the appearance of the oscillatory regime, some additional assumptions are required. According to ref. [160], the activation energies at the step of water production (25C and 25D) depend upon the surface coverage with oxygen, $[O]_{ad}$, *i.e.*:

$$E_C = E_C^0 + \alpha_C[[O]_{ad}]; \qquad E_D = E_D^0 + \alpha_D[[O]_{ad}]$$

In this case the solution of the system of kinetic equations (scheme (25)) for certain parameters leads to self-oscillations. There can be also other qualitative explanations. For example, proceeding from the same scheme [25] and the same assumption of the surface homogeneity, add the buffer step

$$O_2 + 2[\ ] \leftrightarrows 2[O]'_{ad}$$

where $[O]'_{ad}$ is the inert intermediate state of adsorbed oxygen which does not participate in the steps of water formation (25C) and (25D). Then one obtains the complex of kinetic equations whose solution for certain values of parameters leads to self-oscillations in surface concentrations and thus in reaction rates.

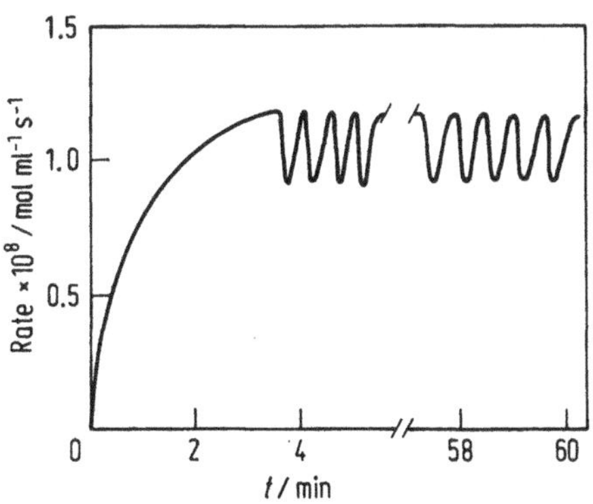

Figure 17. Self-oscillations in the rate of dihydrogen oxidation on nickel (613 K, p_{H_2} = 101 kPa, p_{O_2} = 0.1 kPa). (After [152])

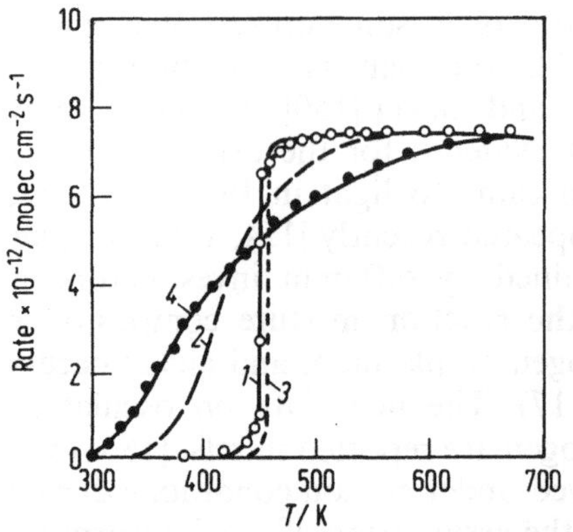

Figure 18. Temperature dependence of dihydrogen oxidation rate on Ni (110) and (111). 1. Experimental curve for Ni (110), 2. Calculated curve based on the assumption of equiprobable distribution of adsorbed oxygen and hydrogen. 3. Calculated curve based on the assumption of the island character of oxygen adsorption. 4. Experimental curve for Ni (111)

4. Effect of Metallic Surface Structure

The surface structure of metallic catalysts may have a pronounced effect on the catalytic activity. Therefore, in order to perform a detailed quantitative analysis, it is necessary to take into account the catalyst surface structure and the topochemical character of the formation of a two-dimensional chemisorbed layer and its interaction with the second reaction component. This can be examplified by the interaction between chemisorbed oxygen and dihydrogen on Ni(110). Figure 18 illustrates an experimental curve (1) of the water formation rate *vs.* temperature. Based on the scheme of the mechanism (24ABDE) the authors have obtained a theoretical curve (2) which is inconsistent with the experimental data. At the same time, if one takes into account the above mentioned constancy of the oxygen adsorption rate, and the topochemical character of the interaction at the boundary of the surface phases, curve (3) will be obtained which reproduces well the experimentally observed sharp increase in the reaction rate at about 450 K [161, 162].

The relative catalytic activity of different planes of platinum was studied with the aid of the Field Emission microscope [163, 164]. After oxygen

Table 11. FEM data on the low-temperature reaction $H_2 + O_{ads}$ on Pt, Rh and Ir

Metal	Conditions of oxygen		adsorption form	p_{H_2}/Pa	T/K	E_{act}/ kJ mol^{-1}	Reaction boundary	Activity of planes
	T_{ads}/ K	exposure/ L						
Pt	78	100	α-molec. β-diss.	5×10^{-4}	105–200	7	observed	(331) > (111) >(110) > $(100)_{step}$ > (210)
Rh	78	300	α-molec. β-diss.	2×10^{-5}	165–180	9	observed	$(100)_{step}$ ~ (110) > (311), (331) > (111) > (210)
Ir	78	150	β-diss.	3×10^{-5}	150–250	8	observed	(321) > (111) > (110) > $(100)_{step}$ > (210)

had been adsorbed on a platinum tip at 78 K (up to complete saturation), hydrogen was introduced at a pressure of 7×10^{-4} Pa. At 135–160 K oxygen was removed which was manifested by a decrease of the work function and in a corresponding increase of emission. This process started on the stepped planes neighboring to plane (111) and then transferred gradually onto the planes (012), (013) and (023). Apparently, oxygen is bound most tightly on these "rough" planes. A similar pattern was also observed for rhodium and iridium in experiments with higher temperatures (150–250 K) (Table 11) [163, 200, 201].

The difference in SCA's of different planes of the same metal is accounted for by the character of oxygen chemisorption thereon. The rate of hydrogen interaction with chemisorbed oxygen on Ni(110), (100) and (111) *vs.* the surface coverage is plotted in Figure 19 [165]. In the range of small coverages, where ideal adsorption takes place, the reaction rate on Ni(100) is some orders of magnitude less than that on the other principal planes. When the surface coverage by oxygen is 0.35 or higher, the chemisorption is of a reconstructive character and the difference in activities is insignificant.

A number of studies are concerned with the effect of the surface structure and dispersity on the SCA. Investigations performed using the gradientless method which excludes the effect of heat and mass transfer processes have indicated that for the surface steady state the SCA of massive platinum (wire, foil), platinum black and silica supported platinum is approximately the same. The platinum surface area was measured by the hydrogen chemisorption method [166]. These data in conjunction with the SCA measurements for other reactions and catalysts allowed formulation of the general rule of an approximate constancy of SCA for the catalysts of constant composition [346]. This rule can be readily explained by the reaction mixture effect on the properties of solid catalysts. Owing to this effect, the catalysts of

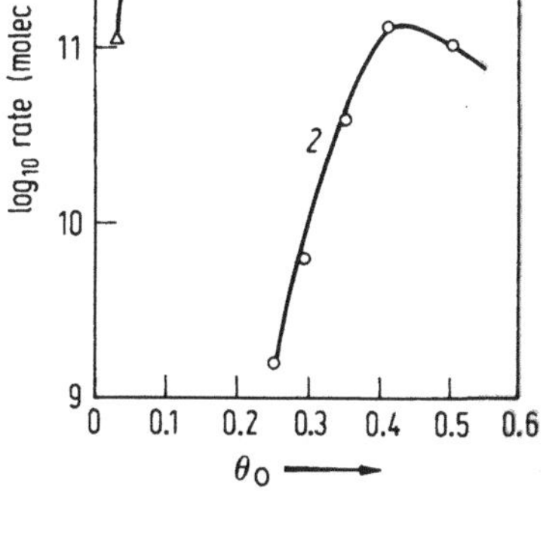

Figure 19. Rate of the interaction of hydrogen with preadsorbed oxygen *vs.* surface coverage ($p_{H_2} = 4.1 \times 10^{-2}$ Pa, 438 K). 1 — Ni (110), 2 — Ni (100), 3 — Ni (111)

the same chemical composition must reach the same steady state independently of the initial state of the surface [347]. Some deviations are possible for very small crystals. For example, by supporting platinum over θ-Al_2O_3 from a solution of bis-π-allyl platinum in pentane and from a solution of chloroplatinic acid, followed by calcination in different conditions it is possible to prepare samples with a platinum crystal size varying from 1.2 to 140 nm at the same total content of platinum. The activity measurements in a flow-circulation system in excess oxygen at 303 K have shown no essential changes of SCA as the crystal size was reduced to 3 nm. However, a subsequent reduction led to a more than one order decrease in the SCA. For silica supported platinum the SCA remains constant upon variation of platinum dispersity over the same intervals [167]. Marshneva *et al.* did not observe any variations in SCA of silica supported platinum when the crystal size was reduced to 1.5 nm (Figure 20) [348]. Variations in the activity of very small platinum crystals are likely to be due to their interaction with a support. In this regard of particular interest are the results of an EXAFS study of disperse platinum on alumina by Katzer and Sayers [168]. According to the authors, the crystals smaller than 2 nm are characterized by a deficiency of electrons which, most probably, are transferred onto the alumina. The electron deficiency increases as the crystal size is reduced. A certain correlation has been found between the electron deficiency and the SCA of platinum in ammonia oxidation.

Poltorak et al. [207] have reported that in the case of silica supported platinum the reactivity of chemisorbed oxygen with respect to dihydrogen substantially decreases when the size of platinum species is diminished.

In contrast, Hanson and Boudart [169] have found for silica supported platinum (dispersity from 0.14 to 1) that in excess hydrogen the SCA increases by a factor of 7 as the dispersity is increased. In oxygen excess the SCA remains approximately constant. The authors explain this by the corrosive adsorption of oxygen in the mixtures with its excess which results in leveling of the surface properties.

The high activity which markedly exceeds an additive one was reported

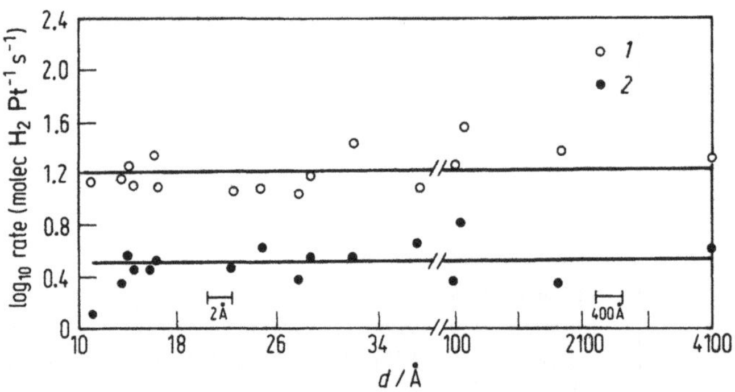

Figure 20. Specific rate of hydrogen oxidation *vs.* particle size of platinum on silica at 293 K. 1. 1% H_2 and 7% O_2, 2. 3% H_2 and 1% O_2 (After [348])

in a number of works concerning hydrogen oxidation on platinum and palladium supported over some oxides [170, 171]. This phenomenon closely relates with the observations of the acceleration of reduction of many oxides in the presence of small amounts of platinum [172–175]. As a possible explanation of these phenomena most of the authors suggest spillover of the dissociatively chemisorbed hydrogen from platinum on the oxide surface whereon it reacts with oxygen of the catalyst. Similar phenomena have also been observed in some other reactions involving hydrogen on platinum catalysts [176]. However, the mechanism of such catalytic events is still unclear.

B. Oxide Catalysts

1. Transition Metal Oxides

Transition metal oxides have lower SCA toward dihydrogen oxidation compared to metals. The oxides of the IV period metals may be arranged in the following series according to the decreasing SCA:

$$Co_3O_4 > MnO_2 > NiO > CuO > Cr_2O_3 > Fe_2O_3 > ZnO > V_2O_5 > TiO_2$$

(see Table 12).

The catalytic activity of the steady state catalysts was measured by the gradientless method in a reaction mixture with excess oxygen so that the possibility of reduction was precluded. The SCA per one cm^2 surface area was determined as the number of oxygen molecules that have reacted per second. In the above series the SCA changes by almost six orders of magnitude, *i.e.*, from 1.7×10^{13} for Co_3O_4 to 3.3×10^7 molecules O_2 s^{-1} cm^{-2} for TiO_2 at 573 K and hydrogen pressure of 1 kPa [185, 186].

The SCA of cadmium and tin oxides is somewhat higher than that of iron oxide whereas that of lead oxide is lower than the latter one; even far less is the SCA of tungsten and molybdenum oxides.

Table 12. Oxidation of dihydrogen in excess oxygen on metal oxides [185, 186]

Catalyst	Temperature range/K	Reaction rate/molec O_2 cm^{-2} s^{-1} at 573 K and conc H_2 0.33 mmol l^{-1}		E/kJ mol^{-1}	Reaction order with respect to H_2		Reaction order with respect to O_2 [186]
		p_{O_2} = 100 kPa [185]	p_{O_2} = 27 kPa [186]		[185]	[186]	
Co_3O_4	323–358	5.50×10^{13}	1.47×10^{14}	50 ± 4	1.0	0.8	0
CuO	358–398	2.20×10^{13}	9.47×10^{13}	59 ± 4	1.0	0.7	0
MnO_2	373–408	1.10×10^{13}	7.57×10^{12}	56 ± 2	1.0	1.0	0
NiO-I	420–470	1.79×10^{12}	7.18×10^{12}	61 ± 2	1.0	0.8	0
Cr_2O_3	458–508	2.00×10^{12}	2.53×10^{12}	77 ± 6	0.5	0.6	0
Fe_2O_3	498–552	2.91×10^{11}	6.82×10^{11}	71 ± 8	0.6	0.7	0
ZnO	573–623	3.85×10^{10}	1.21×10^{10}	96 ± 4	0.7	0.7	0
V_2O_5	693–743	9.10×10^{9}	1.52×10^{9}	82 ± 6	0.8	1.0	0
TiO_2	743–793	8.65×10^{7}	1.85×10^{8}	92 ± 4	0.8	1.0	0

The SCA of the rare earth metal oxides decreases in the following series:

$$Pr_6O_{11} > Nd_2O_3 > La_2O_3 > Ho_2O_3 > Lu_2O_3 > CeO_2 > Yb_2O_3 > Dy_2O_3 > Sm_2O_3 \, .$$

The most active praseodymum oxide is inferior to iron oxide in activity [187]. Magnesium and aluminium oxides are almost inert.

When substituting hydrogen by deuterium the reaction rate decreases: the kinetic isotope effect r_{H_2}/r_{D_2} ranges 1.2 to 1.7 for different oxides [188].

The dissociative adsorption of dihydrogen observed during its oxidation on metals does not occur on oxides. For the stepwise mechanism the following catalytic transformation may occur:

$$H_2 + O_{surf} \rightarrow H_2O_{ad} \rightarrow H_2O \qquad \text{I}$$

$$O_2 \rightarrow 2\,O_{surf} \qquad \text{II} \qquad (26)$$

Most probably, the surface oxygen involved in the reaction is in the form of O^{2-}. Its binding energy with a catalyst and, hence, its reactivity may change depending upon the cation surrounding (coordination). If the desorption of water product is sufficiently rapid, the kinetics of oxidation is first-order with respect to hydrogen and zero order with respect to oxygen:

$$r = Kp_{H_2}$$

The first step of scheme (26) may involve an intermediate formation of surface hydroxyls.

On ZnO, CdO and Cr_2O_3 the low orders (0.5–0.7) with respect to hydrogen were observed.

The above order of the SCA variation of oxides in hydrogen oxidation coincides with that given in the previous section for the isotope exchange of dioxygen following mechanism III. As will be shown below, the same

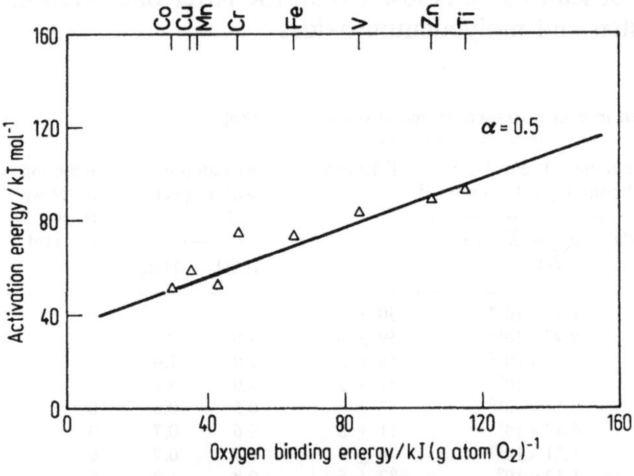

Figure 21. Activation energy of hydrogen oxidation on oxide catalysts *vs.* oxygen binding energy on the catalyst surface. (After [212])

order is valid also for the oxidation of CO, complete oxidation of light hydrocarbons, alcohols and some other compounds. Activation energies and absolute rates of various reactions are considerably different but relative variations in the series of oxides are similar.

The general factor that determines this correlation is the oxygen binding energy on the catalyst surface [189]. The smaller is the oxygen binding energy, the higher the catalytic activity (see Figure 21) [190]. This permits one to propose that the rate-determining step in scheme (26) is step I. At this step, bond breaking occurs between the surface oxygen and the catalyst and between the hydrogen atoms in dihydrogen. Eventually new bonds between hydrogen and oxygen are formed. For the series of oxide catalysts only the energy of oxygen binding to the catalyst is changing. Therefore the energy of the active complex of this step should change by some fraction of the change of the oxygen binding energy when passing from one catalyst to another.

$$\ln r \sim E = E_0 + \alpha q$$

here r is the SCA of oxide catalyst, E is the activation energy, q is the energy of oxygen binding to the catalyst, E_0 and α are the constants $(1 \geqq \alpha \geqq 0)$. From the slope of the straight line in Figure 21 it follows that for hydrogen oxidation α is 0.5 ± 0.1.

The difference in the catalytic properties of oxide catalysts in the oxidation of hydrogen is determined primarily by the energy of oxygen binding to their surfaces. The interaction of an oxidizable substance with a catalyst is of minor importance. Deviations are observed only in the case of high binding energies of oxygen. From all of the above data one may conclude that the interaction of hydrogen with oxygen of the catalyst and the reoxidation of the catalytic surface with dioxygen are two separate steps. This proposal

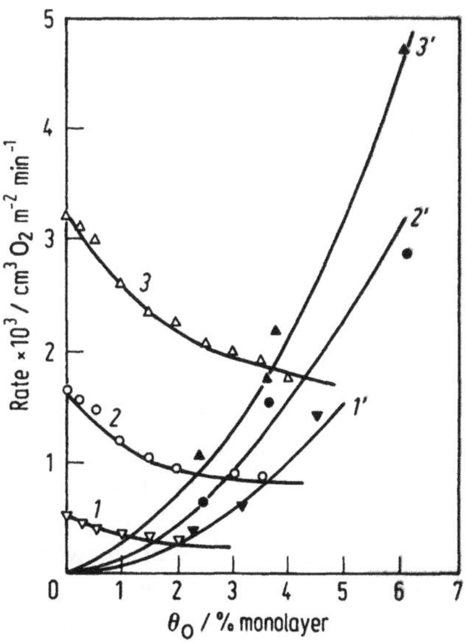

Figure 22. Rates of reduction (1–3) and reoxidation (1′–3′) of manganese dioxide *vs.* oxygen removed from the surface. $p_{H_2} = p_{O_2} = 825$ Pa. (After [186])
1–1′ — 374 K, 2–2′ — 395 K, 3–3′ — 410 K

has been verified experimentally by means of independent measurements of the rates of the catalyst reduction with hydrogen and its reoxidation with oxygen at the catalyst steady state.

For the stepwise mechanism these rates should coincide with the catalytic reaction rates in the presence of the both components. An example of such verification is displayed in Figure 22. As found, at elevated temperatures the reaction of complete oxidation follows the stepwise mechanism of the reactant-catalyst interaction. The step of catalyst reduction with an oxidizable substance occurs more slowly. In this case the concentration of oxygen on the catalyst surface only slightly differs from the equilibrium one under the influence of oxygen solely. At elevated temperatures the rates of the catalytic reaction and of the catalyst reduction with an oxidizable substance are almost the same. However, in many cases as the temperature is decreased the catalytic reaction rate (starting from a certain temperature) tends to exceed the reduction rate. This may be due to the incomplete leveling of the energy distribution that follows the step of catalyst reoxidation with dioxygen. Upon the dissociative adsorption oxygen does not reach immediately a state corresponding to a minimum free energy. It seems natural to propose that the interaction of an oxidizable substance with such oxygen has a lower activation energy and, thus, a higher rate. This corresponds to the special case of interstep compensation when the energy evolved during a facile step (catalyst reoxidation with dioxygen) is used to surmount an activation energy barrier at the rate determining step (withdrawal of oxygen from the catalyst by an oxidizable substance).

The reduction and reoxidation steps of the stepwise mechanism transform to the concerted one when the reaction occurs by a simultaneous interaction of an oxidizable substance and of oxygen with the catalyst.

In all cases, the catalytic activity of oxide catalysts is determined by the oxygen binding energy on the oxide surface. If this energy is not very high, the catalytic activity is independent of the interaction of a catalyst with oxidizable substance. The step of oxygen withdrawal from the catalyst by the oxidizable substance involves electron transfer from the oxygen ions to the metal cations. The electron transfer energy determines the relative catalytic properties of oxides of various metals.

Transition metal oxides are semi-conductors either of n- or p-type. Attempts to correlate the semiconductor and chemisorption properties of solids have been made since the 1930's. The most systematic development of this idea has been performed by Wolkenshtein [202] and Hauffe [203] on the basis of the semiconductor band model. To ascertain the qualitative dependences, Wolkenshtein made an assumption that the catalytic reaction rate depends upon the concentration of chemisorbed intermediates formed with or without the participation of free electrons and holes. The position of the Fermi level affects the heat of adsorption of the first chemisorption form and, thus, its equilibrium surface concentration. The heat of adsorption of the second chemisorption form is independent of the Fermi level: however, at large surface coverages its concentration may vary depending upon the position of the Fermi level as a result of its displacement by the first chemisorption form. These ideas, suggesting clear conclusions and opening a way to control of the catalytic and adsorption properties of semiconductors by shifting the Fermi level by doping with small quantities of altervalent ions (cations of another charge), have enjoyed great popularity. However, the hope of finding a simple correlation between the position of the Fermi level and the chemisorption and catalytic properties of solid catalysts, and thus to control the catalytic properties, did not come true. The reason is that the chemisorption energy of interacting substances (in our case oxygen) with solid catalysts is determined not only by the electron transfer energy but also by parameters independent of the Fermi level. Variation of the catalyst composition aimed at shifting the Fermi level affects these parameters too. The effect of the change of these parameters may be of an opposite character and may be even more important than that of the work function variations [189]. Furthermore, as a result of this effect, the whole energy spectrum of the surface may be modified [204]. Elimination of these shortcomings extremely complicates the model and thus deprives it of its main advantages, *i.e.* clearness of conclusions and possibility of prediction. Therefore, it is more usual to use an approach that takes into account local interactions and is based on experimental evidence.

2. *Catalytic Activity of Mixed Oxides*

The catalytic activity of mixed oxide catalysts in the oxidation of dihydrogen is, as a rule, nonadditive. By way of example consider the binary oxide

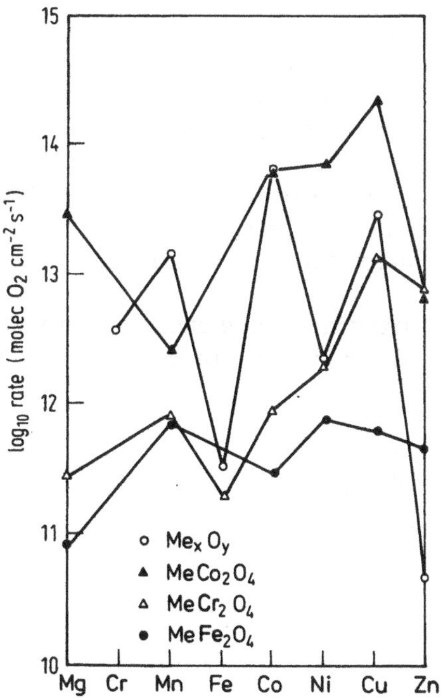

Figure 23. SCA's of cobaltites, ferrites and chromites of the IV period metals in dihydrogen oxidation

compounds with the spinel structure. Figure 23 shows the SCA's of ferrites, chromites and cobaltites in regard to the oxidation of dihydrogen [190, 192, 193]. For comparison, the SCA of simple oxides is also given in the figure. On comparing the given data it is seen that the catalytic properties of transition metal cations incorporated into spinels are nonadditive, the properties of three-valence cations in octahedral coordination play a dominant role whereas the properties of two-valence cations in tetrahedral coordination become of minor importance. This is clearly manifested in the case of ferrites whose catalytic activity is close to the activity of iron oxide. The SCA of cobalt spinel does not differ from that of iron oxide although the activity of Co_3O_4 is 200 times higher than that of Fe_2O_3. This is also the case for manganese and copper ferrites.

In the series of chromites the difference in activities is somewhat larger, but on the average the SCA thereof does not drastically differ from that of chromium oxide with the exception of copper chromite whose SCA is close to SCA of copper oxide. The catalytic activities of cobaltites differ even more, but the most active of cobaltites are close in activity to Co_3O_4.

For all spinels, like for simple oxides, a linear correlation between the logarithm of SCA and the oxygen binding energy is observed. This observation does prove in this case also a predominant role for the step of oxygen withdrawal from the catalyst surface by the oxidizable substance. The difference in the SCA's of transition metal cations in spinels, which depends on the coordination, is attributed to the change of the ligand field. This latter affects the cation ionization potential and hence the energy of the withdrawal of an oxygen ion donating its electrons to a metal cation.

The catalytic activity per one atom (cation), ACA, largely depends upon the type of the metal-oxygen bond. The ACA increases with the bond covalence and decreases with an increase of its ionic character [194]. This becomes evident if we compare the ACA's of oxides and oxygen-containing salts of the same metal. As shown in Table 13, the ACA of copper in the oxide is three orders of magnitude greater than that of copper in the salts. In spinels the movement of electron pairs toward the three-valence cations facilitates electron transfer from oxygen to these cations. As a result, the nature of a three-valence ion becomes a determining factor for the catalytic properties of spinels.

As soon as a new compound with a low oxygen binding energy is formed in mixed catalysts, the ACA may drastically increase. For example, with

Table 13. Atomic catalytic activity of copper cations in different compounds with respect to dihydrogen oxidation

Compound	ACA/molec O_2 s^{-1} (metal atom)$^{-1}$
$CuSO_4$	0.0025
$Cu_3(PO_4)_2$	0.002
$CuCl_2$	0.0003
CuO	0.23

Table 14. Oxidation of dihydrogen in excess oxygen on vanadium oxide catalysts promoted by potassium sulfate [195]

Catalyst	Specific reaction rate/ mol H_2 m^{-2} hr^{-1}; 673 K; conc. H_2 = 10^{-4} mol l^{-1}	E_{act}/ kJ mol^{-1}	Reaction order with respect to hydrogen
V_2O_5	3.3×10^{-6}	88	0.8
$V_2O_5 - 0.1\ K_2SO_4$	7.2×10^{-4}	21	0.9
$V_2O_5 - 0.3\ K_2SO_4$	1.15×10^{-3}	17	0.9
$V_2O_5 - 0.5\ K_2SO_4$	8.1×10^{-5}	67	0.9
$V_2O_5 - K_2SO_4$	2.5×10^{-5}	75	0.9

the addition of potassium sulfate to vanadium pentoxide the SCA increases by three orders of magnitude (Table 14) owing to the formation of potassium polyvanadates and sulfovanadates [195].

The SCA of mixed $V_2O_5 - MoO_3$ catalysts passes through the maximum at 30% of MoO_3 and increases by a factor of 3 compared to pure V_2O_5. A study of the kinetics of hydrogen oxidation on these catalysts has been undertaken by Il'chenko *et al.* [196]. The authors have observed a hetero-homogeneous reaction proceeding with reaction mixture compositions close to the critical limits of ignition. In these conditions the reaction starts on the catalyst surface producing H or OH radicals which desorb to the volume and induce therein the homogeneous chain process. Under certain conditions similar phenomena may take place on other catalysts too.

The influence of the coordination on the catalytic properties becomes even more pronounced when the transition metal cations are introduced into inert matrices. For example, in cobalt — magnesia catalysts which represent solid solutions of substitution of magnesium ions by cobalt ions, two types of cobalt ions have been identified by the ESR method. The ions of the type I are located on the surface in the field of a distorted tetrahedron and remain in the form of Co^{2+} even when exposed to oxygen at 773 K. The ions of type II are in a 6-coordinated position and can be readily reduced and oxidized at low temperatures [197]. Hydrogen oxidation on the type I ions occurs with an activation energy of 127 kJ mol^{-1}, while on the type II ions with an activation energy of 63 kJ mol^{-1}. The proportion of the type II ions is four orders of magnitude smaller than that of the type I ions but due to the low activation energy their participation in the reaction is observed already at 740 K with a 10% cobalt content. A similar state of cobalt ions has been also found by Yurieva and Kuznetsova [198] for solid solutions in alumomagnesium spinels. Figures 24 and 25 show the activation energies and ACA's of solid solutions as a function of cobalt content.

As evidenced by ESR studies of solid solutions of copper oxide in magnesium oxide, at small concentrations copper is predominantly in the form of isolated ions. The growth of copper concentration leads to the formation of associates with a strong interaction between copper ions. ACA measurements in dihydrogen oxidation have shown that over the range of small concentrations the ACA is low (by three orders of magnitude less than that in

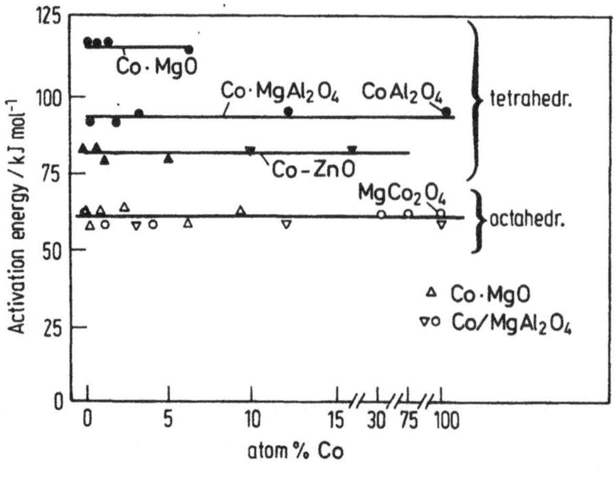

Figure 24. Activation energies of dihydrogen oxidation on the ions of cobalt in tetrahedral and octahedral coordination in various matrices. (After [197])

copper oxide). With the formation of associates the ACA increases and at a 20% concentration (the limit of dissolution) becomes 30 times higher than ACA at low concentrations [199].

C. Metal Carbide Catalysts

Transition metal carbides exhibit relatively high catalytic activity toward the H_2–O_2 interaction. This activity is smaller than that of the corresponding elements in the metal form but is considerably higher than that of their oxides. In the reaction mixtures with excess oxygen the SCA of carbides changes in the following order:

$$WC > Cr_7C_3 > NbC > VC > TiC > Mo_2C > TaC > ZrC > HfC .$$

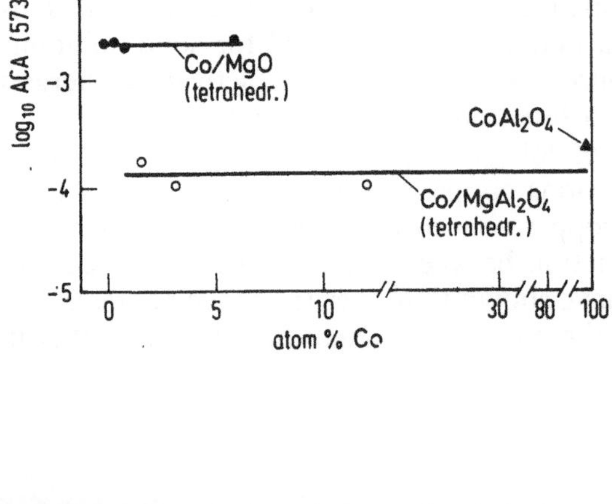

Figure 25. ACA of the ions of cobalt of different coordination in various matrices toward dihydrogen oxidation at 573 K. (After [198])

Table 15. Kinetic characteristics of dihydrogen oxidation on transition metal carbides [206]

Catalyst	Catalysis in excess oxygen		Catalysis in excess hydrogen	
	623 K, p_{H_2} = 5.5 kPa, p_{O_2} = 22 kPa		623 K, p_{H_2} = 22 kPa, p_{O_2} = 2.8 kPa	
	rate × 10^{-13}/ molec cm^{-2} s^{-1}	activation energy/ kJ mol^{-1}	rate × 10^{-13}/ molec cm^{-2} s^{-1}	activation energy/ kJ mol^{-1}
TiC	0.32	113.0 ± 4	0.32	125.5
ZrC	0.14	125.5 ± 4	0.40	79.5 ± 8
HfC	0.10	75.3	0.17	75.3
VC	0.33	96.2	0.37	192.5
NbC	0.41	154.8	0.12	41.8
TaC	0.19	54.4 ± 8	0.32	75.3
Cr_7C_3	2.14	62.8 ± 8	4.27	71.1 ± 8
Mo_2C	0.24	75.3	0.98	125.5
WC	10.0	46.0 ± 8	20.9	29.3

In hydrogen excess in the reaction mixture the above order somewhat changes:

$$WC > Cr_7C_3 > Mo_2C > ZrC > VC > TiC > TaC > HfC > \\ > NbC\,.$$

The SCA of tungsten carbide exceeds that of such active oxide catalysts as MnO_2 and NiO and is lower than the SCA of the most active oxides, Co_3O_4 and CuO, by less than one order of magnitude [205]. The kinetic characteristics of the hydrogen oxidation reaction are summarized in Table 15. The SCA's of metals, carbides and oxides are compared in Table 16 [206].

In a reaction mixture with excess oxygen the surface of carbides is largely covered with atomically adsorbed oxygen which slows down the dissociative adsorption of hydrogen and thus makes more probable the reaction proceeding *via* the interaction of dihydrogen with atomically adsorbed oxygen (see scheme (23)). With hydrogen excess in the gas phase the surface of carbides becomes oxygen-free. Thus dissociative adsorption of hydrogen becomes possible and the reaction proceeds *via* scheme (24) [206]. It should be also noted that carbides are stable in reactions conducted in oxygen or hydrogen excess.

Table 16. Specific catalytic activity of metals, carbides and oxides in dihydrogen oxidation (623 K, p_{H_2} = 22 kPa, p_{O_2} = 5.5 kPa)

Catalyst	rate/molec cm^{-2} s^{-1}	Catalyst	rate/molec cm^{-2} s^{-1}	Catalyst	rate/molec cm^{-2} s^{-1}	Catalyst	rate/molec cm^{-2} s^{-1}
Cr	1.6×10^{17}	Mo	1.6×10^{17}	W	5×10^{16}	V	1.5×10^{16}
Cr_7C_3	4.3×10^{13}	Mo_2C	1.0×10^{13}	WC	2.1×10^{14}	VC	3.7×10^{12}
Cr_2O_3	10^{13}	MoO_3	3.3×10^{10}	WO_3	2.2×10^{12}	V_2O_5	7.3×10^{10}

5. Oxidation of Carbon Monoxide

Reaction $CO + {}^{1}/_{2}\,O_2 \rightarrow CO_2$ is known to be exothermic ($\Delta H^\circ_{298} = -282.6$ kJ mol^{-1}) and practically irreversible upto 1500 K ($\Delta G^\circ_{298} = -256.7$ kJ mol^{-1}; $\Delta S^\circ_{298} = -20.7$ e.u.).

A vast amount of existing catalysis literature is devoted to the interaction of CO with dioxygen. This seemingly simple catalytic reaction has long been a favorite subject to researchers investigating general problems of heterogeneous catalysis involving factors such as catalytic activity dependence on the surface structure and dispersion of active component, type of the catalyst-reagent interaction, energy of surface bonds, etc.

CO oxidation is also an interesting reaction in terms of practical importance, which is associated with CO toxicity and the necessity to purify industrial and automobile exhaust gases.

Homogeneous oxidation of CO follows a mechanism involving branched chains. Since chain propagation occurs with the participation of hydroxyl radicals and hydrogen atoms, the presence of small amounts of hydrogen-containing compounds is indispensable. Catalytic oxidation on solid catalysts does not follow the chain mechanism but in certain conditions the chain reaction may occur in the near-surface volume of the gas phase.

The relative catalytic activity of various substances toward CO oxidation is very close to the activity toward dihydrogen oxidation. Transition metals and their compounds have the highest activity. In practice both pure and supported (most often on alumina) oxide compounds of copper, chromium, cobalt, manganese, etc. are usually used. Catalysts for detoxication of automobile exhaust gases usually contain, along with oxides, small amounts of platinum or palladium.

A. Catalytic Activity of Metals — CO Adsorption on Metals

Numerous studies have appeared recently concerning chemisorption of CO on metals and its interaction with dioxygen. These studies are based on LEED, electron spectroscopy and thermodesorption methods with special emphasis being laid on thorough and controlled cleaning of surfaces in high vacuum (10^{-8} Pa). On clean metal surfaces CO is adsorbed at a high rate with a sticking probability of 0.2–0.6, that is without an activation energy. Adsorbed molecules are located perpendicularly to the metal surface with carbon atoms facing the metal. The interaction involves an acceptor-donor bond with electron transfer from the $5\,\sigma$ orbital of CO into the d_{xy}-orbital of the metal and back-donation of metallic *d*-electrons into the antibonding $2\pi^*$ orbital of CO. This transfer leads to the weakening of the C–O bond and a decrease of its frequency. The shift of the negative charge to the adsorbate increases the electron work function. In Figure 26 there is shown in schematic fashion CO interaction with a transition metal surface.

On a Pt (111) plane during the first steps of adsorption at temperatures below 300 K the diffraction pattern corresponds to the $(\sqrt{3}\times\sqrt{3})$ R 30°

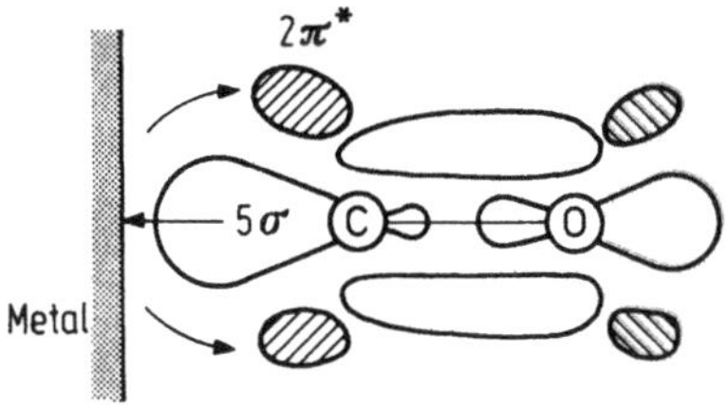

Figure 26. Scheme of the donor-acceptor interaction of CO molecules with the surface of transition metal according to Engel and Ertl. (After [259])

structure[1] at a maximum coverage $\theta = 0.33$ [213, 214]. Upon increasing the exposure up to 2L this structure transforms to c (4×2) that corresponds to a coverage of 0.5. With further coverage increase the diffraction pattern indicates compression of a unit cell along the (110)-surface direction. The maximum coverage is 0.68 (1.02×10^{14} molecules cm^{-2}). Figure 27 displays schematically the distribution of adsorbed molecules. In the $(\sqrt{3} \times \sqrt{3})$ R 30° structure each CO molecule interacts with three metal atoms whereas in the c (4×2) structure bridged bonding to the two metal atoms occurs. The initital adsorption heat equals 138 kJ mol^{-1} and tends to decrease as the coverage increases.

[1] Physical meaning of structural designations see in ref. [292].

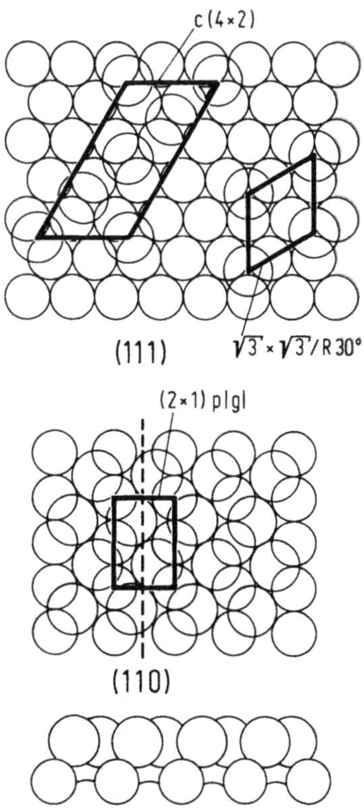

Figure 27. Distribution of adsorbed molecules of CO on Pt (111) and (110). [After [214])

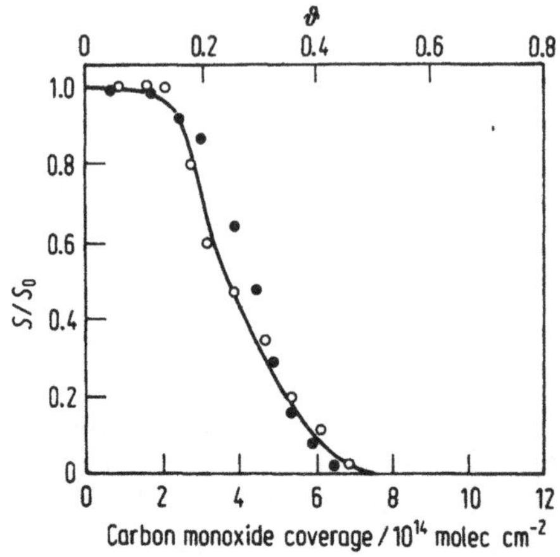

Figure 28. The change of the relative sticking probability S/S_0 with increasing surface coverage with CO on Pt (111) at 300 K. ○ [213], ● [260]

On a Pt (110) plane preheated at 500 K the diffraction pattern corresponding to the (2×1) p |g| structure is observed. The initital adsorption heat is 113 kJ mol^{-1}.

The ordered distribution of adsorbed CO molecules in certain two-dimensional structures has also been found for other transition metals [259].

The rate of adsorption is determined by the sticking probability (*S*) equal to the ratio of the rate of adsorption to the number of molecules incident on the surface. During adsorption of carbon monoxide the sticking probability remains constant upon increasing the surface coverage up to $\theta =$ $= 0.2—0.4$, then starts to sharply decrease with further coverage increase [241]. The sticking probability as a function of surface coverage for CO adsorption on Pt (111) is plotted in Figure 28. This dependency is explained on the assumption of the initial binding of a molecule colliding with the surface and forming a disordered (precursor) state in which it can move along the surface until it finds a vacant site where chemisorption occurs. The bond energy changes insignificantly when the molecule moves within an elementary cell.

Thermal desorption (TD) allows the determination of the number of adsorbed forms of CO having different bond energies. For example, a single form has been found for platinum, two α- and β-forms for palladium, and five forms for rhenium. No reliable data are as yet available on the physical nature of these forms.

The change of the C—O bond energy during adsorption is essential for catalytic transformations. This change depends upon the transfer of metallic *d*-electrons into the antibonding orbital of CO and may lead to the dissociative adsorption. In the case of platinum, this electron donation is insignificant which is reflected by the absence of an increase in the work function. The dissociative adsorption of CO is unobservable even at higher temperatures. Some investigators indicate that for Pd, Ir and Rh dissociative adsorption becomes notable at elevated temperatures. For example, Zhdan et al.. [242] have observed over an Ir (110) plane partial dissociation of CO when heating the metal upto 600 K in the flow of CO. According to these authors, on Ir (111) heating to 700 K does not lead to dissociative adsorption of CO. Gorodetskii and Nieuwenhuys [289] observed no CO dissociation during its adsorption on rhodium at 13 Pa over the temperature range from 300 to 1000 K. Over Ni, Co and Fe [221, 222] dissociative adsorption of CO is observed even at temperatures above 320 K. Some investigators suggest, however, that on these metals also complete dissociative adsorption does not occur. The phenomenon observed must be simply the transformation of the α-form (C atom interacts with the metal) to the β-form (both C and O interact with the metal). For example, during the adsorption of CO in the β-form on tungsten the binding energy in the molecule of CO remains equal to 170 kJ mol^{-1}. During CO adsorption on nickel there was observed the disproportionation reaction:

$$\alpha\text{-CO} + \beta\text{-CO} \rightarrow C_{surf} + CO_2$$

The heat values of CO adsorption for transition metals are comparable (Table 17) and close to the energies of the M-CO bonding in transition metal carbonyls.

1. Kinetics and Mechanism of CO Oxidation on Platinum and Palladium Catalysts

The past decade is characterized by numerous attempts to investigate the reaction of CO oxidation on platinum and palladium catalysts mainly at low pressures ($< 10^{-2}$ Pa) using various physical methods: LEED, electron spectroscopy, modulated molecular beams, TD, etc. Such a vast information even enabled Ertl [259] to assert that CO oxidation is the most comprehensively studied and understood reaction in the field of heterogeneous catalysis. However, the literature contains an abundance of conflicting opinions both on the mechanism of CO oxidation and on the numerical values of the main kinetic characteristics.

Langmuir, who studied the kinetics and mechanism of CO oxidation on Pt catalysts, found that above 500 K the reaction occurs *via* interaction between gaseous CO molecules and atomically adsorbed oxygen [223].

Tretyakov et al. [224, 225] cleaned the platinum surface in vacuum (10^{-8} Pa) and studied the kinetics in a static system at low pressures. Alternatively,

Table 17. CO adsorption heats at small coverages/kJ mol^{-1}

Metal	Plane		
	(111)	(100)	(110)
Ru	121 [217]		104 [261]
Rh	130 [262]	121 [262]	
Pd	143 [218]	152 [263]	167 [216]
Ir	143 [220]	146 [264]	155 [265]
Pt	138 [213]	134 [264]	104 [266]

these authors suggest that both atomically and molecularly adsorbed oxygen takes part in oxidation. Gaseous CO reacts with oxygen whereas adsorbed CO only blocks the active sites on Pt surface. These situations may be described by the following scheme:

1. $O_2 + [\ ] \rightleftarrows [O_2]$
2. $CO + [\ ] \rightleftarrows [CO]$
3. $[O_2] + [\ ] \rightleftarrows 2[O]$
4. $[O_2] + CO \rightarrow CO_2 + [O]$
5. $[O] + CO \rightarrow CO_2 + [\ ]$

Further investigations did not, however, confirm the proposals that form the basis for the above scheme. In the reaction conditions of CO oxidation (200–700 K) at low pressures the oxygen adsorption on platinum and palladium is dissociative and irreversible. No adsorbed dioxygen is detected on the surface. Carbon monoxide is adsorbed reversibly and non-dissociatively.

Investigations of the interaction between chemisorbed CO and gaseous dioxygen have shown that if the platinum surface is completely saturated with CO, the reaction does not occur [225, 230, 231]. However, at a partly CO-free surface which provides sites for the dissociative adsorption of dioxygen, the reaction proceeds at a rather high rate. Thus, on platinum this gives evidence for the possibility of interaction between chemisorbed CO and oxygen atoms.

Application of $^{18}O_2$ to study the isotopic composition of the carbon dioxide formed has indicated that during the reaction carbon monoxide does not decompose and the surface carbonate is not an intermediate reaction product [229].

At complete saturation of a platinum surface with atomically chemisorbed oxygen and with the addition of gaseous CO, the formation of CO_2 is very fast. Simultaneously chemisorption of CO also takes place.

Consequently, when considering the CO oxidation mechanism it is necessary to take into account the following possible elementary steps:

$$O_2 + 2[\ ] \rightarrow 2[O] \quad \text{(a)}$$
$$CO + [\ ] \rightarrow [CO] \quad \text{(b)}$$
$$CO + [O] \rightarrow CO_2 + [\ ] \quad \text{(c)}$$
$$[CO] + [O] \rightarrow CO_2 + 2[\ ] \quad \text{(d)}$$

These steps may form two reaction mechanisms. One of them is an adsorption Langmuir-Hinshelwood (LH) mechanism consisting of the steps (a), (b) and (d), the other one is an "impact" Eley-Rideal (ER) mechanism consisting of the steps (a) and (c).

The most disputable question is whether the ER mechanism is realistic in this case or not. Understanding of this point is of principal significance for the explanation and prediction of the catalytic action. If this mechanism is valid, it would mean that the chemisorbed oxygen is so reactive that an oxidizable substance can interact with it being nonactivated by catalyst. In this case the role of the catalyst is determined by its effect on dioxygen, the bond energy and reactivity of oxygen on the catalyst surface. The impossibility of a reaction path *via* an ER mechanism implies that in addition to oxygen activation, an oxidizable substance should also be catalytically activated. In the situation discussed, CO can chemically interact with the catalyst thus promoting the formation of an active complex. This complex is composed of the catalyst and O and CO particles chemisorbed thereon. In principle it also possible that a situation exists where a strong chemical interaction of an oxidizable substance with a catalyst diminishes its reactivity for oxygen.

An ER mechanism has been taken into account by many investigators to interprete their experimental data. Matsushima et al. [226] have found that upon interaction of CO with oxygen that covers the platinum surface the reaction rate is high and proportional to CO pressure and is independent of the surface coverage with oxygen over the range of large coverages ($1 \geqq \theta_o \geqq 0.4$). The activation energy of this interaction is very small. The maximum probability of the reaction when CO is colliding with the platinum surface is 0.4 for a smooth plane (111) at $\theta_o = 0.03$ and for a stepped plane at $\theta_o = 0.07$. This probability value remains unchanged with further coverage growth [221]. This phenomenon can be explained by the mobility of CO on the surface and its capability to reach an oxygen atom (*prior* to desorption) at a distance three times larger than the lattice constant. On a polycristalline platinum [228] at 300 K the probability of the interaction of CO from the gas phase with adsorbed oxygen reaches a maximum, that is 0.7 ± 0.05, with a surface coverage being of about 3×10^{14} atoms O cm^{-2}. Adsorption of CO occurs simultaneously with the above reaction. According to the authors [228], the reaction occurs between the strongly bound atomic oxygen and weakly bound CO that migrates along the surface.

To ascertain the relative role of ER and LH mechanisms, the response method has been used in ref. [232]. With a sharp variation of CO concentration a rapid change in the reaction rate was followed by some slowdown. It seemed reasonable to attribute this rapid change to the ER (impact) mechanism whereas the slow change to the LH mechanism (*i.e.* to the rate variations of the reaction between adsorbed species). It became therefore possible to measure separately the rates of CO_2 production *via* the above mechanisms and to establish their dependence on the reaction conditions.

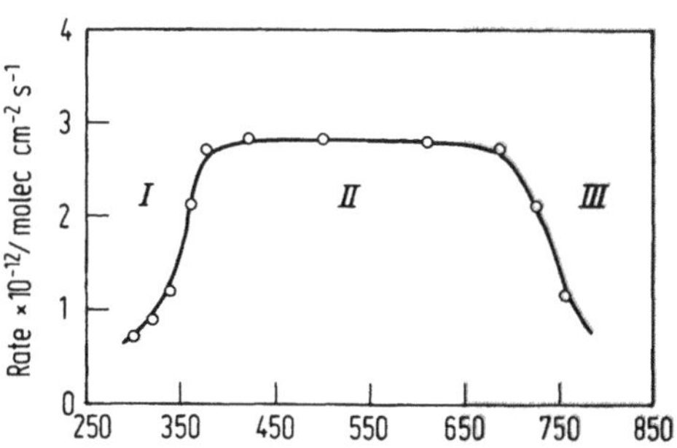

Figure 29. The rate of CO oxidation on polycrystalline platinum *vs.* temperature

As found, the rate of an ER pathway is proportional to CO pressure and at the same time in a complex manner depends upon the oxygen-covered surface (being proportional to this surface coverage only in the range of very small coverages).

The temperature dependence of the rate plotted in Figure 29 is also rather surprising. Over the range of low temperatures (293–470 K) the rate increases with the temperature rise in accord with an activation energy of 30 ± 4 kJ mol^{-1}. At temperatures above 470 K up to 700 K the rate remains constant but tends to decrease upon further heating. The authors think that CO molecules that reach the platinum surface initially are weakly bound and can migrate along the surface. The reaction probability is determined by the capability of a CO molecule to reach an oxygen atom on the surface and react with it *prior* to desorption or strong bonding. This probability is proportional to the rate of surface diffusion. In the light of the above proposal the activation energy observed at low temperatures is thought to be the the diffusion activation energy.

Over the range of moderate temperatures the reaction probability is about unity, that is, each CO molecule colliding with the surface has a chance to react with oxygen. As a result, no appreciable change in the reaction rate is observed with the temperature increase.

Separate measurements of the rates of CO production *via* ER and LM mechanisms on a polycrystalline platinum were conducted by Pacia *et al.* [269] using a molecular beam technique. They have found that the reaction *via* the "adsorption" mechanism shows an activation energy of 92 kJ mol^{-1}. The interaction of adsorbed oxygen with CO from the gas phase or from the weakly bound state (impact mechanism) proceeds without an activation energy and with a probability of about 0.5.

Palladium and platinum catalysts were found to be similar in their catalytic effect on CO oxidation. In the earlier studies concerning the mechanism of CO oxidation on palladium the conclusion was made that the reaction occurs between adsorbed CO and O species (adsorption mechanism) and that carbon monoxide is a strong inhibitor. The following kinetic equations were proposed:

$$r = kp_{O_2}p_{CO}^{-1} \text{ (ref. [233]) } \quad \text{and} \quad r = kp_{O_2}p_{CO}^{-2} \text{ (ref. [234])}$$

However, later more preference was given to the "impact" mechanism. For example, Tretyakov *et al.* [235], who studied the reaction at 1.3×10^{-3} to 13 Pa and 350–770 K over palladium cleaned in high vacuum, have suggested that the reaction is reduced to the interaction of gaseous CO with the atomically adsorbed oxygen. This interaction shows no activation energy and is characterized by a 0.2 probability. This indicates the high reactivity of oxygen atoms adsorbed on palladium. Close and White [238] studied the reaction over Pd treated in oxygen (10^{-4} Pa) and then heated at 900 K. The catalyst so prepared contained oxygen in its subsurface layer, sulphur and possibly other impurities on its surface. The authors distinguished three temperature regions characterized by different orders with respect

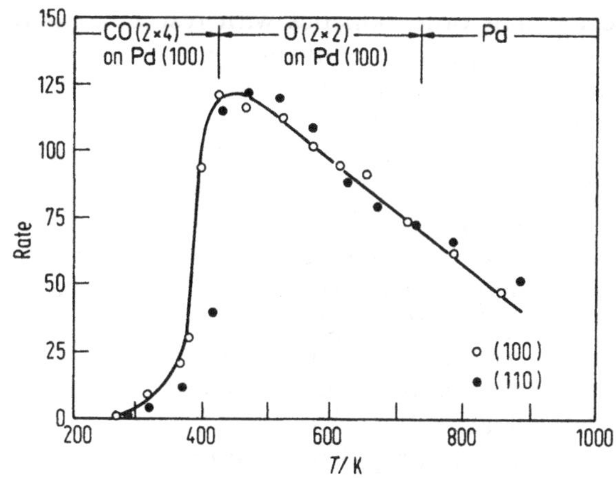

Figure 30. The influence of temperature on the rate of CO oxidation on Pd (100) — ○ and (110) — △. (After [237])

to dioxygen and CO pressures. For the low temperature region (I) first order with respect to dioxygen and negative order (−0.5) with respect to CO were observed. The reaction rate increase with temperature corresponds to an activation energy of 47.7 ± 3.3 kJ mol^{-1}. The reaction in this region is assumed to obey an "impact" mechanism involving inhibition by adsorbed CO. The same mechanism has been found in the region II (473–673 K) when the surface is CO-free and is covered with adsorbed oxygen. Accordingly, kinetic orders of zero and first order with respect to oxygen and CO respectively are observed. The apparent activation energy is −8.84 ± 0.08 kJ mol^{-1}. A small negative activation energy may be attributed to the decrease of the oxygen concentration on the surface with the temperature increase.

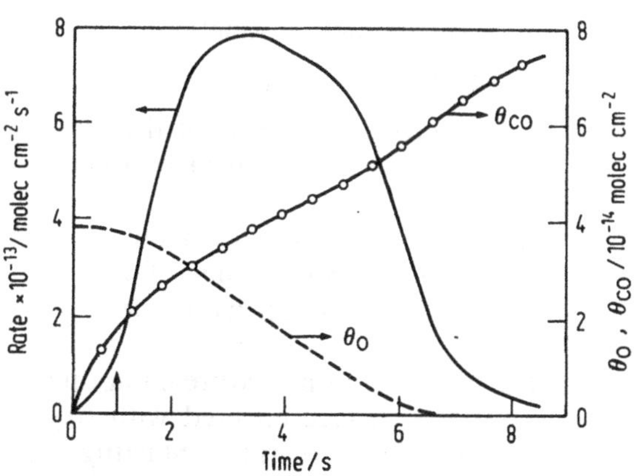

Figure 31. Variations with time of the rate of CO_2 production and of the surface concentrations of O and CO on Pd (111) covered with adsorbed oxygen under the action of CO molecular beam (surface temperature 374 K) (After [267])

In the high-temperature region half order kinetics with respect to both oxygen and CO was observed. The apparent activation energy is -54.0 ± 3.9 kJ mol^{-1}, which is due to a sharp decrease in the concentration of adsorbed oxygen when increasing temperature.

Ertl *et al.* examined CO oxidation on Pd in more detail. In their earlier studies [236, 237] on CO oxidation over low index planes of palladium single crystals an "impact" mechanism was proposed. The temperature dependence they found (Figure 30) shows that chemisorbed CO blocks the surface and thus a fast reaction is possible only after the surface has become partly free to allow the dissociative adsorption of oxygen. The highest reaction rate is observed when the surface is completely covered with oxygen atoms.

The subsequent results led Ertl to question the validity of the experimental evidence for the existence of an ER mechanism. One of the main evidences in favor of this mechanism was a rapid reaction occurence under the action of gaseous CO on the oxygen-covered catalyst surface. More accurate experiments with the use of molecular beams [267] have proved that the reaction needs some induction period to reach its maximum. Figure 31 illustrates the reaction rates and surface concentrations of oxygen atoms and CO molecules measured during the interaction between a CO molecular beam and the oxygen-covered Pd(111) at 374 K [267]. Upon introduction of a CO beam the rate of CO_2 formation is zero but increases rapidly and attains its maximum in three seconds. Hence, all previous conclusions concerning the ER mechanism characterized by the high (close to unity) yield of CO_2 compared to the number of CO molecules colliding with the surface become, indeed, doubtful. The results of this and earlier work may be explained on the basis of the adsorption mechanism if one takes into account that the time resolution in the previous experiments was insufficient to observe an induction period. According to Ertl, this conclusion may be considered to be proved for palladium catalysts and highly probable for platinum catalysts.

The reaction that follows the adsorption mechanism (assuming ideal adsorption) may be described by the following simple kinetic equation:

$$r = k_o \theta_O \theta_{CO} \exp[-E_p/RT] = f(T, p_{O_2}, p_{CO}, p_{CO_2})$$

where k_o is the frequency factor, E_p is the activation energy of the reaction between adsorbed species, θ_O and θ_{CO} are the current surface concentrations of oxygen atoms and CO molecules.

It does not seem possible to describe in any general form the reaction rate as a function of the reactants pressure over a wide range of variations of reaction mixture composition and temperature since, as a rule, θ_O and θ_{CO} are nonequilibrium and depend not only on the pressure of a given reactant but also on the pressure ratio of the both reaction mixture components.

In the steady state reaction conditions the surface covered with oxygen is small and is determined by the reactants pressure ratio. Starting from p_{CO}/p_{O_2} ratio equal to about 0.1, θ_O is much smaller than the equilibrium one with respect to oxygen, and at $p_{CO}/p_{O_2} \geqq 1$, θ_O becomes very small [226].

The adsorption of carbon dioxide is very weak such that its effect on the reaction rate can be neglected.

Consider now the kinetic dependences of CO oxidation on platinum metals for some particular situations assuming that the catalytic surface is uniform. We restrict ourselves to the region irreversible adsorption of oxygen that occurs at low pressures at temperatures below 700 K. CO adsorption is irreversible at temperatures below 500 K and $p_{CO} \leqq 10^{-4}$ Pa. Let us take into account the known experimental data which indicate that when the surface is completely covered with CO, its interaction with oxygen and adsorption of this latter does not take place. When the metal surface is covered with oxygen atoms to form a (2 × 2) O structure CO is adsorbed and enters the reaction at a high rate. During the formation of the (2 × 2) O structure the sticking probability of oxygen falls sharply (by several orders of magnitude) whereas that of CO remains equal to some tenths of unity. Adsorbed CO may react with oxygen but this reaction occurs only after a small induction period (*e.g.* 3 s at 374 K). Yet the nature of the induction period remains in question. It may be associated with the "search" of a CO molecule for an advantageous orientation with respect to oxygen atoms in the (2 × 2) O structure.

As a result of the formation of CO_2, equal amounts of oxygen atoms and CO molecules are removed from the surface. The arrival of oxygen atoms per unit time is $2S_{O_2}Z_{O_2}p_{O_2}$ and of CO molecules is $S_{CO}Z_{CO}p_{CO}$. Here S_i stands for the sticking probability, Z_i stands for the coefficients for the pressure in the expression for the number of impacts on the surface, p_i is the pressure. The difference between Z_{O_2} and Z_{CO} may be neglected. According to Ertl, for the clean surface S_{O_2} is about 0.4 and S_{CO} is about unity. Therefore, for the free surface $2S_{O_2}Z_{O_2} \approx S_{CO}Z_{CO}$ and the advantage to occupy a free surface site is determined by the pressure.

1. CO is adsorbed irreversibly. $p_{O_2} < p_{CO}$.
In this case the free surface is occupied predominantly by CO. Since only the adsorbed oxygen is involved in the reaction, with time the total surface will be covered with CO and the reaction will be thus terminated. For the reaction to occur, the temperature must be raised up to the value at which CO desorption occurs.

2. CO is adsorbed irreversibly. $p_{O_2} > p_{CO}$.
The free surface is occupied predominantly by oxygen. Assuming that only chemisorbed CO is involved in the reaction, the surface in the steady state will be covered with oxygen. Since oxygen does not inhibit CO adsorption, all striking and sorbed CO molecules will be involved in the reaction. The reaction rate is

$$r = S_{CO}Z_{CO}p_{CO}$$

and does not depend upon the oxygen pressure. Here S_{CO} is the sticking probability of CO relative to the oxygencovered surface of (2 × 2) O. The activation energy is determined by the temperature dependence of S_{CO}.

3. CO is adsorbed reversibly. $p_{O_2} < p_{CO}$.
The free surface is covered predominantly with CO but due to the significant desorption, θ_{CO} will be smaller than unity. One may therefore assume that θ_{CO} equals approximately the equilibrium coverage with respect to $p_{CO}(\theta_{CO} = \theta_{CO}^{Eq})$. In excess CO, θ_O is very small and has no effect on the reaction rate. All the molecules of dioxygen adsorbed on the free surface participate in the reaction. The portion of the free surface (θ^o) is determined as

$$\theta^o = (1 - \theta_{CO}) = 1/1 + b_{CO}p_{CO} \approx 1/b_{CO}p_{CO} = 1/b_{CO}^o \exp(q/RT) p_{CO}$$

where b_{CO} is the adsorption coefficient of CO, q is the heat of CO adsorption. The reaction rate is proportional to oxygen pressure and the fraction of the free surface

$$r = S_{O_2} Z_{O_2} P_{O_2} \theta^o \propto p_{O_2} p_{CO}^{-1} .$$

The observed activation energy equals the sum of the activation energy of dioxygen adsorption (determined by the temperature dependence of dioxygen sticking probability, S_{O_2}) and heat of CO adsorption.

4. CO is adsorbed reversibly. $p_{O_2} > p_{CO}$.
The free surface is covered predominantly with oxygen and if CO is reactive only in a chemisorbed state, the steady state surface is covered with oxygen that produces the (2 × 2) O structure. As has been mentioned above, when a CO molecule strikes such a surface, CO can adsorb and participate in the reaction. The reaction rate of CO adsorption on this surface is

$$r = S_{CO} Z_{CO} p_{CO} \propto p_{CO}$$

and is proportional to CO pressure and independent of oxygen pressure. When decreasing the oxygen pressure, some portions of a nonuniform surface may become oxygen-free and the order with respect to oxygen may exceed zero (likewise in the previous case it will become less than unity). The activation energy is determined by the temperature dependence of the CO sticking probability on the oxygencovered surface.

On the basis of the LEED measurements in unsteady state conditions [268] Engel and Ertl propose that in the case under consideration oxygen forms surface "islands" and the reaction occurs by virtue of CO molecules diffusing to the boundaries of these islands [259]. Apparently, in the steady state the oxygen islands should either occupy the whole surface or disappear.

We have considered so far CO_2 production only *via* the adsorption (LH) mechanism. If one adopts the point of view that gaseous CO takes part in the reaction (ER mechnism) it is necessary then to take into account an additional rate of oxygen removal from the surface without the participation of chemisorbed CO. This changes the balance between the production and removal of adsorbed species in favor of CO, and the relative oxygen pressure that determines the boundaries between the cases 1–2 and 3–4 must increase considerably.

Additional complexity is associated with the possible nonuniformity of the metal surface. The structural sensitivity of CO oxidation reaction on platinum and palladium is assumed to be insignificant [259] although it is not as yet unequivocally defined [243]. Hopster, Ibach and Comsa [227] have found that the effect of a stepped structure in the plane (111) is twofold: the oxygen sticking probability increases exponentially with the increase in the number of steps while the probability of CO interaction with adsorbed oxygen decreases. It is interesting that the authors explain this phenomenon by the change of the collective electronic properties (by analogy to the effect of chemisorbed species [227]) leaving aside possible structural effects. It is necessary to take into consideration that the surface coverage also may influence the oxygen bond energy and, consequently, its reactivity. All of the above facts lead to a somewhat pessimistic conclusion about the possibility of constructing kinetic equations on the basis of simple schemes of elementary reactions.

An essential question is whether the use of the principles established for the clean surface of Pt at very low pressures are correct for real catalysts at pressures close to atmospheric. As reported by Hori and Schmidt [243], at high pressures up to 1200 K the platinum surface is covered with a polymolecular film of complexes formed by reactants, and the reaction rate differs from that found for the clean surface of Pt. McCarthy, Zahradnik, Kuczinski and Carberry [244] studied the kinetics of CO oxidation at atmospheric pressure on a catalyst containing 0.035% Pt supported over $\alpha\text{-}Al_2O_3$. Their conclusion is that at a high oxygen content the reaction obeys the impact mechanism and its rate is proportional to p_{CO} whereas at high pressures of CO the oxidation follows the adsorption mechanism and is inhibited by CO. In the cited work, a reaction oscillatory regime has also been established for a certain interval of CO concentrations (Figure 32). The oscillatory regime in CO oxidation has been observed by many investigators [245–248]. The peculiarities and the nature of this phenomenon have been considered in the previous Section on oxidation of dihydrogen. This phenomenon results from the multiplicity of steady states of the catalytic system. For CO oxidation *via* the adsorption mechanism

$$O_2 + 2[\ ] \rightarrow 2[O]$$

$$CO + [\ ] \leftrightarrows [CO]$$

$$[CO] + [O] \rightarrow CO_2 + 2[\ ]$$

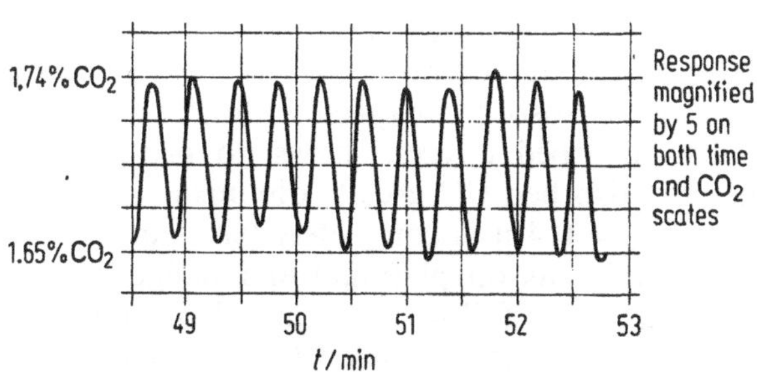

Figure 32. Oscillations in the rate of CO_2 formation on the catalyst surface (After [244])

the kinetic orders with respect to surface concentrations of intermediates [ ], [O] and [CO] differ for individual steps. Here [ ] is the concentration of the free surface sites, and [O] and [CO] are the concentrations of the surface sites occupied by oxygen and CO. As shown in ref. [249], with a certain ratio of the rate constants over a certain pressure interval, a multiplicity of steady states is possible which manifests itself in a sharp change of the reaction rate and in the appearance of hysteresis. Accomplishment of an oscillatory regime requires an additional accumulation of intermediates. If one introduces an additional reversible step of adsorption of one of the components on the inert part of the surface or diffusion of this component in the subsurface layer of the catalyst then the solution of the system of kinetic equations will indicate the possibility of an oscillatory regime at a certain ratio of the rate constants and reactants pressures. Creation of oscillatory regimes may also be related to the dependence of the activation energies of individual steps on the coverage (see also Section on dihydrogen oxidation).

Interesting correlations were observed for CO-dihydrogen co-oxidation. On a platinum wire the oxidation of dihydrogen, CO and CO-H_2 mixture begins at 330 K, 463 K and 425 K, respectively. Thus hydrogen fascilitates CO oxidation. This may be due to the fact that adsorbed CO blocks the platinum surface and thus inhibits oxygen adsorption and terminates both reactions until CO desorption starts (463 K). In the presence of hydrogen the adsorbed CO produces the $Pt = C\langle^{H}_{OH}$ compound which desorbs at lower temperatures (425 K) [240].

2. *Oxidation of CO on Other Metals*

The catalytic activity of iridium has been studied on (111) and (110) planes of an Ir single crystal [250–252, 270, 271] as well as on a polycrystalline metal and that supported over alumina [253]. Adsorption of CO produces the same twodimensional structures as those observed for palladium. The sticking probability is 0.7 for (111) and about unity for (110). The heat of adsorption is 104–140 kJ mol^{-1}. Heating in oxygen changes the structure of (110) which becomes faceted by (111). CO adsorption at low temperatures occurs without the dissociation. Partial dissociation is observed over Ir(110) at above 600 K. Oxygen is dissolved in the subsurface layer of iridium to form, probably, the bulk oxide. The dissolved oxygen does not interact with CO and can be removed when heated to temperatures above 1300 K. However, dissolved oxygen affects the catalytic properties of the surface by notably accelerating the interaction of adsorbed oxygen with CO.

During co-adsorption, the pre-adsorbed oxygen shows no pronounced effect on CO adsorption while pre-adsorbed CO greatly diminishes the rate of oxygen adsorption so that at $p_{CO} = 0.4$ it is completely terminated.

An opinion has been expressed that, as on platinum and palladium on Ir the interaction of oxygen with CO also follows both an adsorption and an impact mechanism upon interaction of gaseous CO molecules with oxygen atoms. For the adsorption mechanism the activation energy is 46 $\pm$ 4 kJ mol^{-1}. The frequency factor is 10^{-11} molec s^{-1} cm^{-2}. Such a low value can be explained on the assumption of an island character of adsorption. For the impact mechanism the activation energy was found to be 4 $\pm$ 2 kJ mol^{-1} [252]. This small value indicates that the atomic oxygen is highly reactive on an irridium surface.

Chemisorbed oxygen in the form of ordered structures and surface phases is less reactive than that randomly distributed on the surface. This is manifested most noticeably in the interaction of CO with chemisorbed oxygen over iridium single crystal. Over Ir(110) with admission of CO the reaction rate is high and proportional to CO pressure and surface oxygen concentration. At the same time it is almost independent of the temperature (Figure 33, curve 1). The case stands otherwise for Ir(111) (Figure 33, curve 2). At low temperatures the reaction occurs very slowly. At 330 K the rate of CO_2 production over Ir(111) is by more than one order of magnitude lower than that over Ir(110). When raising the temperature the reaction rate increases and at 450 K becomes equal to the rate on Ir(110). Disordering of oxygen over Ir(111) is observed in the same temperature region. This may be concluded from the decrease of the intensity of diffraction spots (1/2, 1/2) of the (2 × 1) O structure (see Figure 33, curve 3). One may explain a sharp increase in the rate by assuming that the reaction proceeds through the linear complex O ... C ... O which is parallel to the crystal surface. For the formation of this complex it is necessary that two vacant sites be located nearby the adsorbed oxygen on the same straight line. Such a situation is impossible in the ordered layer, however, its probability grows with the extent of disordering of a two-dimensional phase. On the basis of the above assumption it is also possible to explain the high rate on a microfaceted Ir(110) [271].

When the reaction is conducted in an open system under steady state conditions at temperatures below 450 K the surface is blocked with adsorbed

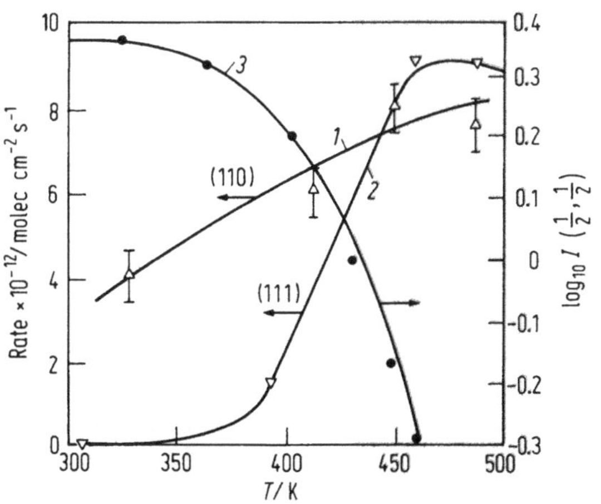

Figure 33. Initial rates of CO_2 formation under the action of CO on the oxygen-covered surface. Curve 1 — plane (110), curve 2 — plane (111), curve 3 — the change of the intensity of spots (1/2, 1/2) corresponding to the (2 × 1) O structure on Ir (111)

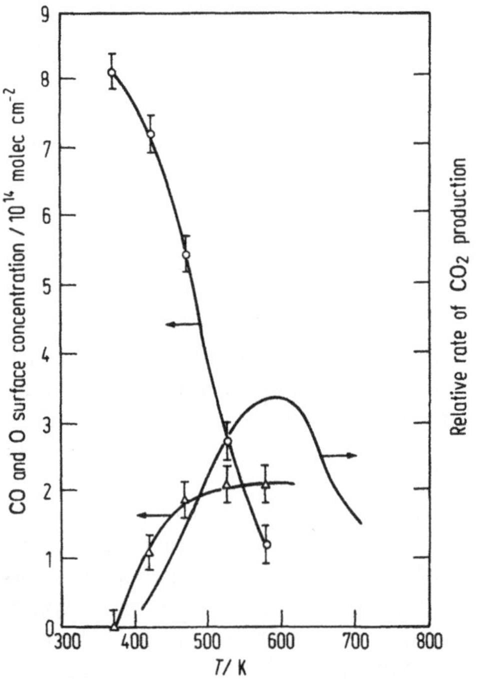

Figure 34. The rate of CO_2 formation and surface concentrations of O (△) and CO (○) in an open system at various temperatures ($p_{O_2} = p_{CO} = 2.4 \times 10^{-5}$ Pa) on Ir (111). [After [259, 293, 251]]

CO molecules, and the reaction starts at higher temperatures after CO has been partially desorbed (Figure 34).

The maximum reaction rate is attained when the extent of CO adsorption significantly decreases and that of oxygen adsorption approaches the maximum. At these temperatures the surface oxygen is disordered and the kinetics can be described in terms of the surface action law. As a result of a detailed kinetic study of the reaction over Ir(110) under the same conditions [270], the following equations have been derived for the temperature region below that of the rate maximum

$$r_1 = \nu_1 p_{O_2}^2 / p_{CO} \exp(-E_1/RT)$$

where E_1 ranges 55 to 65 kJ mol^{-1} and $\nu_1 = 2 \times 10^{15 \pm 1}$ molec cm^{-2} Torr^{-1} $\times$ s^{-1}, and for the temperature region above that of the rate maximum

$$r_2 = \nu_2 p_{CO} \exp(-E_2/RT)$$

where the observed apparent activation energy E_2 ranges -25 to -12.5 kJ $\times$ mol^{-1} and $\nu_2 = 2 \times 10^{19 \pm 1}$ molec cm^{-2} Torr^{-1} s^{-1}.

The authors suggest that the experimentally found second order with respect to oxygen (instead of the calculated first order) is due to pairwise oxygen diffusion which is thought to be the rate determining step. This daring hypothesis of course requires experimental justification.

In the work discussed the rate was measured both upon increasing and decreasing the temperature. The results were found to differ, and a pronounced and well reproducible hysteresis was observed. This hysteresis cannot be attributed only to the formation of the bulk phase of the oxide since it was also was observed for the mixtures with a large excess of CO when the bulk oxide was not produced. Quite probably the hysteresis is the manifestation of the nonuniqueness of steady states of the system that we have considered above in connection with oscillatory regimes in platinum systems.

Far less information is available on CO oxidation over other platinum group metals. Rhodium is known to have a high activity. Campbell *et al.* [254] studied the interaction of gaseous CO with oxygen adsorbed on a rhodium wire at 360–779 K at 5×10^{-6} Pa. The reaction rate is appreciable at 360 K, it attains a maximum at 440 K and then slows down due to acceleration of CO desorption. The authors' statement is such that all the experimental data can be described assuming the existence of only the adsorption mechanism. The activation energy is 104 kJ mol^{-1} and 60 kJ mol^{-1} at temperatures above and below 529 K, respectively. As reported in ref. [289], the interaction of dioxygen with CO chemisorbed on rhodium is observed at 475 K. Madey *et al.* [255] and Reed *et al.* [256] studied CO oxidation over a ruthenium single crystal. Its activity is lower than that of the metals discussed above. In steady state conditions at $p_{O_2} = p_{CO} = 5.5 \times 10^{-5}$ Pa, CO_2 production is observed at 500 K and the formation rate has a maximum at 730 K on Ru(101) and at 950 K on Ru(001). The reaction rate on the former plane is higher by a factor of 5 than that on the latter one. The activation energy of the reaction is 39 kJ mol^{-1}. The rate dependence on the pressure of reactants is described by expressions

$$r = kp_{O_2}^{0.62}p_{CO}^{0.29} \qquad \text{for } p_{CO} < p_{O_2}$$

$$r = kp_{O_2}^{0.62}p_{CO}^{0.55} \qquad \text{for } p_{CO} > p_{O_2}$$

The iron group metals are unstable in conditions of CO oxidation due to a possible dissociative adsorption of CO accompanied by deposition of carbon on the surface or to the oxidation of subsurface metal layers to form bulk oxides. For example, over Ni(111) upon the action of oxygen the (2×2) O structure transforms to the two-dimensional oxide Ni(100) ×c(2×2) O/Ni(111) and further to the bulk oxide NiO [257].

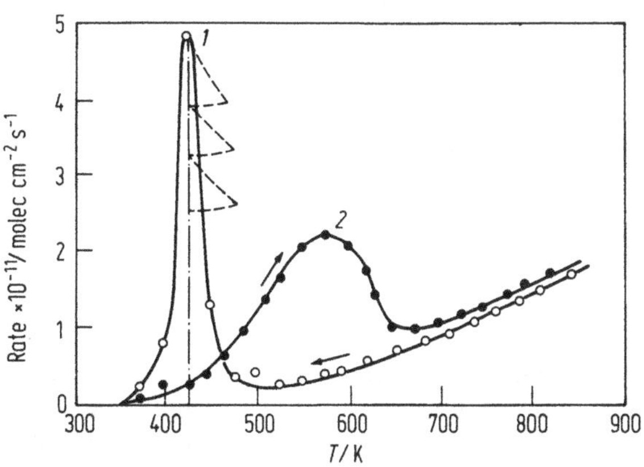

Figure 35. The rate of CO oxidation on Ni (111) upon cooling (1) and heating (2) of the sample ($p_{CO} = 4 \times 10^{-4}$ Pa, $p_{O_2} = 2.6 \times 10^{-5}$ Pa). (After [258])

On a clean metal surface CO oxidation occurs at 500 K at $p_{O_2} = 3 \times 10^{-5}$ Pa and $p_{CO} = 4 \times 10^{-4}$ Pa. When the temperature is raised, the reaction rate tends to increase. Upon returning to lower temperatures, a sharp maximum appears at about 435 K (Figure 35). The maximum activity does not correspond to the catalyst steady state since keeping the catalyst at this temperature leads to a gradual decrease of the catalytic activity. The appearance of the rate maximum coincides with the formation of a two-dimensional structure Ni(100) c(2 × 2) O/Ni(111), *i.e.* a drastic increase in the catalytic activity is, evidently, caused by the rearrangement of the surface structure. Further oxidation produces the oxide bulk phase and is accompanied by a sharp fall of the catalytic activity [258].

B. Oxide Catalysts

1. Simple Oxides

A number of transition metal oxides show high catalytic activity in CO oxidation and have found wide use in respirators and catalytic apparatus for detoxication of exhaust gases. Nevertheless, they are inferior in specific catalytic activity (SCA) to the platinum group metals. Having analyzed numerous literature data, Krylov [272] has found the following order of SCA variations: $MnO_2 > CoO > Co_3O_4 > MnO > CdO > Ag_2O_3 > CuO > NiO > SnO_2 > Cu_2O > Co_2O_3 > ZnO > TiO_2 > Fe_2O_3 > ZrO_2 > Cr_2O_3 > CeO_2 > V_2O_5 > HgO > WO_3 > ThO_2 > BeO > MgO > GeO_2 > Al_2O_3 > SiO_2$.

Measurements of the reaction rate in mixtures with a stoichiometric ratio of components in steady state conditions at 500 K indicate a similar order [273]:

$$Co_3O_4 > CuO > NiO > Mn_2O_3 > Cr_2O_3 > Fe_2O_3 > ZnO > V_2O_5 > TiO_2 .$$

This order is close to that found for SCA's of the same oxides with respect to hydrogen oxidation (cited in the foregoing section). Figure 36 illustrates the SCA's for CO (upper curve) and H_2 (lower curve) oxidation reactions expressed by the number of molecules of oxidation products $cm^{-2}\,s^{-1}$ at atmospheric pressure [273]. The rate of CO oxidation exceeds that of hydrogen by a factor of 10—100. Of particular interest is that the relative activity of oxides with respect to both reactions changes exactly in the same manner. One can therefore anticipate that the mechanisms of CO and dihydrogen oxidation must be similar. At elevated temperatures the oxidation of CO obeys a stepwise mechanism

$$CO + O_{cat} \rightarrow CO_2 + [\ ]$$

$$1/2\, O_2 + [\ ] \rightarrow O_{cat}$$

Here [ ] is an oxygen vacancy on the oxide surface. O_{cat} is most probably O^{2-} although some other forms cannot be ruled out (*vide supra*). Re-oxidation of the catalytic surface is rather fast and the first step of oxygen with-

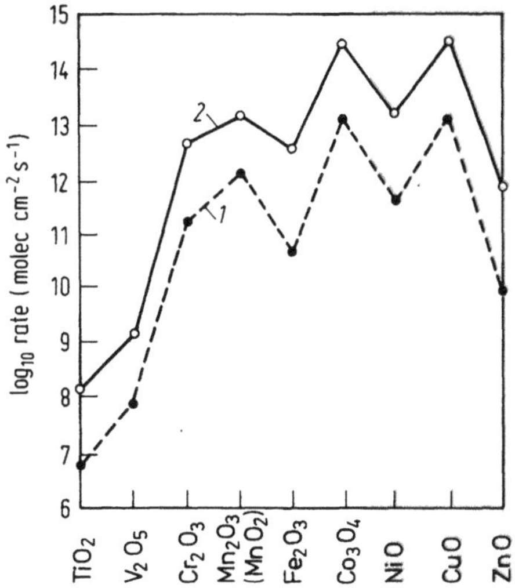

Figure 36. Comparison of the SCA's of oxides of transition metals of the IVth period in oxidation of dihydrogen (curve 1) and carbon monoxide (curve 2) (T = 500 K, pressure of oxidizable substances 270 Pa in the mixture with air, the total pressure 1 atm; SCA's are expressed in molecules of products per cm^2 per s). After [273])

drawal from the catalyst appears to be the rate-determining step. Hence, the smaller the bond energy of oxygen on the surface, the higher is the catalytic activity of oxides. These dependences have been considered in more detail in the foregoing section to describe the mechanism of dihydrogen oxidation.

The idea that the oxidation of CO occurs *via* a stepwise mechanism has been verified by directly comparing the rates of the steps of reduction and

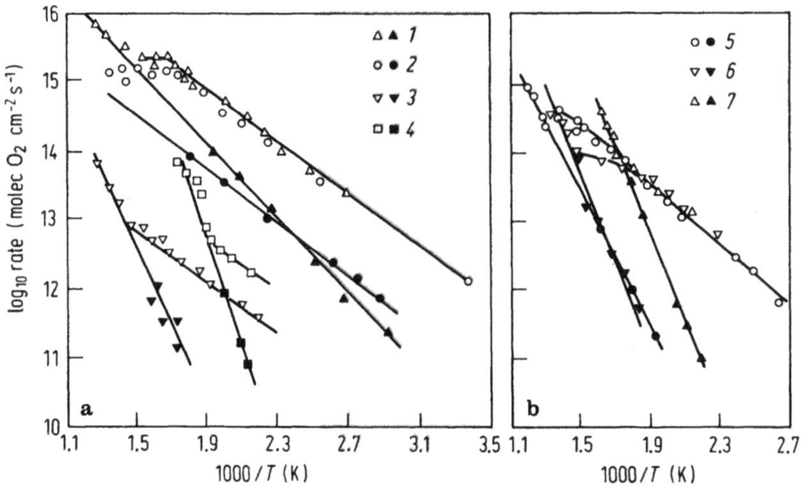

Figure 37. Temperature dependence of the rates of the catalytic oxidation of CO and the steps of oxidation and reduction of oxides (molec CO cm^{-2} s^{-1})
a: 1 — Co_3O_4, 2 — CuO, 3 — ZnO, 4 — Fe_2O_3
b: 5 — Mn_2O_3, 6 — Cr_2O_3, 7 — NiO.
Light points — the rates of catalysis,
dark points — the rates of the steps of oxidation and reduction of oxides

re-oxidation of oxide catalysts with the reaction rate at the catalyst steady state. At high temperatures the rates of these steps coincide with the independently measured rates of the catalytic reaction. The situation significantly changes in the region of lower temperatures. This is illustrated in Figure 37 for oxides of the 4th period: as the temperature is decreased, the catalytic reaction rate decreases to a lesser extent than the rates of reduction and reoxidation [273]. Analogous results were obtained in some other studies too [274–277]. As shown by a detailed study on copper oxide [274–277], at the temperatures above 525 K the reaction follows the stepwise mechanism, however some deviations are observed at lower temperatures. It has been found in this work that the rate of re-oxidation decreases with increasing the time of exposure that follows the reduction. Such dependence is supposed to be caused by the diffusion of oxygen vacancies into the crystal bulk during the exposure.

If the available information is put together, it is possible to state with greater assurance that at high temperatures the catalytic oxidation on transition metal oxides follows the stepwise mechanism. At low temperatures another reaction pathway corresponding to the concerted mechanism may simultaneously occur.

The oxidation of CO on oxides, unlike that on metals, produces surface carbonate structure. The rates of formation and destruction of these structures can be measured from the intensity variations of their absorption bands in IR spectra. The destruction process requires considerable activation energy since it includes the oxygen — oxide bond rupture. At elevated temperatures the destruction of surface carbonate producing CO_2 is sufficiently fast and CO oxidation follows the stepwise mechanism. As the temperature is decreased, the rate of destruction rapidly falls and substantial contribution is made by another pathway, *i.e.* conjugated destruction of surface carbonate and reoxidation of the oxide reported by Sokolovskii [293]. As was found, at low temperatures the production of CO_2 is notably accelerated by dioxygen present in the reaction mixture due to the imposition of the concerted mechanism. For example, CO oxidation on zinc oxide *via* the stepwise and concerted mechanisms can be represented as:

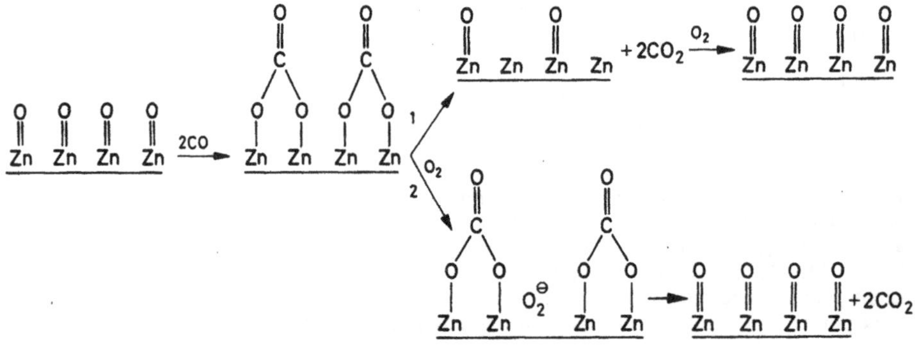

For V_2O_5 the catalytic reaction rate coincides with the rates of reduction and reoxidation steps only at 860 K but at lower temperatures the concerted type pathways play an important role. For silica-supported 1% V_2O_5 the catalytic reaction rate at 733 K was three orders of magnitude greater than the rate of reoxidation of a partly reduced sample [278]. The authors propose the following scheme of elementary reactions involving oxygen anion-radicals O_2^- and O^- observable by ESR and characterized by a high reactivity for CO:

1. $CO + [O^{2-}] \rightarrow CO_2 + 2e$
2. $2e + 2[V^{5+}] \rightarrow 2[V^{4+}]$
3. $[V^{4+}] + O_2 \rightarrow [V^{5+}O_2^-]$
4. $[V^{5+}O_2^-] + [V^{4+}] \rightarrow 2[V^{5+}O^-]$
5. $[V^{5+}O^-] + [V^{4+}] \rightarrow 2[V^{5+}] + [O^{2-}]$
6. $[V^{5+}O_2^-] + CO \rightarrow CO_2 + [V^{5+}O^-]$
7. $[V^{5+}O^-] + CO \rightarrow CO_2 + [V^{4+}]$

According to Golodets [279], this scheme includes the two reaction pathways. Steps 1–5 are reduced to the detailed scheme of the stepwise redox mechanism whereas steps 3, 6 and 7 form the cycle of the associative (concerted) type. To prove the validity of this scheme, it is necessary to measure the rates of individual steps which is rather a difficult problem. The concerted mechanism has been suggested previously for CO oxidation on manganese dioxide [280]. After CO chemisorption on the surface of manganese dioxide CO_2 is thought to be formed by virtue of dioxygen chemisorbed from the gas phase rather than *via* bonding of the catalyst oxygen. Otherwise (when oxygen withdrawal from the gas phase is insufficiently fast) the catalyst is reduced irreversibly, and the catalytic activity falls. In contrast, in a number of works [281] more prefernece was given to the stepwise mechanism. The above consideration strongly suggests that differing interpretations result from different conditions of reaction performance employed by the investigators. As reported by Kabayashi [282], CO oxidation on manganese dioxide at temperatures below 270 K obeys the concerted mechanism. The authors suggest the presence on the catalyst surface of an active form of oxygen, probably, O_2^- or O^-, with which gaseous CO can react without adsorption. The rate determining step is that of the desorption of produced CO_2.

2. *Mixed Oxide Catalysts*

The search for active catalysts for detoxication of CO-containing gases and control of combustion processes has led to the development of numerous mixed oxide catalysts. The most active are the mixtures of those oxides that exhibit a high activity in their pure state. For example, the so-called "hopkalites" prepared by mixing manganese dioxide with copper oxide (60% MnO_2 and 40% copper oxide) or a large number of oxides (50%

MnO_2, 30% CuO, 15% CoO_3 and 5% Ag_2O_3) have found wide application in practice. Many other mixtures and compounds of spinel and perovskites have also been studied. Some of chromites and ferrites exhibit notable activity, however, at higher temperatures than does manganese dioxide. Alumina supported copper oxide, catalysts based on copper chromite and on some of manganites and ferrites have also found practical application.

When transition metal cations are introduced into the zeolite composition to produce isolated cations (low extents of substitution), ACA of copper, nickel, and cobalt is four orders of magnitude less than the ACA of these cations in oxides (Figure 38) [290]. Zeolites exchanged in this way can be regarded as the salts of a strong acid in which the cations are bonded only to the oxygen of the zeolite framework [290]. Oxygen electron pairs are shifted to the anion which retards the electron transfer to the cation upon oxygen withdrawal.

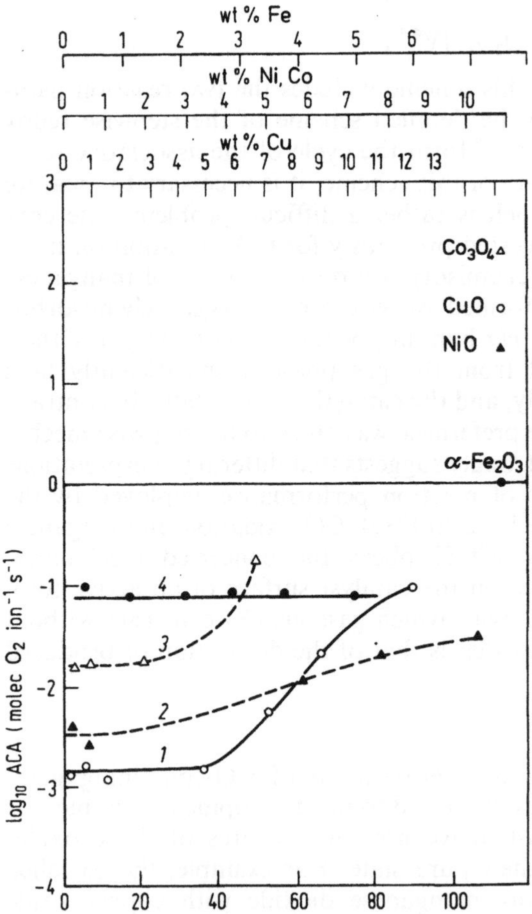

Figure 38. Atomic catalytic activity of transition metal cations in zeolites and oxides, (molec O_2 s^{-1} ion^{-1})
1 — copper zeolite, 2 — nickel zeolite, 3 — cobalt zeolite, 4 — iron zeolite, 5 — Fe_2O_3, 6 — NiO, 7 — CuO, 8 — Co_3O_4

The situation is different when the cations are introduced into zeolites *via* the ion exchange with a partially hydrolized solution containing cation associates. The cation groups, which after dehydration are linked with each other by oxygen bridges, are located in zeolite cavities. In this case the metal-oxygen bond is significantly more covalent which eventually leads to a drastic increase of the ACA. Hydrolysis always takes place in iron salts, therefore, isolated cations are not formed, and the ACA of such cations in zeolites is close to that of oxides [291].

Introduction of additives allows one to increase the activity of oxides which are slightly active in their pure state. For example, MoO_3 or K_2SO_4 additives [283, 284] notably increase the activity of vanadium pentoxide. In the both cases the activity increase is associated with the decrease of the oxygen binding energy.

In order to check the ideas of the so-called "electronic theory" of catalysis [285], in a number of studies the transition metal oxides were modified by additives which changed their semiconductor properties. With this aim, nickel oxide (p-semiconductor) and zinc oxide (n-semiconductor) modified by cations of Li, Ag, Ga, Cr, etc. were most often used [286, 287]. Interesting results have been obtained on the shift of the Fermi level, the change of electrical conductivity and catalytic properties (SCA and activation energy of CO oxidation). However, since usually additives affect not only the semiconductor characteristics but also the chemical properties of catalysts, the investigators met with no success in their attempts to prove the validity of predictions by "electronic theory" [288]. Thus, the most general feature that allows prediction of the catalytic action still remains the oxygen binding energy on the catalyst surface.

6. Catalytic Oxidation of Hydrocarbons and Other Organic Compounds

Catalytic interaction of dioxygen with organic compounds proceeds, as a rule, *via* many pathways producing various compounds. When these reactions are performed on a commercial scale, the end products are usually either those of partial oxidation (aldehydes, ketones, acids) or oxygen-free products of oxidative dehydrogenation. In most cases multicomponent catalysts are used. The only exception is production of ethylene oxide by the oxidation of ethylene which is carried out on metallic silver.

The mechanism of partial oxidation is rather complex and diverse, however, some common principles (at least for oxide catalysts) can be outlined. The surface of catalysts for partial oxidation should not contain weakly bound oxygen, otherwise the reaction will lead to only complete oxidation products, namely, water and carbon dioxide. Products of partial oxidation are formed and retained if the catalyst surface has only strongly bound oxygen with a bond energy no less than 250 kJ mol^{-1}. However, as has been shown hereabove for oxidation of simple molecules, the reactivity of oxygen rapidly decreases with increasing the energy of its bonding to a catalyst. In contrast to reactions involving weakly bound oxygen for which the interaction of an oxidizable substance with a catalyst has no importance, in reactions of partial oxidation the activation of an oxidizable compound strongly influences the rate of conversion. Therefore, selection of catalysts for partial oxidation is a more complicated problem compared to selection of active catalysts for complete oxidation.

In this section we restrict ourselves to reactions of complete oxidation of organic compounds producing water and carbon dioxide. In this case the main requirement of oxide catalysts is also the presence on the surface of weakly bound oxygen (likewise for oxidation of hydrogen and carbon monoxide). This statement is probably valid for metals as well, however, it needs further experimental confirmation.

A. Metallic Catalysts

Reactions of complete oxidation of organic compounds on metallic catalysts have been studied mainly in relation to detoxication of exhaust gases from industry and transportation. For this purpose the main attention has been devoted to catalysts containing platinum and palladium. The apparent lack of interest in catalytic properties of other metals is explained either by their costs or instability in conditions of the process performance.

1. Oxidation of Paraffins

The oxidation of paraffins has been studied mainly on the platinum group metals. The high catalytic activity of platinum for the interaction of methane with dioxygen has been established in a classical work of Davy [295]. The other platinum group metals, *e.g.* Pd, Rh, Ir, and Ru also show fairly high activity. The comparative data on activities are rather contradictory [296 to 298]. The most probable (according to Golodets [299]) values of the activation energy of methane oxidation over a number of alumina supported metals are compiled in Table 18.

The rate of methane oxidation is proportional to the pressure of methane and is independent of the pressure of oxygen. It is likely that when oxygen is present in the reaction mixture the extent of surface coverage with oxygen is close to unity and the reaction rate is determined by the step of methane

Table 18. Activation energies of methane oxidation on alumina supported metals

Metal	Activation energy/ $kJ\,mol^{-1}$	References
Pt	100	[296]
Pd	92	[296]
Ir	71	[298]
Rh	113	[298]
Au	134	[298]

interaction with chemisorbed. oxygen. Whether this interaction occurs by the impact mechanism or by pre-chemisorption of methane has not been so far reliably determined.

The rate of oxidation of methane homologs increases with the molecular weight; however, the relative catalytic activities of the platinum group metals remain unchanged.

2. *Oxidation of Olefins*

The oxidation of olefins is more rapid than that of paraffins on the same metals. According to Kemball [301], the relative activity of metals decreases in the following order:

$$Pt > Pd > Rh \gg Au > W\,.$$

For transition metals the heat of oxygen adsorption increases in the same order. The catalytic activity of silica supported metals decreases in the order Pt > Pd > Cu > Ag [302]. In excess oxygen on palladium, first order with respect to ethylene and zero order with respect to oxygen are observed [303]. On silver ethylene is oxidized producing mainly ethylene oxide.

The catalytic activity of platinum metals in oxidation of propylene changes in the same order as it does in ethylene oxidation [301]. In the course of oxidation on silver only small amounts of propylene oxide are formed, the main products being carbon dioxide and water.

One may propose that upon interaction of olefins with oxygen adsorbed on the metal surface the π-bond has to be broken in the olefin molecule producing the surface complex of the type

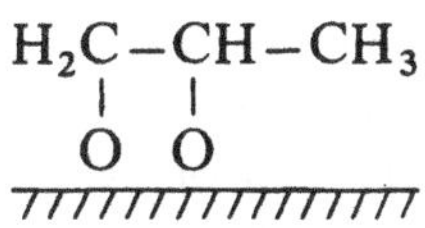

This complex undergoes the fast oxidation with removal of hydrogen and formation of carboxylate complexes that transform to products of complete oxidation. During propylene oxidation small amounts of acetone are formed most likely *via* the parallel pathway. Oxidation of dienes is slightly slower than that of the respective olefins.

B. Oxide Catalysts

As was mentioned above, the complete oxidation of organic compounds requires the presence on the surface of weakly bound oxygen. Correspondingly one may expect the same order of SCA variations in the series of oxides during oxidation of hydrocarbons as was observed during oxidation of dihydrogen and carbon monoxide. Some deviations are, however, possible especially for oxides with a high oxygen binding energy, which can be associated with the specific interaction of an oxidizable substance with a catalyst. In partial oxidation this interaction has a significant influence on the reaction rate.

1. Oxidation of Paraffins

The oxidation of paraffins on oxide catalysts occurs more slowly than that of other hydrocarbons. Among paraffins themselves the least reactive is methane. Figure 39 shows SCA's of oxides of the metals of the 4th period with respect to the oxidation of methane [304]. For comparison, the figure displays the data for the oxidation of other substances. The rate of methane oxidation is 1–2 orders of magnitude less than that of dihydrogen. For oxides with a small oxygen binding energy the order of the activity variations is similar. However, for oxides with a high oxygen binding energy this order is notably variable. For example, on vanadium pentoxide the oxidation of methane does not occur even at 800 K. The relative rate of methane oxidation is substantially slower also on copper oxide. Hence, one may conclude that the SCA depends not only on the energy of bonding of oxygen to the catalyst surface but also on the nature of the interaction of methane with catalysts. Significant differences are observed in the position of the temperature boundary of the stepwise mechanism which is predominant in methane oxidation only at temperature above 650 K. At low temperature the rate of the catalytic oxidation of paraffins is significantly higher than that of reduction of the catalytic surface with an oxidizable substance, which indicates that the concerted mechanism is predominant. As shown by IR spectroscopy study, complexes of the carboxylate structure, most probably formates, are present on the surface [305]. When the temperature increases, thus approaching the temperature boundary of the stepwise mechanism, the concentration of these complexes drastically decreases.

This allows one to propose that on oxide catalysts for complete oxidation having weakly bound oxygen on their surface, the first step of methane oxidation is that of the formation of carboxylate complexes with the participation of the oxygen from the catalyst surface [300]. In the case of the stepwise mechanism (above 650 K) then follows the step of destruction of these complexes producing CO_2 and water. The final step must be the reoxidation of the catalyst with dioxygen from the gas phase which, as was shown for oxidation of H_2 and CO, occurs rather rapidly.

The step of destruction of carboxylate complexes requires a large activation energy since it involves breaking of the oxygen-catalyst bonds. As a result, the rate of this step rapidly decreases with decreasing temperature, and the active catalytic surface becomes covered with carboxylate complexes. In these conditions the destruction is considerably accelerated by dioxygen. One could propose that this acceleration is associated with a subsequent oxidation of carboxylate complexes with dioxygen. However, a study with the use of $^{18}O_2$ has shown [306] that CO_2 formed during the destruction of carboxylate complexes does not contain ^{18}O. From this it follows that this acceleration should be explained by conjugation of the processes of breaking of the oxygen-catalyst bonds during desorption of carboxylate intermediates and reoxidation of the catalyst surface by dioxygen that diminishes the activation energy of destruction of these intermediates. Due to this fact, in the region of low temperature (*e.g.* below 650 K for methane) the concerted mechanism is found to be predominant.

Taking into account all of the above data, the oxidation of methane on oxide catalysts can be described in a schematic fashion as

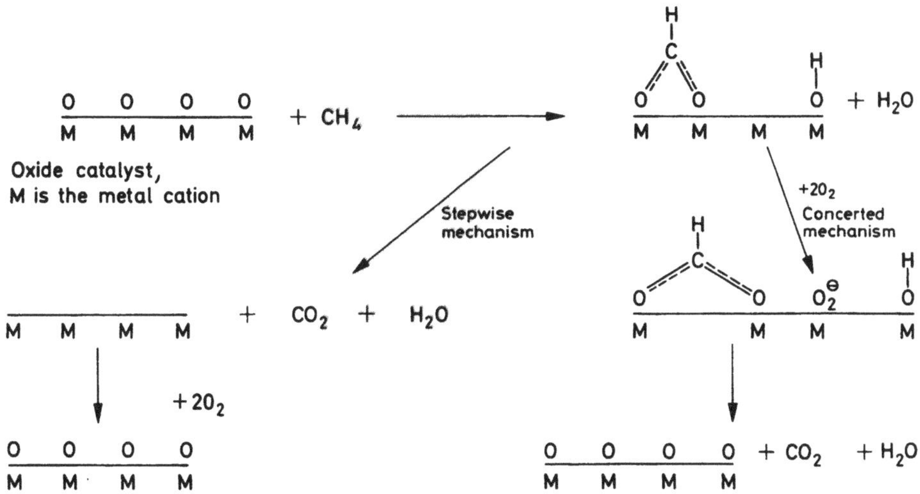

In the course of these transformations, the electron transfer from oxygen on the oxide surface to the metal cations occurs at the first step and back donation from the cations to the atoms of dioxygen that reoxidizes the catalyst occurs at the final step. In the case of the concerted mechanism, in the final step the adsorbed dioxygen participates, probably, in the form of O_2^-. In the temperature range of the stepwise mechanism the primary interaction of hydrocarbons with a catalyst is the rate-determining step, and the reaction rate is proportional to the pressure of paraffin and independent of the pressure of oxygen. In the temperature range of the concerted mechanism the rate-determing step is the destruction of the surface carboxylate intermediates in the presence of dioxygen, and first and zero orders are observed with respect to oxygen and paraffin, respectively.

The relative rates of oxidation of various hydrocarbons on the same catalysts has been measured by many investigators [296, 301, 308, 309]. Popovskii [310] has found that the reactivity of hydrocarbon is higher the smaller the mean energy of one bond is. According to Golodets [311], the rate of oxidation of hydrocarbons is associated with the energy of the rupture of the least strong carbon-carbon bond. Sokolovskii [300] has shown that the relative rate of oxidation of hydrocarbons depends upon the reaction mechanism. For example, for paraffins the reaction rate expressed as the number of CO_2 molecules formed *per* unit time is approximately constant in the temperature range of the concerted mechanism (Table 19). In the temperature range of the stepwise mechanism the rate of oxidation rapidly increases with an increase of the number of carbon atoms in hydrocarbon (Table 20).

The constancy of the rate of oxidation of paraffins in the temperature range of the concerted mechanism is caused by the fragmentation of carboxylate intermediates during the interaction of paraffins with the catalyst.

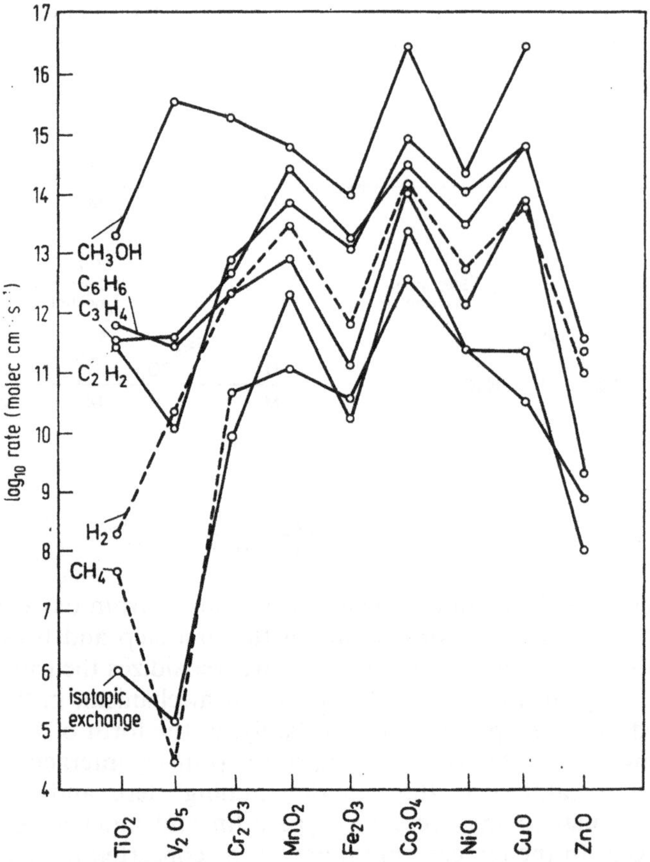

Figure 39. SCA of oxides of the 4th period elements in reactions involving molecular oxygen (excess oxygen, 573 K)

Table 19. Oxidation of *n*-paraffins on cuprous oxide at 473 K (region of the concerted mechanism)

Hydrocarbon	rate × 10^{-10}/molec HC cm^{-2} s^{-1}	rate × 10^{-10}/molec CO_2 cm^{-2} s^{-1}
CH_4	3.85	3.85
C_2H_6	1.85	3.7
C_3H_8	1.1	3.3
C_4H_{10}	0.83	3.3
C_5H_{12}	0.65	3.25
C_6H_{14}	0.6	3.6

Table 20. Oxidation of *n*-paraffins on cuprous oxide [300] at 57 K, $p_{O_2} = 40$ Pa, $p_{HC} = 2.7$ Pa (region of the stepwise mechanism)

Hydrocarbon	rate × 10^{-10}/molec HC cm^{-2} s^{-1}	rate × 10^{-10}/molec CO_2 cm^{-2} s^{-1}
CH_4	6	6
C_2H_6	25	50
C_3H_8	82	246
C_4H_{10}	110	440
C_5H_{12}	160	800
C_6H_{14}	200	1200
C_7H_{16}	210	1470
C_8H_{18}	270	2160

As a result, the surface becomes covered with the same complexes which are very similar to formates. The rate of carbon dioxide production is equal to the rate of destruction of these complexes in the presence of dioxygen which is the same for all paraffins. The high rate of oxidation of C_7–C_8 paraffins, apparently, is due to the ease of their oxidative dehydrogenation. As will be shown below, olefins are oxidized far more readily than paraffins.

At elevated temperatures when the oxidation follows the stepwise mechanism, the destruction of carboxylate compounds is rather fast, and the rate-determining step of the oxidation is that of hydrocarbon bonding to the surface of the oxide catalyst. For paraffins this step is the faster, the easier the removal of one hydrogen atom (see Figure 40). This dependency is confirmed by the high kinetic isotope effect upon substituting hydrogen by deuterium, and also by a significant increase in the reaction rate upon substituting a hydrogen atom by a chlorine or bromine atom whose energy of bonding to carbon is far less than that of hydrogen [300]. As the number of carbon atoms in paraffins increases, the C—H bond energy decreases and, correspondingly, increases the oxidation reaction rate. During the oxidation of *n*-paraffins at the temperatures of the stepwise mechanism the rate of carbon dioxide formation increases by a factor of 360 when passing from methane to octane.

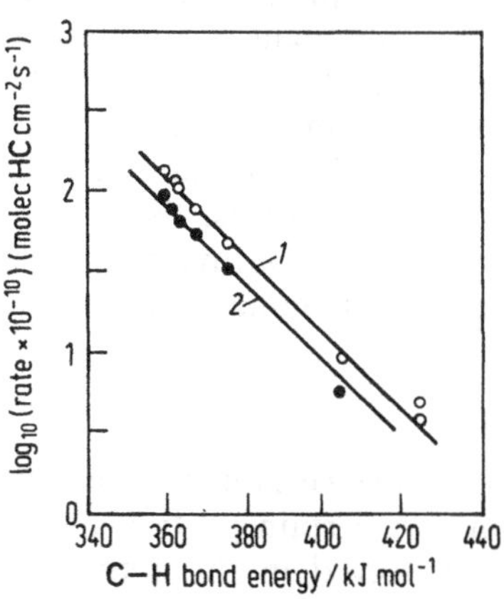

Figure 40. The rates of catalysis and reduction *vs.* the C—H bond energy in complete oxidation of *n*-paraffins on cuprous oxide at 673 K. 1 — catalysis, 2 — reduction.

2. *Oxidation of Unsaturated Compounds*

The oxidation of unsaturated compounds, similar to that of paraffins, obeys the concerted mechanism at low temperatures and the stepwise mechanism at high temperatures. The rate of their oxidation, at least at a small number of carbon atoms, is higher than that of paraffins and changes in the following order:

$$\text{alkynes} > \text{alkenes} > \text{aromatics} > \text{paraffins}\,.$$

For example, in the series of C_6 hydrocarbons the relative rates of oxidation increase in the following order:

n-hexane <	benzene <	1-hexene >	1-hexyne
1	1.8	3.9	5

Cyclization as such shows no appreciable effect on the reactivity.

Oxidation of unsaturated compounds occurs through the formation of carbonate-formate intermediates on the catalyst surface involving fragmentation of the initial hydrocarbon. At low temperatures the rate of oxidation is determined by the decomposition of these intermediates accelerated by dioxygen. In the temperature range of the concerted mechanism the high rate of oxidation of unsaturated hydrocarbons is due to the ease with which the corresponding carboxylate compounds may decompose. Moreover, in the case of unsaturated hydrocarbons these intermediates are formed on the major part of the surface.

At high temperatures the transition to the stepwise mechanism occurs at lower temperatures compared to the oxidation of paraffins. For example, for alkanes it occurs at 625–675 K, for alkenes below 573 K, and for alkynes below 475 K. In the region of the stepwise mechanism the destruction of carboxylate complexes is fast without the participation of dioxygen, and the rate of oxidation is determined by the step of the interaction of an oxidizable substance with a catalyst producing carboxylate complexes involving oxygen of the catalyst surface. The rate of formation of these complexes is independent of the C—H bond energy since no kinetic isotope effect is observed when proceeding to deuterium-containing compounds. Most probably, the interaction with a catalyst involves the rupture of a π-component of the double and triple bonds with the further rapid oxidation and fragmentation. The rupture of π-bonds occurs more easily when passing from aromatic hydrocarbons to alkenes and then to alkynes. The ease of this interaction does not depend upon the number of carbon atoms. Owing to this, the rate of oxidation of unsaturated hydrocarbons, unlike that of paraffins, is almost the same for light and heavy hydrocarbons both for the concerted and the stepwise mechanisms [300].

3. *Strength of the Oxygen Bond to the Catalyst*

The strength of the oxygen bond on the catalyst surface is a determining factor in the complete oxidation of hydrocarbons. A certain portion of the oxygen bond energy on the catalyst surface is involved in the activation energy of both the first step of interaction of hydrocarbons with the catalyst producing carboxylates intermediates and the step of destruction of these compounds either with the participation of dioxygen (concerted mechanism) or without its participation (stepwise mechanism). For this reason the action of all oxide catalysts toward different oxidation reactions is analogous. Figure 39 shows the SCA's of oxides of the 4th period elements toward complete oxidation of H_2, CO, CH_3OH, CH_4, C_2H_4, C_6H_6 and isotope exchange in dioxygen. In all the reactions, oxides in the mid part of the period with a low energy of oxygen bonding on the surface are characterized by the same order of activity variations having the form of a saw-like line with maxima for cobalt, manganese and copper. Hence, for all these oxides the catalytic action depends mainly on the state of oxygen on the catalyst surface. At the same time the effect of the catalyst on oxidizable substances is not too important.

For the elements in the initial part of the period (titanium, vanadium) the oxygen bond energy on oxides is significantly higher and correspondingly the oxygen reactivity is low. The interaction of hydrocarbons with this oxygen is possible only when the catalyst shows a specific effect on the oxidizable substance. This effect manifests itself in different ways on various oxides. For example, on vanadium pentoxide, in spite of a high oxygen binding energy, the oxidation of methyl alcohol occurs at a high rate whereas that of methane is immeasurably slow. Also, on titanium dioxide methyl alcohol is oxidized by almost 5 orders of magnitude faster than is methane. At the same time, on Co_3O_4 this difference is no greater than 2 orders of magnitude.

The specific interaction of an oxidizable substance with a catalyst plays a key role in reactions of partial oxidation. In this case both a strong interaction of an oxidizable substance with the participation of weakly bound oxygen and a weak interaction with the participation of strongly bound oxygen on the catalyst surface are observed. In the former case the desorption yields only the products of complete oxidation, CO_2(CO) and H_2O. In the latter case the initial hydrocarbon and the partial oxidation products are produced. For example, during oxidation of propylene its interaction with a catalyst containing on the surface strongly bound nucleophilic oxygen produces surface allyl complexes which further transform to acrolein. In the light of the above facts the following general scheme of propylene oxidation on oxide catalysts may be proposed:

Pathway 1. Partial oxidation obeys the stepwise mechanism involving strongly bound oxygen with an intermediate formation of allyl compounds and reoxidation of the catalyst surface with dioxygen after desorption of acrolein and water.

Pathway 2. Complete oxidation occurs in the presence of weakly bound oxygen with an intermediate formation of hydroxyl and carboxylate intermediates. Further reaction may proceed *via* two paths.

Pathway 3. Desorption of products and reoxidation of catalysts by dioxygen occur separately with a relatively high activation energy at high temperatures.

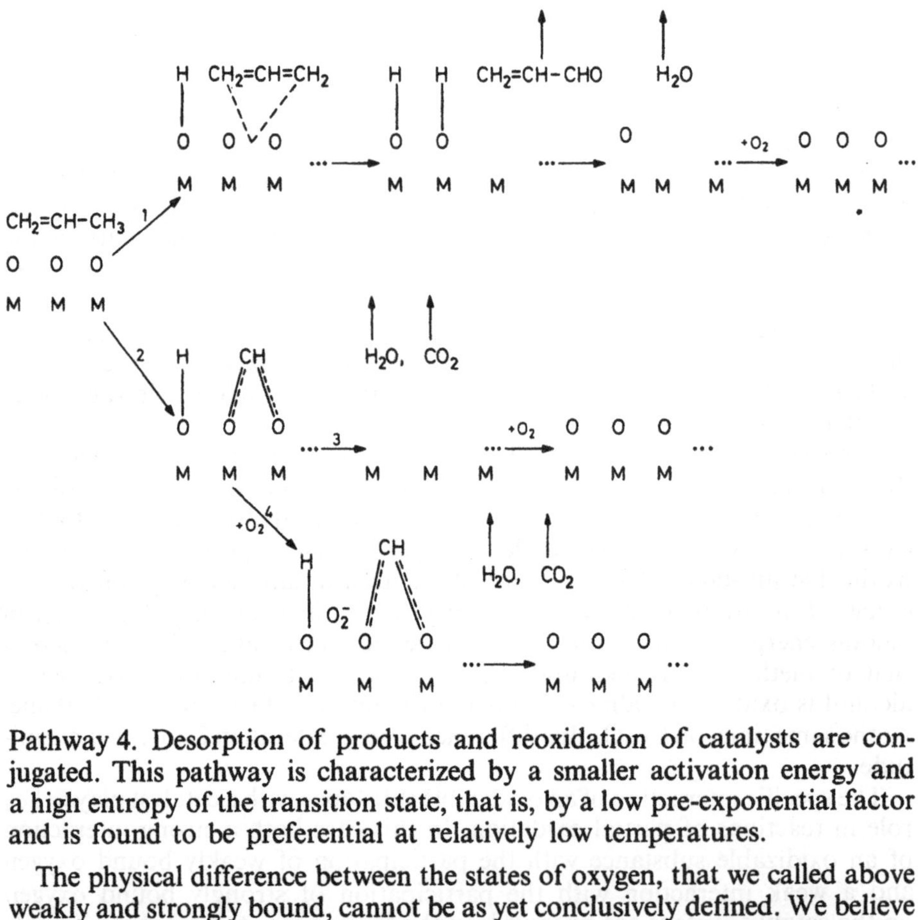

Pathway 4. Desorption of products and reoxidation of catalysts are conjugated. This pathway is characterized by a smaller activation energy and a high entropy of the transition state, that is, by a low pre-exponential factor and is found to be preferential at relatively low temperatures.

The physical difference between the states of oxygen, that we called above weakly and strongly bound, cannot be as yet conclusively defined. We believe that most probably these are O^{2-} ions which differ in the extent of the shift of an electron pair to the catalyst bulk and in their affinity for an electron of neighbouring cations. We have also mentioned an alternative hypothesis, which implies that the weakly bound state of oxygen on the surface of transition metal oxides is a singly-charged oxygen ion, O^-.

The SCA's of the metals most active in hydrocarbon oxidation (Pt, Pd) are somewhat higher than those of the most active oxides (Co_3O_4, MnO_2). In the oxidation of acetylene [308] and methane [310] this difference in SCA's is insignificant whereas in the oxidation of olefins [313] it reaches two orders of magnitude.

7. Interaction of Dioxygen with Sulfur Dioxide

The reaction $SO_2 + \frac{1}{2}O_2 = SO_3$, which is used extensively for production of sulfuric acid, is exothermic and reversible ($\Delta H_{278} = -96.6\ \text{kJ mol}^{-1}$, $\Delta G_{298} = -142.2\ \text{kJ mol}^{-1}$).

A. Oxidation on Platinum Catalysts

In 1831 Phillips had already a patent granted on the use of platinum catalysts for sulfur dioxide oxidation [314]. It took, however, many years to accomplish this process on a large scale because of the problems created by platinum poisoning mainly with arsenic compounds in the process gases. To diminish the consumption of platinum, its support first on asbestos (6–10 %) and later on magnesium sulfate or silica (0.05–0.3 %) was proposed. The specific catalytic activity per unit surface of platinum is known to be constant [315].

Dioxygen chemisorbs dissociatively on platinum at a very high rate *vide supra*. Assuming that SO_2 oxidation on platinum follows a stepwise mechanism, one may propose that interaction of sulfur dioxide with the dissociatively adsorbed oxygen is the rate-determining step of the process. This proposal is confirmed by the value of the reaction molecularity[1] with respect to dioxygen equal to 1/2 [341] determined by the Horiuti method [342] using sulfur and oxygen isotopes. Suppose that adsorption of oxygen is strong, that of sulfur trioxide is weak whereas adsorption of sulfur dioxide is moderate and obeys the laws of adsorption on evenly nonuniform surfaces. Then the rates of direct and back reactions can be written as

$$r_1 = K_1 p_{O_2}^{0.5\alpha} p_{SO_2} p_{SO_3}^{-\alpha} = K_1 p_{O_2}^{0.25} p_{SO_2} p_{SO_3}^{-0.5} \tag{27}$$

$$r_2 = K_2 p_{O_2}^{-0.5(1-\alpha)} p_{SO_3}^{(1-\alpha)} = K_2 p_{O_2}^{-0.25} p_{SO_3}^{0.5} \tag{28}$$

The experimental kinetic studies of this reaction [316–318] led to somewhat differing results, which nevertheless can be described satisfactorily by equations (27) and (28) provided that α is 0.5 [319].

All other metals are not so active toward the reaction under consideration.

B. Catalytic Activity of Oxides

The catalytic activity of simple oxides is restricted by their low temperature transformation into inert sulfates. In this respect some mixed oxides, *e.g.*, the mixture of tin and chromium oxides [320, 321] and some spinels [322] are more advantageous. However, they are not active enough to be used in industry.

The only element whose compounds can replace platinum in catalysts for sulfuric acid production is vanadium. The catalytic activity of pure vanadium pentoxide is very low. However, it can be increased markedly by adding small amounts of alkali such as Na_2O, K_2O, Rb_2O and Cs_2O. An interesting feature of these catalysts is that the active component is in the liquid state on the carrier surface under the reaction conditions.

The composition of the active catalyst is $V_2O_5 \cdot nK_2O \cdot mSO_3$ where n ranges 2 to 4, and m depends upon the reaction conditions and approximates

[1] By molecularity [340] is meant the number of molecules of a given substance that have reacted *per* transformation of one active complex. A stoichiometric number is the reciprocal of molecularity.

2*n* [323]. Porous silica, sometimes with a small alumina addition, is used as a support. Under the influence of the reaction mixture potassium pyrosulfate is formed (its melting point is 688 K) which reacts further with vanadium pentoxide to produce potassium sulfovanadate. In the indicated interval of K_2O/V_2O_5 ratios the melting point of resulting compounds is 627–653 K.

The detailed derivatographic, thermo-optical, X-ray diffraction and IR-spectroscopic analyses of the $V_2O_5 \cdot nK_2S_2O_7$ system have shown that crystallization yields the following compounds: A — $6\,K_2O \cdot V_2O_5\ 12\,SO_3$, B — $K_2O \cdot V_2O_5 \cdot 2\,SO_3$, C — $K_2O \cdot V_2O_5 \cdot 4\,SO_3$, and D — $K_2O \cdot V_2O_4\ 3\,SO_3$. Complex C is formed at low temperature treatment (623 K), complex D — at high concentrations of SO_2 in the reaction mixture at 693 K [324, 325]. Being in the molten state, the mixture may also contain some other compounds.

Similar results have been obtained for the $V_2O_5 \cdot nM_2S_2O_7$ system, where M is Na, Rb, Cs. In the series Na, K, Rb, Cs the melting point decreases when passing from Na to Cs. The catalytic activity significantly increases when proceeding from Li to Na and then to K. However, passing to Rb and Cs leads to a slight decrease [326]. In the commercial process, the temperature is from 680 to 870 K. In the catalytic reactor the active component in the liquid state impregnates the porous carrier. Consequently, special attention should be paid to a selection of the carrier pore structure. This latter must provide sufficient development of the surface of a molten active component located on the pore walls. At the same time the pores should not be very small, otherwise they will be completely filled with the active component, which thus will have no contact with the reaction mixture [321]. The most advantageous is a bidisperse structure which combines average size pores creating a large area of active component with large transport pores whose size exceeds the length of a free run of reactant molecules [328, 329, 344].

It has been suggested that the oxidation of sulfur dioxide on vanadium catalysts follows a stepwise mechanism:

$$2\,V^{5+} + SO_2 + O^{2-} \rightleftarrows 2\,V^{4+} + SO_3 \tag{29}$$

$$2\,V^{4+} + 1/2\,O_2 \rightarrow 2\,V^{5+} + O^{2-} \tag{30}$$

The first step occurs at a high rate and, according to Mars and Maessen [330], equilibrium between the reaction participants is reached under the reaction conditions, and the observed rate of SO_2 oxidation is determined by that of step (30) [331]. Further investigations have proved that in reality the situation is much more complex. Individual measurements of reduction and oxidation rates of the steady state catalyst by means of determination of V^{4+} content from the ESR signal intensities indicate that above 720 K the rates of these supposed steps are far lower than the rate of the catalytic reaction [332]. This leads one to conclude that the catalytic reaction proceeds by another pathway which is independent of the vanadium valence state. Probably, in potassium sulfovanadate either V^{5+} or a binuclear complex composed of two vanadium atoms promotes the bonding of the two SO_2

molecules in the coordination sphere (without the formation of V^{4+} ions), and dioxygen reacts with these SO_2 molecules to produce SO_3. The precise structure of this complex is as yet unknown. If we designate this complex by $[V_2^{5+}]$, the reaction proceeding *via* the concerted mechanism can be described by

$$[V_2^{5+}] \xrightarrow{+SO_2} [V_2^{5+}]SO_2 \xrightarrow{+SO_2} [V_2^{5+}]2SO_2 \xrightarrow{+O_2} [V_2^{5+}] + 2SO_3 \quad (A)$$

$$[V_2^{5+}]SO_2 \xrightarrow{+O^{2-}} [V_2^{4+}]SO_3 \xrightarrow{+1/2\,O_2} [V_2^{5+}] + O^{2-} + SO_3 \quad (B)$$

$$(31)$$

Pathway (A) prevails over (B) at a large concentration of SO_3 in the reaction mixture, that is, at moderate and high conversion degrees. The case stands otherwise with low concentrations of SO_3 in the reaction mixture, that is, at small conversion degrees. In this latter case the reaction rate of pathway (B) involving intermediate reduction and oxidation of vanadium ions in the catalyst drastically increases. Studies on the oxidation by dioxygen of reduced catalysts and catalytic reactions in the absence of SO_3 prove the rate of these processes to be almost the same. Therefore at low conversion degrees the redox mechanism of SO_2 oxidation (pathway B) appears to be predominant.

C. The Kinetics of SO_2 Oxidation on Vanadium Catalysts

The kinetics of SO_2 oxidation on vanadium catalysts has been comprehensively studied by many investigators. As a consequence, different kinetic equations were proposed to describe the process. The most essential results are shown in Table 21. The multiplier in square brackets indicates the effect of the reaction reversibility on its rate, its exponential equals the reaction molecularity, that is, the number of SO_2 molecules oxidized upon transformation of active complex [340]. Equation 7 in Table 21 is consistent with the above scheme of elementary reactions. According to this equation, the reaction molecularity is 2. This implies that during the transformation of one active complex two molecules of sulfur dioxide and one molecule of dioxygen enter the reaction. It is possible to conclude, therefore, that the rate determining step of SO_2 oxidation on vanadium catalysts is that of dioxygen bonding to the catalyst. The other steps, namely, SO_2 bonding or SO_3 removal cannot be rate-determining since the reaction molecularity in this case must be equal to unity.

With high concentrations of SO_3 in the reaction mixture, that is at moderate and high conversion degrees, the reaction obeys the concerted mechanism (A). Let us assume now that the pseudo-equilibrium ratios of the component concentrations are reached in all the steps but not in the rate-determining one. Suppose that binuclear vanadium complexes can combine with SO_2 and SO_3 molecules which have the ability to remove each other. Taking into account the experimental data, it is necessary to assume that one molecule

Table 21. Kinetic equations describing oxidation of SO_2 on vanadium catalysts

No of equation	Kinetic equation	Reference
1	$r = Kp_{O_2}(p_{SO_2}/p_{SO_3})^{0.8}[1 - (p_{SO_3}/p_{SO_2}p_{O_2}^{0.5}K_p)^2]$	[333]
2	$r = Kp_{O_2}p_{SO_2}^{0.5}[1 - p_{SO_3}/p_{SO_2}p_{O_2}^{0.5}K_p]$	[334]
3	$r = Kp_{O_2}(p_{SO_2}/p_{SO_3})^{0.5}[1 - (p_{SO_2}p_{O_2}^{0.5}K_p)^2]$	[335]
4	$\vec{r} = Kp_{O_2}(p_{SO_2}/p_{SO_3})^{0.4}$	[336]
5	$r = \dfrac{Kp_{SO_2}p_{O_2}^{0.5}[1 - p_{SO_3}/p_{SO_2}p_{O_2}^{0.5}Kp]}{(1 + b_{SO_2}p_{SO_2} + \sqrt{b_{O_2}p_{O_2}} + b_{SO_3}p_{SO_3} + b_{N_2}p_{N_2})^2}$	[337]
6	$\vec{r} = Kp_{O_2}(\sqrt{bp_{SO_2}}/\sqrt{p_{SO_3}} + \sqrt{bp_{SO_2}})^2$	[330]
7	$r = Kp_{O_2}\dfrac{1}{1 + Ap_{SO_3}/p_{SO_2}}[1 - (p_{SO_3}/p_{SO_2}p_{O_2}^{0.5}K_p)^2]$	[338] [339]
8	$r = kKp_{SO_2}p_{O_2}/[p_{SO_3}^{0.5} + (Kp_{SO_2})^{0.5}]^2\left\{1 - \left(\dfrac{p_{SO_3}}{p_{SO_3}p_{O_2}^{0.5}K_p}\right)^2\right\}$	[343]

of SO_2 always enters the composition of the complex independently of the p_{SO_2} to p_{SO_3} ratio, and that the portion of complexes containing the second molecule is proportional to this ratio. Then the main steps of the catalytic reaction can be described by equations

$$V_2^V nSO_3SO_2 + SO_2 \rightleftarrows V_2^V(n-1)SO_3\,2\,SO_2 + SO_3 \quad (32a)$$
$$V_2^V(n-1)SO_3\,2\,SO_2 + O_2 \rightarrow V_2^V(n+1)SO_3 \quad (32b)$$
$$V_2^V(n+1)SO_3 + SO_2 \rightarrow V_2^V nSO_3SO_2 + SO_3 \quad (32c)$$
$$V_2^V(n-1)SO_3\,2\,SO_2 + O^{2-} \rightleftarrows V_2^{IV}nSO_3SO_2 \quad (32d)$$
$$2\,V_2^{IV}nSO_3SO_2 + O_2 \rightarrow 2\,V_2^V nSO_3SO_2 + 2\,O^{2-} \quad (32e)$$

Due to the pseudo-equilibrium of the steps (32a) and (32d) we have

$$[V_2^V(n-1)SO_3 2SO_2] = K_1\frac{p_{SO_2}}{p_{SO_3}}[V_2^V nSO_3SO_2]$$

$$[V_2^V(n-1)SO_3 2SO_2] = K_4^{-1}[V_2^{IV}nSO_3SO_2]$$

where K_1 and K_4 are the equilibrium constants of these reactions. Then

$$[V_2^V(n-1)SO_3 2SO_2] = [V°]\frac{1}{1 + K_1^{-1}p_{SO_3}/p_{SO_2} + K_4}$$

where [V°] is the total concentration of vanadium complexes in the catalyst.

The reaction rate will be

$$r = K^1[V_2^V(n-1)SO_3 2SO_2]p_{O_2} = \frac{K^1[V°]p_{O_2}}{1 + K_1^{-1}p_{SO_3}/p_{SO_2} + K_4}$$

$$= \frac{K^1[V°]}{1 + K_4} p_{O_2} \frac{1}{1 + [K_1(1 + K_4)]^{-1} p_{SO_3}/p_{SO_2}}$$

This is consistent with the experimentally derived kinetic expressions. If we define the parameters K and A by:

$$K = K^1[V°]/(1 + K_4) \qquad \text{and} \qquad A = 1/K_1(1 + K_4)$$

then, according to the relevant experimental data, the parameter A is almost independent of temperature. In turn, this suggests that K_1 only slightly depends upon the temperature. As proved by direct measurements at high temperatures, the equilibrium constant K_4 is small. Consequently, at temperatures above 700 K the observed activation energy is determined only by the temperature dependence of the rate constant K^1 corresponding to an activation energy of 85 kJ mol^{-1}. At low temperatures K_4 becomes much greater than unity and the observed activation energy increases by the value of the reaction heat of step (32d). As was shown by some experiments, at temperatures above 733 K the reaction rate remains constant. This effect is due to as yet non-specific transformations of the catalyst active component, quite probably to the decomposition of vanadium complexes resulting from SO_3 elimination. The typical temperature dependences of the rate constant K are shown in Figure 41.

The composition and properties of vanadium catalysts change substantially under the influence of the reaction medium. They also depend sensitively on the temperature and chemical reactions which are not the steps of the catalytic process and usually occur more slowly. V^{5+} is readily reduced to

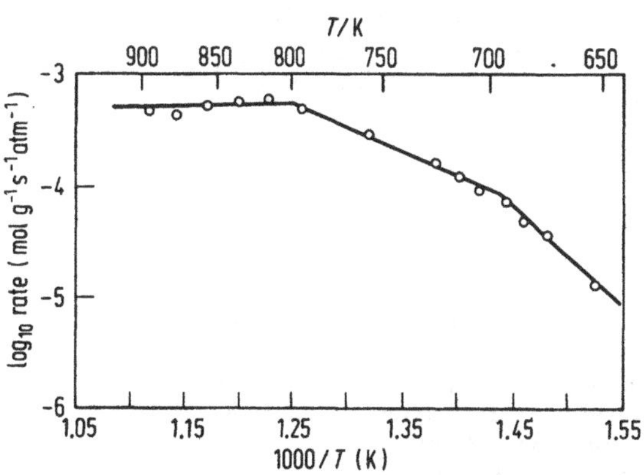

Figure 41. The activity of vanadium catalyst as a function of temperature.

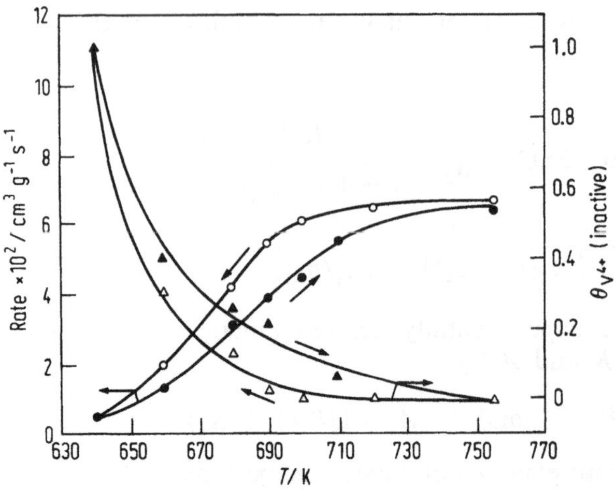

Figure 42. Changes in the composition and activity of the vanadium catalyst during oxidation of SO_2

V^{4+} as the concentration of SO_2 in the reaction mixture is increased and the temperature is decreased. The changes are reversible and are characterized by small relaxation times. Moreover, when the catalyst is subjected to the prolonged action of the reaction mixture containing large concentrations of SO_2 at low temperatures, the ESR spectrum shows a signal which is attributed to some specific state of the 4-valence vanadium inactive in the catalytic reaction. The concentration of this state correlates with the decrease in the catalytic activity (Figure 42). The changes of the catalytic activity and of the ESR signal intensity are characterized by large relaxation times and by the appearance of hysteresis. As a result, the catalytic reaction can be performed on catalysts of the unsteady state composition which slowly transforms to the steady state one. This fact should be taken into account when varying the temperature and reaction mixture composition. As is widely recognized now, performance of some chemical processes in unsteady state regimes is more advantageous compared to that in steady state regimes. So, at the moment creation of artificial unsteady state regimes is quite expedient. To solve the problems of simulation and control of such processes, we need to have an intimate knowledge of the unsteady state kinetics [345].

References

1. Hagward, D. O., Trapnell, B. M. W.: Chemisorption, London: Butterworths 1964
2. Lang, B., Joyner, R. W., Somorjai, G. A.: Surf. Sci. **30**, 454 (1972)
3. Carriere, B., Legare, P., Maire, G.: J. Chim. Phys., Phys.-Chim. Biolog. **71**, 355 (1974)
4. Ivanov, V. P., Savchenko, V. I., Boreskov, G. K., Taylor, K.: Kinet. Katal. **19**, 210 (1978)
5. Morgan, A. E., Somorjai, G. A.: Surf. Sci. **12**, 405 (1968)
6. Lang, B., Legare, P., Maire, G.: Surf. Sci. **47**, 89 (1975)
7. Kneringer, G., Netzer, F. P.: Surf. Sci. **49**, 125 (1975)

8. Helms, C. R., Bonzel, H. P., Kelemen, S.: J. Chem. Phys. **65**, 1773 (1976)
9. Engelhardt, H. A., Menzel, D.: Surf. Sci. **57**, 591 (1976)
10. Hall, P. G., King, D. A.: Surf. Sci. **36**, 810 (1973)
11. Chesters, M. A., Somorjai, G. A.: Surf. Sci. **52**, 21 (1975)
12. Bazhutin, N. B., Boreskov, G. K., Savchenko, V. I.: React Kinet. Catal. Lett. **10**, 337 (1979)
13. Pentenero, A., Pacia, N., Weber, B.: J. de Chim. Phys., Phys.-Chim. Biol. **72**, 941 (1975)
14. Peng, Y. K., Dawson, P. T.: Can. J. Chem. **52**, 3507 (1974)
15. Sobyanin, V. A.: Thesis, Inst. of Catalysis, Siberian Branch of the USSR Acad. of Sciences, Novosibirsk, USSR, 1978
16. Alnot, M., Fusy, I., Cassuto, A.: Surf. Sci. **72**, 467 (1978)
17. Bajer, E., Shretuman, P.: Structure and Bonding, Vol. 2. Berlin, Heidelberg, New York: Springer 1967
18. Clarkson, R. B., Cirillo, A. C.: J. Catal. **33**, 392 (1974).
19. Madey, T. E., Engelhardt, H. A., Menzel, D.: Surf. Sci. **48**, 304 (1975)
20. Rovida, G., Pratesi, F.: Surf. Sci. **52**, 542 (1975)
21. Conrad, H., Ertl, G., Küppers, J., Latta, E. E.: Surf. Sci. **65**, 245 (1977).
22. Ivanov, V. P.: Thesis, Inst. of Catalysis, Siberian Branch of the USSR Acad. of Sciences, Novosibirsk, USSR, 1978
23. Boreskov, G. K., Savchenko, V. I.: Prepr. 7-th Intern. Congr. Catalysis, Tokyo, 1980, A-46
24. Ivanov, V. P., Boreskov, G. K., Savchenko, V. I., Egelhoff, W. T., Weinberg, W. H.: Surf. Sci. **61**, 207 (1976)
25. Thiel, P. A., Yates, J. T., Weinberg, W. H.: Surf. Sci. **82**, 22 (1979)
26. Klein, R., Shih, A.: Surf. Sci. **69**, 403 (1977)
27. Doyen, G., Ertl, G.: J. Chem. Phys. **68**, 5417 (1978)
28. Chan, C. M., Luke, M. A., Weinberg, W. H., Van-Hove, M. A., Withrow, S. P.: Surf. Sci. **78**, 386 (1978)
29. Toyoshima, J., Somorjai, G. A.: Catal. Rev. **19**, 105 (1979)
30. Taylor, I. L., Ibbotson, D. E., Weinberg, W. H.: Surf. Sci. **79**, 349 (1979)
31. Sazonov, V. A., Popovskii, V. V., Boreskov, G. K.: Kinet. Katal. **9**, 307, 312 (1968)
32. Boreskov, G. K., Sazonov, V. A., Popovskii, V. V.: Dokl. Akad. Nauk SSSR **176**, 1331 (1967)
33. Sazonov, V. A.: Thesis, Inst. of Catalysis, Siberian Branch of the USSR Acad. of Sciences, Novosibirsk, 1968
34. Joly, J. P.: J. Chim. Phys. 1013, 1019 (1975)
35. Pankratiev, Yu. D., Boreskov, G. K., Soloviev, V. I., Popovskii, V. V., Sazonov, V. A.: Dokl. Akad. Nauk SSSR **184**, 611 (1969)
36. Marshneva, V. I., Boreskov, G. K.: React. Kinet. Catal. Lett. **1**, 15 (1974)
37. Marshneva, V. I., Boreskov, G. K., Sokolovskii, V. D.: Kinet. Katal. **14**, 210 (1973)
38. Halpern, B., Germain, J. E.: J. Catal. **37**, 44 (1975)
39. Halpern, B., Germain, J. E.: Compt. Rend. Acad. Sci. Paris **277**, Ser.-C., 1287 (1973)
40. Bielanski, A., Haber, J.: Catal. Rev. **19**, 1 (1979)
41. Bielanski, A., Najbar, M.: J. Catal. **25**, 398 (1972)
42. Dyrek, K.: Bull. Acad. Pol. Sci., Ser. Sci. Chem. **20**, 57 (1972)
43. Kokes, R. J.: J. Phys. Chem. **66**, 566 (1962)
44. Glemza, R., Kokes, R. J.: J. Phys. Chem. **66**, 576 (1962); J. Phys. Chem. **69**, 3254 (1965)
45. Solonitsyn, Yu. P.: Kinet. Katal. **6**, 433 (1965); **7**, 328 (1966)
46. Kejer, N. P.: IV-th Intern. Congr. on Catalysis, 1970, 133
47. Volkenshtein, C. F.: Elektronnaya Teoria Kataliza na Poluprovodnikah, M., 1960 Spiridonov, K. H., Krylov, O. V.: Sb. Poverkhnostnye Soedineniya v Geterogennom Katalize, Problemy Kinetiki i Kataliza, M., **16**, 7 (1975)
48. Lunsford, J. H., Jayne, J. P.: J. Chem. Phys. **44**, 1487 (1966)
49. Mikheikin, N. D., Matschenko, A. I., Kazanskii, V. B.: Kinet. Katal. **8**, 1363 (1967)
50. Shwets, V. A., Sarichev, M. E., Kazanskii, V. B.: J. Catal. **11**, 378 (1968)
51. Miller, D. J., Naneman, D.: Phys. Rev. (B) **3**, 2918 (1971)
52. Naccache, C., Meriandean, P., Tench, A. J.: Trans. Faraday Soc. **67**, 606 (1971)

53. Meriandean, P., Naccache, C., Tench, A. J.: J. Catal. **21**, 208 (1971)
54. Setaka, M., Kwan, T.: Bull. Chem. Soc., Japan **43**, 2727 (1970)
55. Matschenko, A. I., Kazanskii, V. B., Sharapov, V. M.: Kinet. Katal. **8**, 853 (1967)
56. Horiguchi, H., Setaka, M., Sancier, K. M., Kwan, T.: IV-th Intern. Congr. on Catalysis 1970, 102
57. Shwets, V. A., Vorotynzev, V. M., Kazanskii, V. B.: Kinet. Katal. **10**, 356 (1969)
58. Shwets, V. A., Kazanskii, V. B.: J. Catal. **25**, 123 (1972)
59. Naccache, C.: Chem. Phys. Lett. **11**, 323 (1971)
60. Davydov, A. A., Konareva, M. P., Anufrienko, V. F., Maksimov, N. G.: Kinet. Katal. **14**, 1519 (1973)
61. Davydov, A. A., Tschekochikhin, Yu. M., Kejer, N. P.: Kinet. Katal. **13**, 1081 (1972)
62. Tsydanenko, A. A., Rodionova, T. A., Filimonov, V. H.: React. Kinet. Catal. Lett. **11**, 113 (1979)
63. Davydov, A. A.: Kinet. Katal. **20**, 1506–1512 (1979)
64. Estrup, P. J.: Structure and Chemistry of Solid Surfaces (Somorjai, G. A. ed.), New York: John Wiley & Sons 1969
65. Gorodetskii, D. A., Melnik, Yu. P.: Izvest. Akad. Nauk SSSR, ser. phys. **35**, 1064 (1971)
66. MacRae, A. V.: Surf. Sci. **1**, 319 (1964)
67. Dadayan, K. A., Boreskov, G. K., Savchenko, V. I.: Kinet. Katal. **18**, 189 (1977)
68. Dadayan, K. A., Boreskov, G. K., Savchenko, V. I., Bulgakov, N. N.: Dokl. Akad. Nauk SSSR **239**, 356 (1978)
69. Dadayan, K. A., Boreskov, G. K., Savchenko, V. I.: Kinet. Katal. **19**, 699 (1978)
70. Niewenhuys, B. E., Somorjai, G. A.: Surf. Sci. **72**, 8 (1978)
71. Ivanov, V. P., Boreskov, G. K., Savchenko, V. I., Tataurov, V. L., Egelhoff, V. F., Weinberg, W. H.: Dokl. Akad. Nauk SSSR **249**, 3, 642 (1979)
72. Conrad, H., Ertl, G., Küppers, J., Latta, E. E.: Surf. Sci. **65**, 245 (1977)
73. Collins, D. M., Spicer, W. E.: Surf. Sci. **69**, 85 (1977)
74. Weinberg, W. H., Lambert, R. M., Comrie, R. M., Linnett, J. W.: V-th Intern. Congr. on Catalysis, 1973, 964
75. Alberts, H., Wan der Wal, W. J. J., Bootsma, G. A.: Surf. Sci. **68**, 47 (1977)
76. Rovida, G., Pratesi, F., Maglietta, M., Ferroni, E.: Surf. Sci. **43**, 230 (1974)
77. Boreskov, G. K., Gruver, V. Sh., Pankratiev, Yu. D., Turkov, V. M., Khasin, A. V.: Dokl. Akad. Nauk SSSR **215**, 359 (1974)
78. Yass, M. J., Budrugeas, P.: J. Catal. **64**, 68 (1980)
79. Somorjai, G. A.: Principles of Surface Chemistry, Englewood Cliffs, New Jersey: Prentice-Hall, 1972
80. Gorodetskii, V. V., Nieuwenhuys, B. E., Sachtler, W. M. H., and Boreskov, G. K., Applications of Surface Science **7**, 355–371 (1981)
81. Gorodetskii, V. V., Sobyanin, V. A., Bulgakov, N. N. and Z. Knor, Surface Sci. **82**, 120 (1979)
82. Boreskov, G. K.: Adv. Catal. **15**, 285 (1964)
83. Boreskov, G. K., Muzykantov, V. S.: Ann. N.Y. Acad. Sci. **213**, 137 (1973)
84. Muzykantov, V. S., Popovskii, V. V., Boreskov, G. K.: Kinet. Katal. **5**, 624 (1964)
85. Klier, K., Novakova, J., Jiru, P.: J. Catal. **2**, 479 (1963)
86. Novakova, J.: Catal. Rev. **4**, 77 (1970)
87. Sutula, V. D., Zeif. A. P.: Dokl. Akad. Nauk SSSR **212**, 1393 (1973)
88. Winter, E. R. S.: J. Chem. Soc. 1522 (1954)
89. Barry, F. J., Stone, F. S.: Proc. Royal Soc. **256**, 124 (1960)
90. Gorgoraki, V. I., Boreskov, G. K., Kasatkina, L. A., Sokolovskii, V. D.: Kinet. Katal. **5**, 120 (1964)
91. Boreskov, G. K., Kasatkina, L. A.: Uspekhi Khimii **37**, 1462 (1968)
92. Sazonov, L. A., Sokolovskii, V. D., Boreskov, G. K.: Kinet. Katal. **7**, 284 (1966)
93. Gorgoraki, V. I., Boreskov, G. K., Kasatkina, L. A.: Kinet. Katal. **7**, 266 (1966)
94. Sokolovskii, V. D., Sazonov, L. A., Boreskov, G. K., Moskvina, Z. V.: Kinet. Katal. **9**, 130, 784 (1968)
95. Nikisha, V. V., Shelimov, B. N., Shvets, V. A., Griva, A. P., Kazansky, V. B.: J. Catal. **28**, 321, (1975)

96. Winter, E. R. S.; J. Chem. Soc. (A), 2889 (1968)
97. Popovskii, V. V., Boreskov, G. K., Muzykantov, V. S., Sazonov, V. A., Panov, G. I., Roshchin, V. A., Plyasova, L. M., Malakhov, V. V.: Kinet. Katal. **13**, 727 (1972)
98. Boreskov, G. K., Muzykantov, V. S., Popovskii, V. V., Goldshtein, N. D.: Dokl. Akad. Nauk SSSR **159**, 1354 (1964)
99. Dzisyak, A. P., Boreskov, G. K., Kasatkina, L. A.: Kinet. Katal. **3**, 81 (1962)
100. Boreskov, G. K., Dzisyak, A. P., Kasatkina, L. A.: Kinet. Katal. **4**, 388 (1963)
101. Muzykantov, V. S., Popovskii, V. V., Boreskov, G. K., Mikigur, N. N.: Kinet. Katal. **5**, 745 (1964)
102. Evald, G., Muzykantov, V. S., Boreskov, G. K.: Kinet. Katal. **14**, 627 (1973)
103. Kasatkina, L. A., Zujev, A. P.: Kinet. Katal. **6**, 478 (1965)
104. Muzykantov, V. S., Boreskov, G. K., Popovskii, V. V., Panov, G. I., Shkrabina, R. A.: Kinet. Katal. **13**, 385 (1972)
105. Boreskov, G. K., Muzykantov, V. S., Panov, G. I., Popovskii, V. V.: Kinet. Katal. 1043 (1969)
106. Semin, G. L., Cherkashin, A. E., Kejer, N. P., Muzykantov, V. S.: Dokl. Akad. Nauk SSSR **203** 391 (1972)
107. Semin, G. L., Kejer, N. P., Cherkashin, A. E.: Kinet. Katal. **13**, 1221 (1972)
108. Gorgoraki, V. I., Kasatkina, L. A., Levin, V. Yu.: Kinet. Katal. **4**, 422 (1969)
109. Blanchard, M., Longnet, G., Boreskov, G. K., Muzykantov, V. S., Panov, G. I.: Bull. Soc. Chem. France 814 (1971)
110. Popovskii, V. V., Boreskov, G. K., Muzykantov, V. S., Sazonov, V. A., Shubnikov, S. G.: Kinet. Katal. **10**, 786 (1969)
111. Muzykantov, V. S., Panov, G. I., Boreskov, G. K.: Kinet. Katal. **10**, 1270 (1969)
112. Winter, E. R. S.: J. Chem. Soc. (A) 1832 (1969)
113. Sazonov, L. A., Sokolovskii, V. D., Boreskov, G. K.: Kinet. Katal. **7**, 521 (1966)
114. Novakova, J., Jiru, P.: Coll. Czech. Chem. Commun. **29**, 1114 (1964)
115. Sazonov, L. A., Mitrofanova, G. N., Preobrazhenskaya, L. V., Moskvina, Z. V.: Kinet. Katal. **13**, 789 (1972)
116. Borisov, Yu. A., Bulgakov, N. N., Boreskov, G. K., Muzykantov, V. S., Panov, G. I., Zyrulnikov, P. G.: Kinet. Katal. **16**, 1246 (1975)
117. Boreskov, G. K., Popovskii, V. V., Sazonov, V. A.: Sb. Osnovy Predvidenija Kataliticheskogo Deistviya, v. 1, 343, Izd. Nauka, Moscow, 1970
118. Aleksandrov, V. Yu., Popovskii, V. V., Bulgakov, N. N., Muzykantov, V. S., Skorobogatova, I. A., Ikorskii, V. N., Zahrov, V. I.: Kinet. Katal. **14**, 390 (1973)
119. Andrushkevich, T. V., Boreskov, G. K., Popovskii, V. V., Muzykantov, V. S., Kimkhai, O. N., Sazonov, V. A.: Kinet. Katal. **9**, 595 (1968)
120. Klier, K., Kucera, E.: J. Phys. Chem. Solids **27**, 1087 (1966)
121. Klier, K.: Coll. Czech. Chem. Commun. **31**, 3820 (1966)
122. Muzykantov, V. S., Panov, G. I.: Kinet. Katal. **13**, 350 (1972)
123. Khasin, A. V., Boreskov, G. K.: Dokl. Akad. Nauk SSSR **152**, 1387 (1963)
124. Boreskov, G. K., Khasin, A. V.: Kinet. Katal. **5**, 956 (1964)
125. Boreskov, G. K., Khasin, A. V., Starostina, T. S.: Dokl. Akad. Nauk SSSR **164**, 606 (1965)
126. Khasin, A. V., Boreskov, G. K., Starostina, T. S.: Sb. Metody Issledovaniya Katalizatorov i Kataliticheskikh Reaktsii, Inst. of Catalysis, Siberian Branch of the USSR Acad of Sci., Novosibirsk, 1965, v. 1, 342
127. Boreskov, G. K.: Discuss. Farad. Soc. **41**, 263 (1966)
128. Starostina, T. S., Khasin, A. V., Boreskov, G. K.: Kinet. Katal. **8**, 942 (1967)
129. Starostina, T. S., Khasin, A. V., Boreskov, G. K., Plyasova, L. M.: Dokl. Akad. Nauk SSR **190**, 394 (1970)
130. Starostina, T. S.: Thesis. Inst. of Catalysis, Siberian Branch of the USSR Acad. of Sci., Novosibirsk, 1975
131. Sandler, Y. L., Hickam, W. M.: Rep. 3rd Intern. Congr. Catalysis, Amsterdam, 1964
132. Sandler, Y. L., Durigon, D. D.: J. Phys. Chem. **69**, 4201 (1965)
133. Sandler, Y. L., Beer, S. L., Durigon, D. D.: J. Phys. Chem. **70**, 3881 (1966)
134. Sandler, Y. L., Durigon, D. D.: J. Phys. Chem. **72**, 1051 (1968).

135. Sandler, Y. L., Durigon, D. D.: J. Phys. Chem. **73**, 2392 (1969)
136. Kajumov, R. P., Kulkova, N. V., Temkin, M. I.: React. Kinet. Catal. Lett. **1**, 29 (1974)
137. Kajumov, R. P., Kulkova, N. V., Temkin, M. I.: Kinet. Katal. **15**, 157 (1974)
138. Boreskov, G. K.: J. Phys. Chem. (USSR) **31**, 937 (1957)
139. Boreskov, G. K., Slin'ko, M. G., Filippova, A. G.: Dokl. Akad. Nauk SSSR **92**, 353 (1953)
140. Boreskov, G. K., Slin'ko, M. G., Filippova, A. G., Guryanova, R. N.: Dokl. Akad. Nauk SSSR **94**, 713 (1954)
141. Davy, H.: Ann. Chim. Phys. **4** (2), 260, 347 (1817)
142. Döbereiner, J.: Schweigg. J. **39**, 1 (1923)
143. Faraday, M.: Pogg. Ann. **33**, 149 (1834).
144. Harkovskaya, E. N., Boreskov, G. K., Slin'ko, M. G.: Dokl. Akad. Nauk SSSR **127**, 145 (1959)
145. Kuchaev, V. L., Temkin, M. I.: Kinet. Katal. **13**, 719, 1024 (1972)
146. Balovnev, Yu. A., Roginskii, S. Z., Tretyakov, I. I.: Dokl. Akad. Nauk SSR **158**, 929 (1964); **163**, 394 (1965)
147. Boreskov, G. K., Khasin, A. V.: J. Res. Inst. Catalysis, Hokkaido Univ. **16**, 477 (1968)
148. Benton, A. F., Eldgin, J. C.: J. Am. Chem. Soc. **48**, 3027 (1926); **51**, 7 (1929)
149. Psezhezkii, S. Ya., Vlodavez, M. L.: J. Phys. Chem. (USSR) **24**, 354 (1950)
150. Boreskov, G. K.: J. Chim. Physique **51**, 759 (1954)
151. Beusch, H., Fieguth, P., Wicke, E.: Chem.-Ing.-Techn. **44**, 445 (1972); Chem. React. Eng., First Intern. Symp., 1970, Washington, D. C., 1972, p. 615
152. Belyaev, V. D., Slin'ko, M. M., Slin'ko, M. G., Timoshenko, V. I.: Kinet. Katal. **14**, 810 (1973)
153. Belyaev, V. D., Slin'ko, M. M., Slin'ko, M. G.: Proc. Sixth Intern. Congr. Catalysis, London, 1976, Preprint B-15
154. Sheintuch, M., Schmitz, R. A.: Catal. Rev. **15**, 107 (1977)
155. Slin'ko, M. G., Slin'ko, M. M.: Catal. Reviews **17**, 119 (1978)
156. Schmitz, R. A., Renola, G. T., Gannigan, P. C.: Ann. N.Y. Acad. Sci. **316**, 638 (1979)
157. Schmitz, R. A., Renola, G. T., Gannigan, P. C.: Kinetics of Physiochem. Oscil., Sept. 1979, **1**, 221, (1979)
158. Kurtanjen, Z., Scheituch, M., Luss, D.: ibid. **1**, 191 (1979)
159. Zuniga, J. E., Luss, D.: J. Catal. **53**, 312 (1978)
160. Zelenyak, T. I., Slin'ko, M. G.: Kinet. Katal. **18**, 1235 (1977)
161. Savchenko, V. I., Boreskov, G. K., Dadayan, K. A.: Kinet. Katal. **20**, 741 (1979)
162. Dadayan, K. A., Boreskov, G. K., Savchenko, V. I., Sadovskaya, E. M., Yablonskii, G. S.: Kinet. Katal. **20**, 795 (1979)
163. Gorodetskii, V. V., Sobjanin, V. A., Bulgakov, N. N., Knor, Z.: Surface Sci. **82**, 120 (1979)
164. Gorodetskii, V. V., Savchenko, V. I.: Proc. Fifth Congr. on Catalysis (J. W. Hightower, ed.), North-Holland, Amsterdam, 1973, p. 527
165. Boreskov, G. K., Savchenko, V. I.: Seventh Intern. Congr. on Catalysis, Preprints, A-46, 1980
166. Karnaukhov, A. P., Boreskov, G. K.: J. Phys. Chem. (USSR) **26**, 1814 (1952)
167. Kimkhai, O. N., Kuznetsov, B. N., Pashkovskaja, N. A., Moroz, E. M., Boreskov, G. K.: React. Kinet. Catal. Lett. **6**, 393 (1977)
168. Katzer, Y. K., Sayers, D. E.: Rep. Symp. Catal. Materials, Boston, 1978
169. Hanson, F. V., Boudart, M.: J. Catal. **53**, 56 (1978)
170. Il'chenko, N. I., Yuza, V. A., Roiter, V. A.: Dokl. Akad. Nauk SSSR **172**, 133 (1967)
171. Kuzmina, G. M., Popovskii, V. V., Boreskov, G. K.: React. Kinet. Catal. Lett. **1**, 351 (1974)
172. Il'chenko, N. I.: Uspekhi Khimii **41**, 84 (1972)
173. Verhoeven, W., Delmon, B.: C. R. Ser. C **262**, 33 (1966)
174. Hegedüs, A. J., Millner, T., Neugebauer, J., Sasvari, K.: Z. f. anorg. u. allg. Chemie **281**, 64 (1955)
175. Novak, E. J., Koros, R. M.: J. Catal. **7**, 50 (1967)
176. Sancier, K. M.: J. Catal. **20**, 106 (1971)

177. Golodez, G. I., Pyatnizkii, Yu. I., Goncharuk, V. V.: Theor. and Experim. Chemistry (USSR) **4**, 53 (1968)
178. Avdeenko, M. A., Boreskov, G. K., Slin'ko, M. G.: Probl. of Kinet. Katal. (USSR) **9**, 61 (1957)
179. Smith, J. N., Palmer, R. L.: J. Chem. Phys. **56**, 13 (1972)
180. Pacia, N., Dumesic, J. A.: J. Catal. **41**, 155 (1976).
181. Norton, P. R.: J. Catal. **36**, 211 (1975)
182. Collins, D. M., Lu, J. B., Spicer, W. E.: J. Vac. Sci. Technol. **13**, 266 (1976)
183. Bernase, S. L., Somorjai, G. A.: Surface Sci. **48**, 204 (1975)
184. Boudart, M., Collins, D. M., Hanson, F. V., Spicer, W. E.: J. Vac. Sci. Technol. **14**, 441 (1977)
185. Popovskii, V. V., Boreskov, G. K.: Probl. Kinet. Katal. (USSR) **10**, 67 (1960)
186. Mamedov, E. A., Popovskii, V. V., Boreskov, G. K.: Kinet. Katal. **10**, 852 (1969); **11**, 969 (1970); **11**, 979 (1970)
187. Bakumenko, T. T.: Kinet. Katal. **6**, 74 (1965)
188. Bulgakov, N. N., Ismagilov, Z. R., Popovskii, V. V., Boreskov, G. K.: Kinet. Katal. **11**, 638 (1970).
189. Boreskov, G. K.: Kinet. Katal. **8**, 1020 (1967)
190. Boreskov, G. K., Popovskii, V. V., Sazonov, V. A.: Fourth Intern. Congr. Catalysis, Moscow, 1968
191. Netzer, F. P., Kneringer, G.: Surface Sci. **51**, 526 (1975)
192. Andrushkevich, T. V., Boreskov, G. K., Popovskii, V. V., Muzykantov, V. S., Kimkhai, O. N., Sazonov, V. A.: Kinet. Katal. **9**, 595 (1968)
193. Popovskii, V. V., Boreskov, G. K., Dzevenzki, Z., Muzykantov, V. S., Shulmeister, T. T.: Kinet. Katal. **12**, 979 (1971)
194. Boreskov, G. K.: Proc. Fifth Intern. Congr. Catalysis, (H. Hightower, ed.) **2**, 1973, p. 981.
195. Boreskov, G. K., Kasatkina, L. A., Popovskii, V. V., Balovnev, Yu. A.: Kinet. Katal. **1**, 229 (1960)
196. Il'chenko, N. I., Golodez, G. I., Pyatnizkii, Yu. I.: Kinet. Katal. **14**, 372 (1973)
197. Kuznetsova, L. I., Boreskov, G. K., Yurieva, T. M., Anufrienko, V. F., Maksimov, N. G.; Dokl. Akad. Nauk SSSR, **216**, 1323 (1974)
198. Kuznetsova, L. I., Yurieva, T. M., Boreskov, G. K.: Kinet. Katal. (in press)
199. Davydova, L. I., Popovskii, V. V., Boreskov, G. K., Yurieva, T. M.: React. Kinet. Catal. Lett. **1**, 175 (1974)
200. Gorodetskii, V. V., Panov, G. I., Sobyanin, V. A., Bulgakov, N. N.: React. Kinet. Catal. Lett. **9**, 239 (1978)
201. Gorodetskii, V. V., Nuwenhuys, B. E., Sachtler, W. M., Boreskov, G. K.: Appl. Surface Sci. (1981), in press
202. Wolkenstein, T.: Elektronentheorie der Katalyse an Halbleitern, Berlin: Deutscher Verlag der Wissenschaften, 1964
203. Hauffe, K.: Reaktionen in und an festen Stoffen. Berlin, Heidelberg, New York: Springer 1966
204. Kiselev, V. F., Krylov, O. V.: Electronic phenomena in adsorption and catalysis on semiconductors and dielectrons (in Russ.), Izd. Nauka, Moscow, 1979, p. 103
205. Il'chenko, N. I.: Kinet. Katal. **14**, 976 (1973)
206. Chebotareva, N. P., Ilchenko, N. I., Golodez, G. I.: Theor. and Experim. Chemistry (USSR) **12**, 202, 361 (1976)
207. Dus, R., Tompkins, F. C.: Proc. R. Soc. London **A343**, 477 (1975)
208. Abasov, S. I., Smirnova, N. N., Boronin, V. S., Poltorak, O. M.: J. Phys. Chem. (USSR) **54**, 1003 (1980)
209. Gentry, S. J., Firth, J. G., Jones, A.: J. Chem. Soc., Faraday Trans. I, J. Phys. Chem. **3**, 600 (1974)
210. Weinberg, W. H., Merrill, R. P.: J. Catal. **40**, 268 (1975)
211. Golodez, G. I.: Heterogeneous Catalytic Reactions Involving Molecular Oxygen (in Russ.), Izd. Naukova Dumka, Kiev, 1977, p. 225
212. Panov, G. I., Kinet. Katal. **1**, 202–207, (1981)

213. Ertl, G., Neumann, M., Streit, R. M.: Surface Sci. **64**, 393 (1977)
214. McCabe, R. W., Schmidt, L. D.: Surface Sci. **66**, 101 (1977)
215. Comrie, C. M., Lambert, R. M.: J. Chem. Soc. Faraday Transactions I, **72**, 1659 (1976)
216. Conrad, H., Ertl, G., Koch, J., Latta, E. E.: Surface Sci. **43**, 462 (1974)
217. Madey, T. E., Menzel, D.: J. Appl. Phys. Suppl. (Tokyo) **2**, Pt. 2, 229 (1974)
218. Ertl, G., Koch, J.: Z. Naturforsch. A25, 1906 (1970)
219. Comrie, C. M., Weinberg, W. H.: J. Chem. Phys. **64**, 250 (1976).
220. Rüppers, J., Plagge, A.: J. Vacuum Sci. Technol. **13**, 259 (1976)
221. Joiner, R. W., Roberts, M. W.: J. Chem. Soc. Faraday Transactions **70**, 1819 (1974)
222. Atkinson, S. J., Brundel, C. R., Roberts, M. W.: Chem. Phys. Lett. **24**, 175 (1974)
223. Langmuir, I.: Trans. Faraday Soc. **17**, 621 (1922)
224. Sklyarov, A. V., Tretyakov, I. I., Shub, B. P., Roginskii, S. Z.: Dokl. Akad. Nauk SSSR **189**, 1302 (1969)
225. Tretyakov, I. I., Sklyarov, A. V., Shub, B. P.: Kinet. Katal. **11**, 166, 479 (1970)
226. Matsushima, T., Almy, D. V., White, J. M.: Surface Sci. **67**, 89 (1977)
227. Hopster, H., Ibach, H., Comsa, G.: J. Catal. **46**, 37 (1977)
228. Alnot, M., Fusy, J., Cassuto, A.: Surface Sci. **57**, 651 (1976)
229. Kuchaev, V. L., Nikitushina, L. M., Temkin, M. I.: Kinet. Katal. **15**, 1202 (1974)
230. Winterbottom, W. L.: Surface Sci. **36**, 205 (1973)
231. Bonzel, H. P., Ku, R.: J. Vacuum Sci. Technol. **9**, 663 (1972)
232. Malakhov, V. F., Shmachkov, V. A., Kolchin, A. M.: Kinet. Katal **18**, 579 (1977)
233. Badour, R. F., Madell, M., Heusser, H. R.: J. Phys. Chem. **72**, 3621 (1968)
234. Daglish, A. G., Eley, D. D.: Proc. Second Intern. Congr. on Catalysis, Paris, 1961, **2**, 1615.
235. Tretyakov, I. I., Sklyarov, A. V., Shub, B. P. Kinet. Katal. **12**, 996 (1971)
236. Ertl, G., Rau, P.: Surface Sci. **15**, 443 (1969)
237. Ertl, G., Koch, J. In: Catalysis, vol. 2. Hightower, J. (ed.). North-Holland, Amsterdam 1973, p. 969; Z. Phys. Chem., (N. F.) **69**, 323 (1970)
238. Close, J. S., White, J. M.: J. Catal. **36**, 185 (1975).
239. Halsey, G. D.: Surface Sci. **64**, 681 (1977)
240. Dabill, D. W., Gentry, S. J., Holland, H. B., Jones, A.: J. Catal. **53**, 164 (1978)
241. McCabe, R. W., Schmidt, L. D.: Surface Sci. **65**, 189 (1977)
242. Zhdan, P. A., Boronin, A. I.: XPS and UPS Study of Adsorption and Reactions on Iridium. In: Proc. Conf. on "Surface Analysis — 79", Karlovy Vary, Czechoslovakia 1979, p. 33
243. Hori, G. R., Schmidt, L. D.: J. Catal. **38**, 335 (1975)
244. McCarthy, E., Zahradnik, J., Ruczgnski, G. C., Carberry, J. J.: J. Catal. **39**, 29 (1975)
245. Hugo, P., Jakubith, M.: Chem.-Ing.-Tech. **44**, 383 (1972)
246. Dauchot, J. P., Cakenberghe, J. Van: Nature (London), Phys. Sci. **249**, 61 (1973)
247. Bench, H., Fieguth, P., Wicke, E.: Chem.-Ing.-Tech. **44**, 445 (1972).
248. Bench, H., Fieguth, P., Wicke, E.: Chem. React. Eng. First International Symposium 1970, Washington, D. C. 1972, p. 615
249. Bykov, V. I., Elokhin, V. I., Yablonskii, G. S.: React. Kinet. Catal. Lett. **4**, 191 (1976)
250. Ivanov, V. P., Savchenko, V. I., Tataurov, V. L., Boreskov, G. K., Egelhoff, W. F., Weinberg, W. H.: Surface Sci. **61**, 207 (1976)
251. Ivanov, V. P., Boreskov, G. K., Savchenko, V. I., Egelhoff, W. F., Weinberg, W. H.: J. Catal. **48**, 269 (1977)
252. Ivanov, V. P., Savchenko, V. I., Tataurov, V. L., Boreskov, G. K., Egelhoff, W. F., Weinberg, W. H.: Dokl. Akad. Nauk SSSR **249**, 642 (1979)
253. Falconer, I. L., Wentrick, P. R., Wise, H.: J. Catal. **45**, 248 (1976)
254. Cambell, Charles T., Shi, Shei-Kung, White, I. M.: J. Phys. Chem. **83**, 2253 (1979)
255. Madey, T. E., Engelhardt, H. A., Menzel, D.: Surface Sci. **48**, 304 (1975)
256. Reed, P. D., Comrie, C. M., Lambert, R. M., Surface Sci. **64**, 603 (1977)
257. Dadayan, K, A., Boreskov, G. K., Savchenko, V. I., Bulgakov, N. N., Smolikov, M. D.: Dokl. Acad. Nauk SSSR **239**, 356 (1978)
258. Kalinkin, A. V., Boreskov, G. K., Savchenko, V. I., Dadayan, K. A.: React. Kinet. Catal. Lett. **13**, 111 (1980)

259. Engel, T., Ertl, G.: Adv. Catal. **28**, 1 (1979)
260. Shigeishi, R. A., King, D. A.: Surface Sci. **58**, 379 (1976)
261. Reed, P. D., Comrie, C. M., Lambert, R. M.: Surface Sci. **59**, 33 (1976)
262. Castner, D. G., Sexton, B. A., Somorjai, G. A.: Surface Sci. **71**, 519 (1978)
263. Tracy, I. C., Palmberg, P. W.: J. Chem. Phys. **51**, 4852 (1969)
264. Küppers, J., Michel, H.: J. Vac. Sci. Technol. **13**, 259 (1976)
265. Christmann, R., Ertl, G.: Z. Naturforsch. **A 28**, 1144 (1973)
266. McCabe, R. W., Schmidt, L. D.: Surface Sci. **60**, 85 (1976)
267. Engle, T., Ertl, G.: J. Chem. Phys. **69**, 1267 (1978)
268. Conrad, H., Ertl, G., Küppers, J.: Surface Sci. **76**, 323 (1978).
269. Pacia, N., Cassuto, A., Pentenero, A., Weber, B.: J. Catal. **41**, 455 (1976).
270. Taylor, J. L., Ibbotson, D. E., Weinberg, W. H.: Surface Sci. **79**, 349 (1979); J. Catal. **62**, 1 (1980)
271. Boreskov, G. K., Savchenko, V. I.: Seventh Intern. Congr. Catalysis, Prepr. A 46, Tokyo, 1980
272. Krylov, O. V.: Kinet. Katal. **3**, 502 (1962)
273. Boreskov, G. K., Marshneva, V. I.: Dokl. Akad. Nauk SSSR **213**, 112 (1973)
274. Garner, W. E., Gray, T. I., Stone, F. S.: Proc. Roy. Soc. **A 197**, 294, 314 (1949); Disc. Faraday Soc. **8**, 246 (1950)
275. Garner, W. E., Stone, F. S., Tiley, P. F.: Proc. Roy. Soc. **A2**, 472 (1952)
276. Praliand, H., Ronsseau, J., Figueras, F., Mathien, M. V.: J. Chim. Phys., Phys.-Chim. Biol. **70**, 1053 (1979)
277. Fesenko, A. V., Korneichuk, G. P.: All-Union Conf. on Kinetics of Catalytic Reactions. Processes of complete oxidation. Inst. of Catalysis of the Siberin Branch of the USSR Acad. of Sciences, Novosibirsk, 1973, p. 66
278. Kon', M. Ya., Shvetz, V. A., Kazanskii, V. B.: Dokl. Akad. Nauk SSSR **203**, 624 (1972)
279. Golodetz, G. I.: Heterogeneous catalytic reactions involving molecular oxygen. Kiev: Naukova Dumka, 1977, p. 284
280. Roginsky, S. Z., Zeldovich, Ya. B.: Acta Phys. Chim. (USSR) **1**, 554, 595 (1934).
281. Bruns, B. P.: J. Phys. Chem. (USSR) **21**, 1011 (1947)
282. Kabayashi, M., Kabayashi, U.: J. Catal. **27**, 100, 108 (1972)
283. Kakioka, H., Dukari, V., Teshner, S. G.: Kinet. Katal. **14**, 78 (1973)
284. Tarama, K., Teranishi, S., Yoshida, S., Tamura, N.: Proc. Third Intern. Congr. on Catalysis, North-Holland Amsterdam, 1965, p. 282
285. Hauffe, K., Wolkenstein, Th.: Symposium on Electronic Phenomena in Chemisorption and Catalysis on Semiconductors, Berlin: de Gruyter 1969
286. Kejer, N. P., Roginski, S. Z., Sazonova, I. S.: Dokl. Akad. Nauk SSSR **106**, 859 (1956)
287. Kejer, N. P., Chizhikova, G. I.: Dokl. Akad. Nauk SSSR **120 4**, 8, 30 (1958)
288. Kiselev, V. F., Krylov, O. V.: Electronic phenomena in Adsorption and Catalysis on Semiconductors and Dielectrics Moscow, Nauka 1979
289. Gorodetskii, V. V., Nieuwenhuys, B. E.: Surface Sci. Lett. (1981) in press
290. Boreskov, G. K., Bobrov, N. N., Maksimov, N. G., Anufrienko, V. F., Ione, K. G., Shestakova, N. A.: Dokl. Akad. Nauk SSSR, Ser. khim. **201**, 887 (1971)
291. Ione, K. G., Kuznetsov, P. N., Romannikov, V. N. In: Application of Zeolites in Catalysis (ed. Boreskov, G. K., Minachev, Kh. M.), Academiai Kiadþ Budapest, p. 87, 1979
292. Somorjai, G. A.: Principles of Surface Chemistry, Englewood Cliffs, New Jersey: Prentice-Hall, 1972
293. Sokolovskii, V. D.: Study of the mechanism of heterogeneous catalytic oxidation and search for catalysts and novel reactions of selective oxidative transformations of low paraffins. Thesis. Novosibirsk, Inst. Catalysis, 1980
294. Zhdan, P. A., Boreskov, G. K., Egelhoff, W. F., Weinberg, W. H.: Surface Sci. **61**, 377 to 390 (1976)
295. Mittasch, A., Theiss, E.: Von Davy und Döbereiner bis Deacon, ein halbes Jahrhundert Grenzflächen-Katalyse (Russ. transl.), ONTI, Ukr. Scientific-Techn. Publ., Kharkov, 1934
296. Anderson, R. B., Stein, K. C., Feenan, J. J., Hofer, L. J. E.: Ind. Engng. Chem. **53**, 809 (1961)

297. Kainz, G., Horwatitsch: Mikrochim. Acta (1962) p. 7
298. First, J. G., Holland, H. B.: Trans. Faraday Soc. **65**, 11 (1969)
299. Golodets, G. I.: Heterogeneous Catalytic Oxidation of Organic Substances, izd. Naukova Dumka, Kiev, 1978, p. 42
300. Sokolovskii, V. D.: Study of the mechanism of heterogeneous catalytic oxidation and search for catalysts for novel reactions of selective oxidative transformations of low paraffins. Thesis, Novosibirsk, Inst. Catalysis, 1980
301. Patterson, W. R., Kemball, C.: J. Catal. **2**, 465 (1969)
302. Reyeeson, L., Swaringen, A. L.: J. Am. Chem. Soc. **50**, 2872 (1928)
303. Kemball, C., Patterson, W. R.: Proc. Roy. Soc. **A 270**, 219 (1962)
304. Andrushkevich, T. V., Popovskii, V. V., Boreskov, G. K.: Kinet. Katal. **6**, 860 (1965)
305. Anshiz, A. G., Sokolovskii, V. D., Boreskov, G. K., Davydov, A. A., Budneva, A. A., Avdeev, V. I., Zakharov, I. I.: Kinet. Katal. **16**, 95 (1975)
306. Anshiz, A. G., Sokolovskii, V. D., Boreskov, G. K., Boronin, A. I.: React. Kinet. Catal. Lett. **7**, 87 (1977)
307. Stein, K. C., Feenan, J. J., Thompson, G. P., Schultz, J. F., Hofer, L. J. E., Anderson, R. B.: Ind. Engng. Chem. **52**, 671 (1960)
308. Moro-Oka, Y., Morikawa, Y., Ozaki, A.: J. Catal. **7**, 23 (1967)
309. Accomazzo, M. A., Nobe, K.: Ind. Engng. Chem. Process Res. and Develop. **4**, 425 (1965)
310. Popovskii, V. V.: Kinet. Katal. **13**, 1190 (1972)
311. Golodets, G. I., Pyatnitskii, Yu. I., Goncharuk, V. V.: Theor. and Experim. Chemistry (USSR) **3**, 830 (1967)
312. Bielanski, A., Haber, J.: Catal. Reviews **19**, 1 (1979).
313. Moro-Oka, Y., Ozaki, A.: J. Catal. **5**, 116 (1966)
314. Peregrine Phillips, United Kingdom Pat. 6091 (1831)
315. Boreskov, G. K., Chesalova, V. S.: Dokl. Akad. Nauk SSSR **85**, 378 (1952)
316. Bodenstein, M., Fink, F.: Z. Phys. Chem. **60**, 1 (1907)
317. Taylor, G. B., Lenher, S.: Z. Phys. Chem. Bodenstein-Festband, p. 30 (1931)
318. Uyehara, O., Watson, K. M.: Ind. Engng. Chem. **35**, 541 (1943)
319. Boreskov, G. K.: Catalysis in Production of Sulfuric Acid (in Russ.) Goskhimizdat, M.-L. 1954, pp. 101, 183
320. Neumann, B., Goevel, E.: Z. Elektrochem. **34**, 734 (1928)
321. Adadurov, I. E.: J. Appl. Chemistry (USSR) **7**, 1355 (1934). J. Phys. Chem. (USSR) **7**, 45 (1936)
322. Certificate of autorship 691185 (USSR). Catalyst for gas purification from sulfuric anhydride. B. I. 1979, No 38.
323. Boreskov, G. K.: Les oxides du soufre dans la chimie Moderne. Compte-rendu du Congrès Intern. du Soufre (Edonard Privat, ed.), Toulouse, 1968, p. 221
324. Bazarova, Zh. G., Boreskov, G. K., Kefeli, L. M., Karakchiev, L. G., Ostankovich, A. A.: Dokl. Akad. Nauk SSSR **180**, 1132 (1968)
325. Bazarova, Zh. G., Boreskov, G. K., Ivanov, A. A., Karakchiev, L. G., Kochkina, L. D.: Kinet. Katal. **12**, 949 (1971)
326. Milisavlivich, B. S., Ivanov, A. A., Polyakova, G. M., Serzhantova, V. V.: Kinet. Katal. **16**, 103 (1975).
327. Villadsen, J., Livbjerg: Catal. Rev. **17**, 203 (1978)
328. Boreskov, G. K.: 4th Intern. Congr. on Catalysis, Symposium III, Akademiai Riado, Budapest, 1972, p. 1
329. Ivanov, A. A., Boreskov, G. K., Beskov, V. S.: ibid. p. 383
330. Mars, P., Maessen, J. G. H.: Proc. Intern. Congr. on Catalysis, 3rd, Amsterdam, 1964, v. 1, p. 266
331. Happel, J., Hnatov, M. A., Rodriguez, A.: Chem. Eng. Progr. Symp. Ser. **68**, 155 (1972)
332. Boreskov, G. K., Polyakova, G. M., Ivanov, A. A., Mastikhin, V. M.: Dokl. Akad. Nauk SSSR **210**, 626 (1973)
333. Boreskov, G. K., Sokolova, T. N.: J. Appl. Chemistry (USSR) **14**, 1240 (1937)
334. Calderbank, P. H.: Chem. Engng. Prog. **49**, 585 (1953)
335. Eklund, R. B.: The Rate of Oxidation of Sulfur Dioxide with a Commercial Vanadium Catalyst, Stockholm, 1956

336. Rzaev, P. B., Roiter, V. A., Korneichuk, G. P.: Ukrainian Chem. J. **26**, 161 (1960)
337. Davidson, B., Thodos, G.: A. J. Ch. E. Journal **10**, 568 (1964)
338. Boreskov, G. K., Buyanov, R. A., Ivanov, A. A.: Kinet. Katal. **8**, 153 (1967)
339. Ivanov, A. A., Boreskov, G. K., Buyanov, R. A., Polyakova, G. P., Davydova, L. P., Kochkina, L. D.: Kinet. Katal. **9**, 560 (1968)
340. Boreskov, G. K.: J. Phys. Chem. (USSR) **19**, 92 (1945)
341. Odanaka, H.: Kinet. Katal. **7**, 571 (1966)
342. Horiuti, J.: Ann. New York Acad. Sci. **213**, 5 (1973)
343. Mezaki, R., Kadlec, B.: J. Catal. **25**, 454 (1972)
344. Happel, J.: Oxidation Communications **1**, 15 (1979)
345. Boreskov, G. K., Ivanov, A. A., Balzhinimaev, B. S., Karanatovskaya, L. M.: React. Kinet. Catal. Lett. **14**, 25 (1980)
346. Boreskov, G. K.: Heterogenious Catalysis in the Chemical Industry, Goskhimizdat, Moscow, 1955, p. 5
347. Boreskov, G. K.: Kinet. Katal. **21**, 5 (1980)
348. Marshneva, V. J., Boreskov, G. K., Pankratova, G. H., Solomennikov, H. A.: Kinet. Katal. (1982), in press

Chapter 3

Catalytic Activation of Carbon Monoxide on Metal Surfaces

M. A Vannice

College of Engineering
Dept. of Chemical Engineering
The Pennsylvania State University
133 Fenske Laboratory
University Park, PA 16802, USA

Contents

1. Introduction

Interest in carbon monoxide has increased dramatically during the past decade. This is due primarily to a new interest in energy resources other than natural gas and petroleum resources which include coal, oil shale, and heavy residua. In any process which involves gasification to convert these hydrogen-deficient materials to hydrocarbons or other organic compounds, CO is one of the principal products of the gasification step, and its subsequent hydrogenation to form the required final products is of extreme importance. It is thermodynamically possible to produce methane as SNG, hydrocarbon liquids as fuels, and alcohols and olefins as chemical intermediates — the major problem is the *selective* production of the desired product. During the overall conversion process, the reaction between CO and water is important, *via* the water gas shift reaction, not only because it is used to adjust the H_2/CO ratio in the gas stream between the gasifier and the reactor, but also because H_2O is a by-product of the CO hydrogenation reaction and can react with unconsumed CO in the syngas reactor to form CO_2.

This paper will address the interaction of activated CO with these two molecules — H_2 and H_2O — and will emphasize the current status of the subject. Numerous review articles exist which describe earlier work [1–11]. In this chapter the discussion will be restricted to the adsorption and reaction of CO on reduced metal surfaces since these systems, as both unsupported and supported catalysts, are those typically used to alter selectivities in the first reaction and enhance the rates of both reactions. It is not the intention of this chapter to discuss in detail the Fischer-Tropsch reaction as an established process: this has been dealt with in Chapter 4, Volume 1 of this series. The present purpose is rather to discuss the important basic aspects of carbon monoxide catalytic chemistry. The two chapters, that is the present one and the previous one on the Fischer-Tropsch process, should be regarded as complementary.

2. The CO Molecule and Chemisorption

Althought the CO molecule is isoelectronic with the N_2 molecule, significant changes occur in the relative energy levels of the molecular orbitals because of the difference in electronegativity between the carbon atom and the oxygen atom. Localization occurs with the orbital of lower energy existing on the more electronegative atom, oxygen, and since the average electronegativity of CO is similar to that of N_2, the lowering of one orbital results in the raising of another, shown schematically in Figure 1.

The lower, symmetric combination of inner-shell, $1s$ type orbitals of N_2 localizes on the O atom in CO and becomes essentially the oxygen $1s$ orbital, while the higher, anti-symmetric combination localizes around the carbon atom. The symmetric 2σ orbital and the two 1π orbitals of N_2 are lowered in CO and both types localize on the O atom. The antisymmetric 2σ N_2

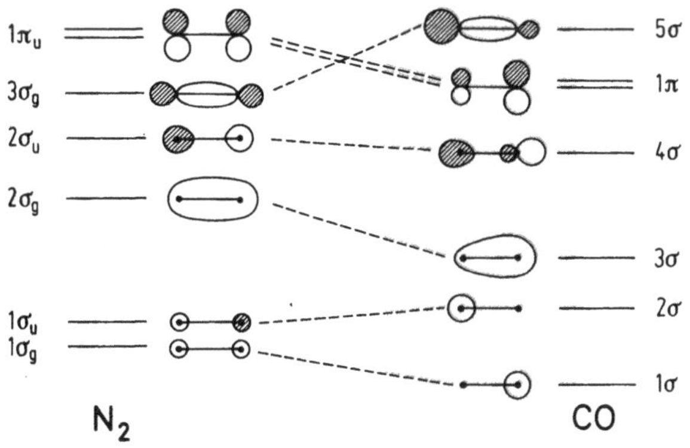

Figure 1. A Comparison of N_2 and CO Molecular Orbital Energy Levels. (After [12])

orbital correlates with a 4π CO orbital which results in some poorly localized oxygen "lone-pair" character, while the symmetric 3σ N_2 orbital becomes a high "lone-pair" orbital localized on the carbon atom. It is this latter labile orbital which results in the small dipole moment of CO, and which is responsible for allowing a strong interaction between CO and metal surfaces since it has the high energy and the requisite directional character to donate electrons into empty metal orbitals. The antibonding 2π orbitals, which represent the next higher level and tend to be localized on the carbon atom, facilitate back donation from the metal to the CO molecule [12]. Figure 2 provides a perspective of the shape of these various molecular orbitals of CO.

Any catalytic event on a solid surface must be preceded by an adsorption step which significantly perturbs the electronic structure of the adsorbate — a requirement associated with a chemisorption process. Therefore, variations in the chemisorbed state might be expected to have marked effects on the reactivity of the surface species, and knowledge about the chemical state of adsorbed molecules on different metal surfaces is of great importance. The formation of a chemisorption bond between CO and a transition metal surface is generally thought to occur according to the Blyholder model [13], which invokes a donor-acceptor mechanism qualitatively similar to the bonding for metal-carbonyl clusters [14]. In this model the bond occurs through a concerted electron transfer from the highest filled (5σ) molecular orbital of CO to unoccupied metal orbitals (essentially *d* orbitals), with back-donation occurring from occupied metal orbitals to the lowest unfilled (2π) orbital of CO. Figure 3 provides a visualization of the configuration of adsorbed CO. As summarized by Brodèn et al. [15], the 4σ and 1π orbitals are not thought to participate appreciably in bond formation, and the 5σ orbital in gaseous CO is considered to be essentially non-bonding with respect to C and O atoms. To a first approximation, then, transfer of electrons from the 5σ orbital will not markedly affect the C—O bond strength in the chemisorbed state. This conclusion has been verified by the work of Doyen and Ertl [16]. However, the 2π orbitals are anti-bonding, and back-donation into these orbitals will tend to weaken the C—O bond. Therefore, the net

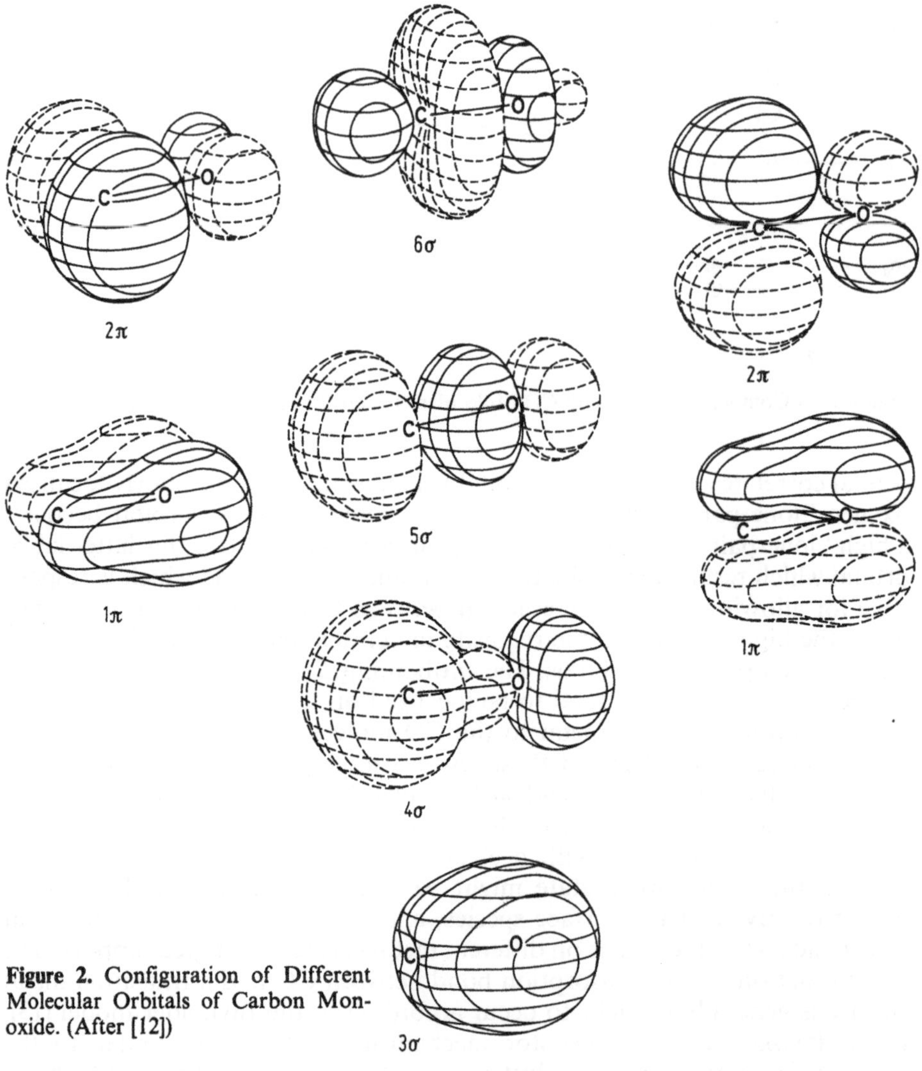

Figure 2. Configuration of Different Molecular Orbitals of Carbon Monoxide. (After [12])

result of the chemisorption process is expected to be a weakened carbon-oxygen bond.

This consequence is observed for carbonyls and chemisorbed CO where infrared spectra, and more recently electron energy loss spectra (EELS), have shown decreases in the C—O stretching frequency from 2143 cm^{-1} for gas phase CO to less than 1800 cm^{-1} in some carbonyls and chemisorbed CO species. This topic will be discussed later in greater detail.

The strength of the carbon-metal adsorption bond (and also the C—O bond strength, which is presumed to weaken as the C-metal bond becomes stronger) might be expected to be dependent upon: 1) the nature of the adsorbent metal, 2) the crystallographic orientation of the surface, and 3) the geometric location of the adsorbed molecule on a given single crystal

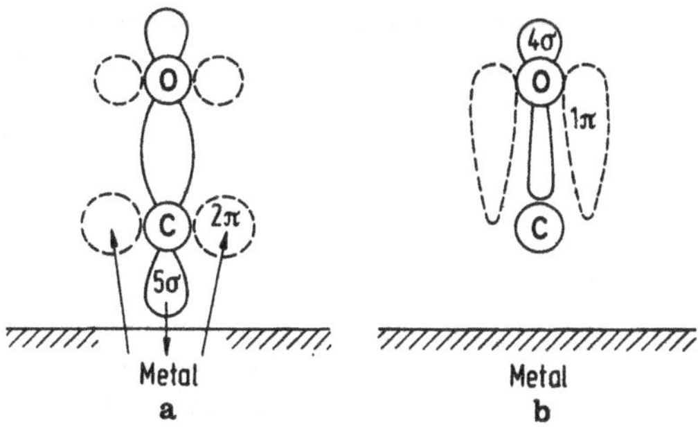

Figure 3. The chemisorption of CO on a metal surface is generally assumed to involve the 5σ and 2π orbitals, shown in (*a*), with insignificant participation of the 4σ and 1π orbitals, shown in (**b**). (After [15])

plane. These three variables can induce changes which are attributable to geometric or electronic factors, terms which have recently been better defined as metallic structure and metallic nature by Boudart [17] or the "ensemble" effect and "ligand" effect by Sachtler [18]. Using a model with one adjustable parameter, Doyen and Ertl [16] calculated the variation of adsorption energy on Ni, Cu, and Pd as the chemisorbed CO molecule was moved to sites of different coordination on five different crystallographic planes. The calculated ranges of initial adsorption energies were 105–126 kJ mol^{-1}, 67.8–82.0 kJ mol^{-1}, and 142–167 kJ mol^{-1} for Ni, Cu, and Pd, respectively, and these values were in very good agreement with experimental values. For the Pd (111) surface, the adsorption energy varied by less than 10.5 kJ × mol^{-1} for any site on the entire surface. Therefore, metallic nature is expected to have the largest effect on the chemisorption bond while geometric location should have the smallest influence.

The relative influence on the C—O bond strength of *d*-band electron concentration (ligand effect) versus the coordination number of the adsorption site, *i.e.*, the number of surface metal atoms with which the CO molecule interacts (ensemble effect), has been a debate of considerable interest [19, 20]. Generally, high frequency (HF) bands above 2000 cm^{-1} are associated with CO coordinated with a single metal atom in a linear or terminal manner, whereas 2-fold (bridged) and other forms of multiply-bonded CO results in lower frequency (LF) bands below 2000 cm^{-1} [21, 22].

Chini and coworkers [23] have reported the synthesis and characterization of a series of platinum carbonyl cluster dianions of general formula $[Pt_3(CO)_3(\mu_2\text{-}CO)_3]_n^{2-}$ (n = 2–5). These clusters have equal numbers of bridged-bonded and linearly-bonded CO ligands with well-defined positions. Since the charge on each cluster remains constant as n increases, the electron density associated with each Pt atom decreases and the effect on CO bond frequencies can be observed. For $n = 2$, the two major IR bands obtained in THF solution are 1990 cm^{-1} and 1775 cm^{-1}; both values increase regularly with n until at $n = 5$ the HF band has shifted to 2055 cm^{-1} and the LF band to 1870 cm^{-1}. The frequency difference between the singly- and doubly-coordinated species is always much greater than the shifts in frequency brought about in each type of bond by the decrease in electron density. Therefore, electronic changes alone in the linear adsorption bond do not appear to be sufficient to account for the range of frequencies observed on metal surfaces. Such results support the original assignments proposed by Eischens et al. [24], which were also based upon IR bands obtained from carbonyl clusters.

3. CO Adsorbed on Metal Surfaces

During the past few years, a tremendous research effort has been devoted to the study of CO adsorption and reactivity on transition metals. In particular, new spectroscopic techniques have been developed in the field of surface science which have been applied to studies of CO adsorption. These investigations have supplemented the large number of recent kinetic studies which have typically been conducted at much higher pressures, and an earlier review has attempted to unite results from these two fields based on literature available through 1975 [11]. This section will emphasize work reported since then, with each of the Group VIII metals being discussed separately as much as possible. Studies of the catalytic properties of these metals will be discussed in Sections 5 and 6.

Spectroscopy of surface vibrations is one of the most widely applied techniques to characterize adsorbed species, and many of the recent surface characterization studies utilize one of the following three techniques: infrared spectroscopy (IR), electron-energy-loss spectroscopy (EELS), and inelastic electron tunneling spectroscopy (IETS).

Transmission infrared spectroscopy was first applied in the 1950's to study adsorbed CO [24], and for supported catalysts it combines high resolution with sensitivity; it is also the only one of these techniques applicable under high pressures. However, the structure of surfaces of supported metal particles is generally not known and, in addition, wavenumbers below $\approx 1000 \times cm^{-1}$, which include the carbon-metal frequencies, are not usually observed due to absorption by the support. Also, when applied in a reflection mode to CO adsorbed on single crystal surfaces, IR sensitivity decreases more than an order of magnitude [25].

IETS is a recently developed technique which can allow the observation of both C—O and C-metal frequencies of CO adsorbed on oxide-supported metals [26]. The necessity of covering the metal surface and the adsorbate with an additional electrode complicates the catalyst system, however.

EELS is yet another new technique which is very appropriate for the study of adsorbed CO on well-defined, single-crystal surfaces [27] and is now beginning to be used for this purpose. EELS has somewhat poorer resolution, but a wide spectral range can be easily scanned and high sensitivity is achieved. A range of 200–3600 cm^{-1} can be scanned in 20 minutes and 0.001 monolayer of CO can be detected. Obviously, all these techniques are complementary and their application should lead to a better understanding of the adsorbed state of CO.

Two recent reviews have discussed the adsorbed state of CO from different perspectives. One, by Joyner [28], reviews photoelectron spectroscopy and includes a discussion of its application to CO while the other, by Bradshaw [29], briefly discusses the interaction of CO with metal surfaces.

A. Nickel

In regard to studies of CO adsorption, nickel has probably received more attention than any other Group VIII metal. Undoubtedly this is not only because it has been considered the best methanation catalyst, but also due to the ease with which Ni surfaces can be cleaned and studied in ultra high vacuum (UHV) systems. Nickel also serves well as a model metal system for theoretical approaches which describe electronic behavior and chemisorption properties.

Bertolini and coworkers [30, 31] have characterized CO adsorbed on a (111) single-crystal surface using electron energy loss spectroscopy (EELS), Augèr electron spectroscopy (AES), low energy electron diffraction (LEED), temperature programmed desorption (TPD) and work function measurements. An increase in the work function, $\Delta\varphi$, as θ increased from 0 to 0.41 coincided with a shift in C—O frequency from 1814 cm^{-1} to 1911 cm^{-1}. These results indicate that adsorbed CO has a net withdrawing effect on the electron concentration of nickel and that back-donation of nickel electrons into the 2π antibonding orbital is the major reason for the frequency shift. Coadsorption of electron-donating adsorbates, such as ethylene and benzene, lowered the C—O stretching frequency [30]. In a later study, it was shown by TPD curves and EELS spectra that the presence of "surface carbide" resulted in a more weakly bound form of CO [31]. Coadsorption of CO and H_2 at 413 K followed by cooling to 300 K led to the appearance of three surface species: weakly-bonded CO, an oxyhydrocarbonated compound, and a CH_x species. The formation of surface carbon *via* CO disproportionation (the Boudouard reaction) at 413 K appeared to facilitate the formation of these species — this was further demonstrated by the appearance of the first two aforementioned species when CO and H_2 were coadsorbed at 300 K on a carburized Ni (111) surface. However, in this case the CH_x species was not observed. More recently, Bertolini and Tardy have studied CO adsorption on clean and carbided Ni (111), (100) and (110) surfaces using EELS [32]. On the clean surfaces at 300 K a single CO species with 2-fold coordination was found while both singly and doubly coordinated CO existed on the other two surfaces. A small band at 1565 cm^{-1} on the (110) crystal may be indicative of defect sites which provide very strong C-metal bonding. The presence of surface carbon markedly lowered the heats of adsorption, and no adsorption was visible on the carbided (100) plane. These results, as well as others described later, point to the significant influence of surface carbon in CO hydrogenation reactions.

Andersson [33] was one of the first to apply EELS to CO adsorption when he studied the behavior of this gas on a Ni (100) surface. One of the advantages of EELS in these systems is the capability of resolving both the C-metal and the C—O bond frequencies in the adsorbed molecule. At low CO coverages at 173 K, peaks were observed at 359 cm^{-1}, 657 cm^{-1}, and 1932 cm^{-1} —

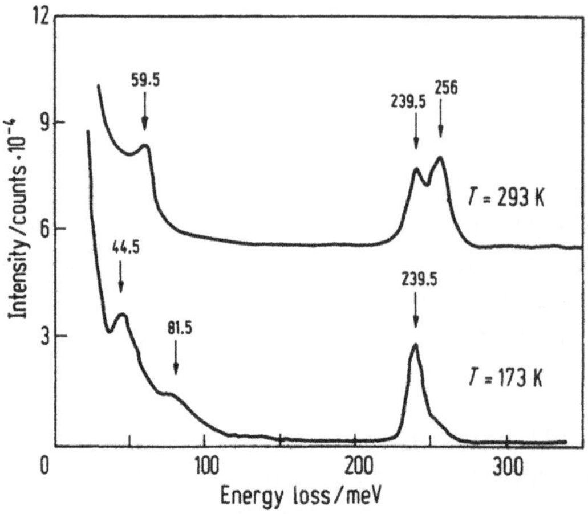

Figure 4. Electron energy loss spectra at a low CO coverage (0.3 L exposure) on the Ni (100) surface at 173 K and 293 K. (After [33])

the first two of which were attributed to Ni—C stretching vibrations; these results indicated bridged-bonded CO. At 293 K, additional vibrations at 480 cm^{-1} and 2065 cm^{-1} occured which indicated linearly-bonded CO was also present. These spectra are shown in Figure 4. At higher coverages a c(2 × 2) LEED pattern was observed, which was assumed to correspond to a coverage of $\theta = 0.5$, and only the last two IR peaks were observed. At θ values above 0.5, the adsorbed CO layer lost registry with the Ni (100) matrix and developed a quasihexagonal structure, which gave two shifted and broadened CO peaks, as shown in Figure 4. Later, Andersson and Pendry [34] calculated the intensity spectra of various LEED beams to determine the c(2 × 2) configuration of CO chemisorbed on Ni (100). The best agreement was attained when CO molecules were placed directly above the Ni atoms, and these authors initially proposed that the CO molecule was tipped over at an angle of 34° from the surface normal so that a C—O bond length close to that in $Ni(CO)_4$ is achieved. However, later work showed that the CO molecule was essentially perpendicular to the nickel surface [35].

X-ray photoelectron spectroscopy (XPS) and another new surface-sensitive technique, SIMS (secondary ion mass spectroscopy) have recently been applied to CO adsorption on a Ni (100) surface by Fleisch, Ott, Delgass and Winograd [36]. Analysis of the positive ion spectra showed that the $Ni_2CO^+/NiCO^+$ ratio was higher at low coverages than at saturation coverage of $\theta = 0.5$; this is indirect evidence that a high percentage of bridged-bonded CO may be present during initial stages of adsorption. Evidence was found for CO disproportionation at 523 K, and CO adsorption on these carbon-covered surfaces was greatly diminished. In an earlier SIMS study of CO on polycrystalline Ni foils, Barber, Vickerman, and Wolstenholme [37] also observed both $NiCO_2^+$ and $NiCO^+$ peaks, and they also associated these species with bridged- and linearly-bonded CO , respectively. They found that prolonged heating at 390 K gave evidence for the formation of surface carbon.

Erley and Wagner [38] observed that CO dissociation essentially does

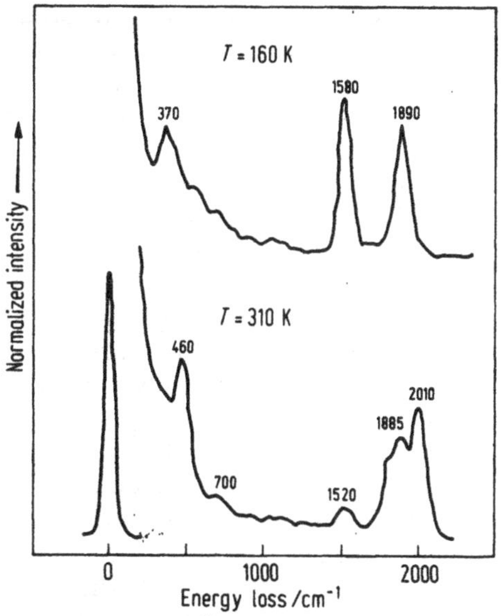

Figure 5. Energy loss spectra at 160 and 310 K for CO on a stepped Ni[5(111) × (1$\bar{1}$0)] surface after CO exposure of 0.2 L at 310 K. (After [39])

not occur on low index Ni faces up to temperatures at which molecular desorption occurs, whereas the dissociation process occurs more readily on films and roughened surfaces. Surmising that surface defects such as steps and kinks might facilitate CO dissociation, they compared the stepped Ni(S) [5(111) × (110)] surface with a Ni (111) plane [38]. Additional TPD peaks were observed from the stepped surface which were associated with an associative desorption process involving surface carbon and oxygen. They concluded that steps lower the activation barrier for CO dissociation, compared to Ni (111), and increase the activation energy for associative desorption. More recently, Erley, Ibach, Lehwald, and Wagner [39] have applied EELS to further investigate CO adsorption on this stepped Ni surface. The adsorption of CO at 150 K and low coverages produced an intense 1520 cm^{-1} peak, similar to that reported by Bertolini and Tardy for the (110) plane [32], as well as a 1900 cm^{-1} peak. The unusually low C—O stretching frequency was associated with CO adsorbed on special step sites while the second peak was considered to indicate bridged-bonded CO close to the edge of the steps. At higher coverages and higher temperatures, a band near 2000 cm^{-1} predominated, as indicated in Figure 5. The low frequency peak is indicative of a weakened C—O bond, which supports their previous conclusion that special surface sites enhance CO dissociation. Doering, Poppa and Dickinson [40] have reported that CO dissociation occurs more rapidly on small Ni and Pd crystallites, which is consistent with these findings because special surface sites (with low coordination numbers) occur in higher concentrations on small metal particles.

Finally, CO species adsorbed on the Ni (111) and Ni (100) single crystal surfaces have similar dipole moments, μ, at surface coverages of $\theta = 0.5$ or lower, with $\mu = 0.28$ Debye on the former surface and $\mu = 0.33$ Debye on the latter [41, 42].

Van Dijk, Groenewegen, and Ponec [43] used IR spectroscopy to study the effect of surface carbon on CO adsorption on silica-supported nickel. The intensity of the LF band near 1925 cm^{-1} decreased continuously with increasing carbon coverage whereas the intensity of the HF band around 2055 cm^{-1} went through a maximum before declining. These results were interpreted to mean that surface carbon is preferentially formed at sites which provide multiple coordination for adsorbed CO. Carbon blocks these sites, thus reducing the LF band intensity, but the linearly-bonded CO increases because either IR-inactive CO is removed to free new sites or additional adsorption occurs on top of surface Ni atoms. It is clear that significant changes in CO adsorption can occur as a consequence of surface carbon.

A very informative IR spectroscopic and magnetization study was conducted by Primet, Dalmon, and Martin [44] on a series of Ni/SiO_2 catalysts. The average Ni crystallite sizes were determined by hydrogen chemisorption and saturation magnetization measurements, and very good agreement was obtained between the two techniques as long as the nickel was completely reduced in the samples. The ratio of integrated optical densities for the HF and LF peaks was approximately constant at 0.3 to 0.4 for all the well-reduced samples, indicating a preponderance of multiply-bonded CO. Magnetization measurements made at various CO coverages indicated a bond number, n, of 1.85, which was independent of CO coverage, particle sizes in the 2.5–9.5 nm range, and temperature between 293–373 K. Incomplete nickel reduction resulted in a larger HF infrared peak and a lower bond number. The consistency of these results strongly supports the hypothesis that linear ($n = 1$) and multicentered ($n \geqq 2$) forms of adsorbed CO exist [23].

Additional studies of H_2 and CO coadsorption on Ni/SiO_2 catalysts by these same investigators led them to conclude that, although a weak interaction occurred between these two adsorbates, no evidence of a -CHOH complex was observed at 300 K [45]. These results agree with earlier work which also indicated an interaction between adsorbed CO and hydrogen but no formation of a detectable CH_x complex on nickel [11].

B. Iron

Studies on iron surfaces in UHV systems have been relatively few in number, due to the difficulty in removing impurities such as oxygen to obtain a clean, well-defined surface. Recently, Wedler and coworkers [46, 47] have prepared iron films under UHV conditions and studied H_2 and CO adsorption at 77 and 273 K. Hydrogen monolayer coverages at 273 K appear to reach values near saturation at pressures as low as 10^{-3} Pa, and initial heats of adsorption for H_2, determined calorimetrically, were 98 ± 3 kJ mol^{-1} [46]. A similar study with CO showed that the iron surface is essentially saturated at 10^{-3} Pa, and the heats of adsorption indicated an initial value of 155 $\pm$ 5 kJ mol^{-1} with three general regions of adsorption visible up to coverages of 2.2×10^{15} molecules cm^{-2} [47]. TPD spectra produced peaks at 4000 K and 700 K, with the former attributed to molecular desorption and the latter to a recombination process of C and O, since a slow dissociation process appeared to exist at 273 K. Evidence from UPS data indicating CO dissociation at room temperature on Fe films was also reported by Spicer and coworkers [48].

CO adsorption on three low-index, single-crystal planes of Fe has been studied by three different groups of investigators. Brodèn, Gafner, and Bonzel [49] found that CO adsorption is nondissociative on the close-packed, Fe(110) surface at 320 K, but a slow dissociation process occurs at this temperature; however, this dissociation process is very rapid at 385 K. On an Fe(100) plane, Rhodin and Brucker [50] observed that CO adsorption was molecular at 123 K, but warming the surface to 300 K began to induce dissociation. They concluded from LEED patterns that after dissociation both O and C atoms resided in hole positions with a 4-fold coordination. Interestingly, preadsorbed sulfur inhibited dissociation and produced molecular desorption upon warming. Finally, Textor, Gay, and Mason [51] determined that at 300 K only dissociative CO adsorption occurs at low and medium coverages on the rough Fe(111) surface, but that at saturation coverage molecular adsorption also exists. Exposure of this CO-saturated surface to hydrogen produced no reaction at 300 K, but at 470 K there was a preferential reaction with the adsorbed CO molecules which removed them from the surface. The carbide layer remained and was stable toward hydrogen for extended periods. From integrated XPS intensities, they calculated that 15% of the surface was covered with CO molecules, and the estimated coverages were: 7.2×10^{14} atoms C cm^{-2}, 7.8×10^{14} atoms O cm^{-2}, and 2.4×10^{14} molecules CO cm^{-2}. These studies indicate that structural effects may play an important role in determining the catalytic properties of iron for CO activation. These last authors mention that hydrogen preadsorbed on this surface results in an enhancement of molecularly adsorbed CO, similar to the effect produced by preadsorbed sulfur. Since chain-growth properties and alcohol production on a metal surface may be very dependent on the relative amounts of dissociated and molecular CO present, as discussed later, these results for iron are particularly intriguing.

Kroeker, Hansma and Kaska [52] have utilized IETS to study the low energy vibrational modes of CO on very small (2–3 nm) iron particles dispersed on silica. Only a linearly adsorbed species was found whose three LF bands at 436, 519, and 569 cm^{-1} were attributed to 2 C-Fe bending and 1 C-Fe stretching modes. The dipole derivative of this CO species was calculated to be $\sim$10 Debye $Å^{-1}$.

Benzinger and Madix [53] have examined the effect of surface carbon, oxygen, sulfur, and potassium on CO adsorption on the Fe(100) plane. XPS and TPD showed that C, O, and especially S adlayers reduced CO and H_2 binding energies and inhibited CO dissociation. However, K enhanced both H and CO binding energies, and the stronger CO bond was attributed to an interaction between K (4s) and CO ($2\pi^*$) orbitals. The activation energy for CO dissociation was estimated to be near 105 kJ mol^{-1}. This direct interaction of CO and H with K clusters to increase surface coverages of CO and hydrogen is important because it does not support the old explanation of higher activity and selectivity shifts for promoted Fe via electron transfer from K to the *d*-band of the iron. A more localized effect of K is a distinctly new perspective on the role of alkali metal promoters.

C. Cobalt

In addition to the problem of surface cleanliness, surface studies of cobalt are further complicated by the fact that an hcp → fcc transition occurs at temperatures near 700 K. However, CO adsorption on clean Co surfaces, as determined by AES, has recently been studied by Lambert et al. [54, 55]. Adsorption at 300 K on Co foil and on the close-packed Co(0001) plane produced a single TPD peak at ~430 K on both surfaces and a $(\sqrt{3} \times \sqrt{3})$ R30° LEED pattern from the single crystal, which was associated with a coverage of one-third [54]. At higher CO pressures ($>5 \times 10^{-9}$ torr), compression occured in the adsorbed layer as a coverage of 0.6 was approached. The desorption energy was very dependent on coverage and at $\theta = 0.022$, $E_d = 103 \pm 8$ kJ mol^{-1}. In a study of CO adsorption on the stepped Co(1012) surface, a similar E_d value of 106 kJ mol^{-1} was estimated from a TPD peak at 370 K, and again rapid, nondissociative adsorption occured at 300 K [55]. However, on this surface, TPD runs at 5 °K s^{-1} produced substantial CO dissociation whereas runs at 10 °K s^{-1} did not. Since no oxygen Augér signal was observed, it was concluded that no oxygen resided near the surface and O atoms rapidly diffuse into the bulk. An important result of this study was that the surface carbide can still chemisorb substantial amounts of CO (~20% of the saturation value on clean Co), but cannot dissociate it.

This last result infers that the working surface of a cobalt catalyst under reaction conditions is likely to possess very different properties than the initial reduced Co surface. Even metal surfaces which readily dissociate CO at low temperatures can be modified to allow substantial nondissociated CO adsorption to occur. From these studies of the first-row Group VIII metals, the possible importance of rough or stepped surfaces (structural effects) is evident — and a general pattern emerges which indicates that these surfaces, which contain step or kink sites, facilitate CO dissociation.

D. Palladium

Substantial efforts have been devoted to the characterization of palladium surfaces, and studies continue to provide additional information about the adsorbed state of CO on this metal. Hoffmann and Bradshaw [56] utilized IR reflection spectroscopy to examine low coverages of CO on a Pd film at 300 K. The only observable peak occurred below 2000 cm^{-1} and was initially very broad and weak although it sharpened somewhat at higher coverages, which were achieved at pressures near 1.3×10^{-4} Pa. Assuming that low-index planes predominate, especially the (111) face, the authors concluded that the electronic nature of the chemisorption bond appeared

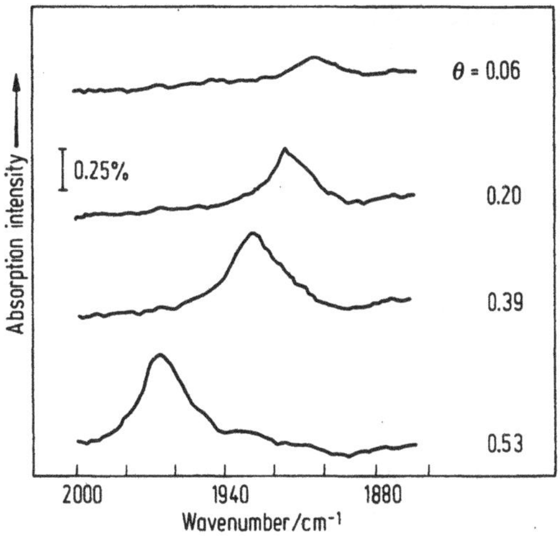

Figure 6. IR reflection spectra for CO adsorbed on the Pd (100) surface at 300 K and at different coverages. After [57]

more important than the geometry of the adsorption site thereby tending to support Blyholder's [20] model invoking only linear adsorption.

However, in a subsequent study of the (100), (111), and (210) single-crystal surfaces of Pd using both LEED and IR reflection spectroscopy, Bradshaw and Hoffmann [57] observed both HF and LF CO bands. On the (100) face the IR frequency was 1895 cm^{-1} at low coverage, and this shifted continuously to 1930 cm^{-1} at $\theta = 0.4$. A much sharper increase occurred above half coverage which was attributed to compression of the surface layer. Only sites of 2-fold coordination were assumed to exist on this face. The IR spectra for this surface are shown in Figure 6, and can be compared with the EELS spectra in Figure 4 for resolution and sensitivity. The same general trend occurred on the Pd(111) plane, with the frequency shifting from 1823 cm^{-1} at low coverage to 1936 cm^{-1} at $\theta = 0.5$. However, the sharpest increase began at $\theta = 1/3$, and a 2092 cm^{-1} band was formed at coverages above one-half. On the open (210) surface, the frequency increased more rapidly with coverage than on the other two planes, and it varied from 1880 cm^{-1} to 1996 cm^{-1}. As a consequence of this study, these authors made the following assignments for adsorbed CO, which are quite similar to those proposed by Hulse and Moskovits [21]: bonding which is single-fold — 2050–2120 cm^{-1}, 2-fold — 1880–2000 cm^{-1}, and 3-fold — 1800–1880 cm^{-1}.

CO adsorption on Pd and Pd-Ag films was studied by Stephan, Franke, and Ponec [58]. At 295 K and higher pressures (2.7×10^{-1} Pa), TPD spectra indicated two low-temperature peaks (γ_1 at 100 K and γ_2 at 150 K), and two broad maxima at $\approx$300 and $\approx$500 K. They associated the γ_2 state, which is not observed for pure Pd and is very sensitive to small amounts of Ag, to a binding state requiring an ensemble of Pd atoms. The 300 K peak was attributed to linearly-bonded CO, but the assignment of the 500 K peak to multiply-bonded CO, although plausible, was much less certain. The different binding energies appear to correlate with the different states of adsorption observed by Bradshaw and Hoffmann [57]. In contrast to Ni, Co and Fe, dissociative CO adsorption at 300–400 K on Pd does not happen, even on open faces such as the (210) and on rough surfaces such as the stepped (001) surface, as reported by Bader et al. [59].

E. Rhodium

Castner, Sexton, and Somorjai [60] investigated CO adsorption on the (111) face, where $(\sqrt{3} \times \sqrt{3})$ R30° LEED patterns indicated 3-fold coordination corresponding to $\theta = 1/3$, while at higher coverages a hexagonal overlayer formed which resulted in a maximum coverage of 3/4. A c(2 × 2) structure initially occurred on the (100) face up to $\theta = 0.5$, which is indicative of 4-fold coordination, then this structure also was compressed into a hexagonal monolayer at higher pressures, reaching a maximum θ of 0.83. TPD showed that a low-temperature peak began forming simultaneously with the onset of the compression of the CO overlayer. This peak was particularly apparent for the (100) face and indicated a significant weakening of the adsorption bond at coverages above 1/2. The study of CO adsorption on Rh (111) by Thiel et al. [61] confirmed the results of Somorjai and coworkers [60]. They measured a desorption energy of 132 kJ mol^{-1} compared to 130 kJ mol^{-1} in reference [60] and concluded that CO adsorbs *via* a mobile precursor state with an initial sticking probability of 0.76. Two possible orientations on the surface were proposed for the CO overlayer, each of which possessed two types of binding states. Further studies by Yates et al. [62] have provided strong evidence that CO dissociation does not occur on Rh surfaces. The probability of dissociation between 300 and 870 K $(< 10^{-4})$ was much smaller than that for desorption. The absence of CO dissociation conflicts with the earlier work of Castner et al. [60].

Dubois and Somorjai [63] have shown that linearly adsorbed CO first populates sites on top of Rh (111) surface atoms, then bridged-bonded CO occurs later. The binding energy of the latter species was calculated to be about 17 kJ mol^{-1} lower. This behavior is similar to that of CO on Pt (111) but different from Ni and Pd surfaces. Surface carbon and oxygen inhibited adsorption and weakened the C-metal bond, but adsorbed hydrogen had no effect.

The role that the support material plays in the preparation of supported metal catalysts is commonly viewed as nothing more than providing a solid matrix on which small metal crystallites are formed and maintained. However, the possibility exists that metal-support interactions can occur which alter the electronic properties of the crystallites or provide surface structures normally not present, for instance, *via* expitaxial stabilization. Such interactions can alter the adsorption and catalytic properties of the metal component.

An example of stabilization of atypical surface structures is provided by rhodium dispersed on γ-Al_2O_3. Yates, Murrell, and Prestridge [64] have employed IR spectroscopy, H_2 and CO chemisorption, and high-resolution transmission electron microscopy (TEM) to show that Rh crystallites can exist on an alumina surface as small (1.5 nm), 2-dimensional rafts which appear to be only 1 atom thick. Very high $CO_{(ad)}$/Rh ratios near 1.5 correlated with an IR spectrum dominated by symmetric peaks at 2100 cm^{-1} and 2030 cm^{-1}. These peaks have been associated with the symmetric and asymmetric vibration modes of two CO molecules bound to a single Rh atom [65]. Three-dimensional crystallites with an average size near 2.6 nm provided an adsorbed CO species with one dominant band lying between 2055–2075 cm^{-1}, which is typically associated with one linearly-adsorbed CO molecule per Rh atom. The electron opacity of the particles during TEM measurements was used to determine whether the particles were 2- or 3-dimensional. Yao and Rothchild [66] had also studied a $Rh/\gamma\text{-}Al_2O_3$ catalyst and observed the aforementioned doublet, but found additional peaks at 2000 cm^{-1} and 1850–1900 cm^{-1} in their sample. However, this catalyst had been calcined at 574 K prior to reduction at 673 K. An additional air calcination step at 773 K enhanced the doublet formation; therefore, these two studies show the significant influence that pretreatment can have on the state of CO adsorption. Regardless, Yao and coworkers [66] also concluded that Rh can exist in a two-dimensional phase. The effect of these changes in particle morphology on the catalytic activity of CO over Rh has not yet been reported, but it would seem to be a study worthy of attention.

An IETS study of adsorbed CO has been conducted by Hansma et al. [67] on alumina-supported Rh particles. A spectrum obtained from this technique is shown in Figure 7, and can be compared with previous IR and EELS spectra in Figures 4 and 6. Two CO stretching modes at 1721 cm^{-1} and 1942 cm^{-1} were observed on relatively uniform 2.5 nm Rh particles. Carbon-metal bending modes at 413 cm^{-1} and 465 cm^{-1} were associated with the species having the latter C–O stretch frequency, and a peak at 600 cm^{-1} was associated with the HF CO species. The resolution of both C–O and C-metal bond frequencies is very useful; however, the C–O stretching peaks were downshifted approximately 120 cm^{-1} compared to IR results, and this shift was attributed to a dipole imaging effect produced by the top metal electrode.

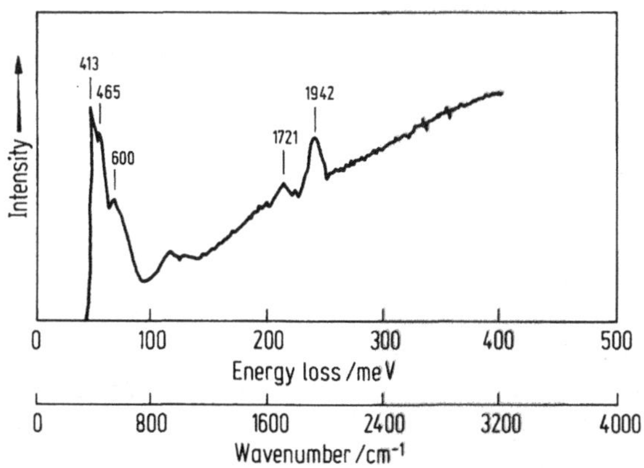

Figure 7. Differential inelastic electron tunneling spectra of CO adsorbed on alumina-supported Rh. After [67]

F. Ruthenium

The close-packed Ru (001) plane has recently been studied in some depth. The LEED study by Williams and Weinberg [68] showed a $(\sqrt{3} \times \sqrt{3})$ R30° pattern, optimized at $\theta = 1/3$, which shifted to a pattern indicative of a compressed overlayer at a coverage of 0.65. This behavior is quite similar to that discussed previously for other Group VIII metals.

Pfnur et al. [69] found that a single CO band occurred on the Ru (001) surface whose frequency shifted from 1984 to 2061 cm^{-1} as the coverage increased to $\theta = 0.68$. The integrated absorbance reached a maximum at $\theta = 1/3$. A linear CO species occupying on-top sites was assumed up to $\theta = 1/3$, then repulsive interactions shifted this linearly bound CO away from on-top sites, and dipole-dipole coupling increased the band frequency. Madey [70] has obtained electron stimulated desorption ion angular distributions on this surface which verify that CO molecules are bonded to the metal surface via the carbon atom with the CO axis perpendicular to the surface. This orientation on the clean surface was independent of coverage and temperature in the range of 90–350 K, and still occurred on an oxygen-covered Ru surface.

An earlier study had shown that the CO sticking coefficient on the Ru (001) plane was proportional to $(1-\theta)$ and had a high initial value of 0.85 [71]. This same study by Fuggle et al. [71] also showed that CO adsorption onto an oxygen-covered Ru (001) plane displaced 15–20% of the oxygen monolayer, presumably by reaction to form CO_2, and deposited 25–30% of a CO monolayer thereby increasing the total oxygen content on the surface. These results are of interest because significant differences appear to exist between the IR spectra of CO adsorbed on reduced and oxidized Ru crystallites. For example, Brown and Gonzalez [72] found that the major band on a reduced 6% Ru/SiO_2 catalyst was at 2030 cm^{-1} whereas a strong 2080 cm^{-1} band and two medium intensity bands at 2135 and 2030 cm^{-1} were observed on the oxidized sample. The 2080 cm^{-1} band was assigned to linearly-adsorbed CO on Ru, and the last two bands were assigned to CO adsorbed on a surface oxide and on Rh atoms perturbed by neighboring oxygen atoms, respectively. An earlier explanation by Dalla Betta [73] that multiple CO bands on Ru were a function of particle size is contradicted by the interpretation of this study, which suggests that highly dispersed Ru may be in a state of partial oxidation.

Using TPD, McCarty and Wise [74] investigated the isotopic exchange of $^{13}C^{16}O$ and $^{12}C^{18}O$ at 101 kPa on two Ru/Al_2O_3 catalysts. Exchange began near 375 K and a maximum rate of 0.01 molecule $site^{-1}$ s^{-1} was attained at 427 K. Formation of CO_2 during these runs was low. The maximum rate of this reaction occurred 45 K lower than that of the methanation reaction on these catalysts, and strongly indicated that methane can be formed via a sequence involving CO dissociation prior to any hydrogenation step. Bossi et al. [75] studied this exchange reaction on Ru/Al_2O_3 at 373 K and found it was strongly inhibited by hydrogen. The inhibition was much greater than that attributable to competitive adsorption, so they proposed that hydrogen donated electrons to increase CO backbonding and reduce exchange lability.

G. Platinum

The Pt (111) plane has been thoroughly studied by a number of investigators [76–82]. Ertl, Neumann, and Streit [76] observed a $(\sqrt{3} \times \sqrt{3})$ R30° pattern up to $\theta = 1/3$, which then transformed into a c(4 × 2) structure at $\theta = 1/2$. Finally a hexagonal close-packed layer formed as a saturation value of $\theta = 0.68$ was approached. This seems to be characteristic behavior of Group VIII metals. The initial isosteric heat of adsorption was 138 kJ × mol^{-1} while TPD data provided a desorption energy of $E_d = 117$ kJ mol^{-1} assuming $\nu = 10^{13}\ s^{-1}$. Since desorption energies must equal or exceed heats of adsorption, a value of $\nu = 10^{13}\ s^{-1}$ was required to provide equilivancy. As will be seen, E_d values for CO which appear to be too small are not uncommon. The variation of the work function, $\Delta\varphi$, resulted in a sharp minimum at $\theta = 1/3$, followed by a sharp maximum at $\theta = 1/2$, before finally decreasing again and levelling out near saturation coverage. From an analysis of this behavior, Ertl and coworkers [76] concluded that CO on the surface was mobile, and less than 2 kJ mol^{-1} separated adsorption energies between sites.

Shigeishi and King [77] employed reflection IR spectroscopy and TPD to examine CO adsorption on a (111)-oriented Pt ribbon. The initial sticking probability at 300 K was high, 0.67, and a saturation coverage of 7×10^{14} molecules cm^{-2} was determined. One desorption peak at 650 K occurred for $\theta > 0.14$, which was 200 K higher than that obtained by Ertl et al. [76]. Above $\theta = 0.3$, the integrated IR intensity leveled out, and at values above 0.4 the desorption energy remained constant at 135 kJ mol^{-1}. Only one IR band, which occurred between 2065 cm^{-1} and 2100 cm^{-1}, was observed. A possible IR-inactive state or in-phase vibrations of adsorbed islands of CO on the Pt surface were suggested as explanations for the independence of the IR intensity on coverages above $\theta = 1/3$.

However, Krebs and Lüth [78] used a more sensitive double-beam reflection IR spectroscopic technique and, in contrast to the previous authors, they

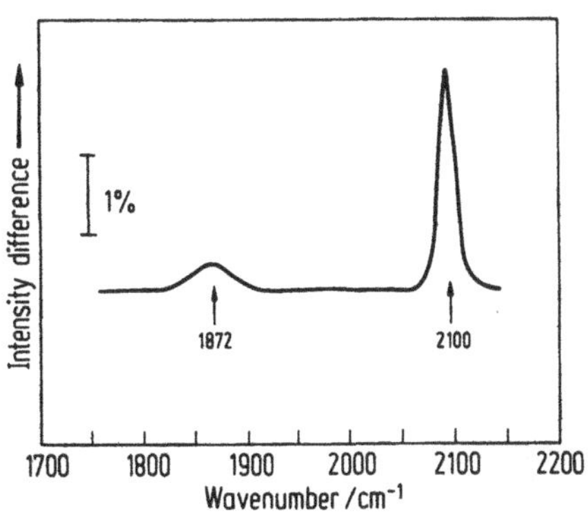

Figure 8. Double-beam IR reflection spectra of CO adsorbed on a Pt (111) surface at 300 K and saturation coverage. After [78]

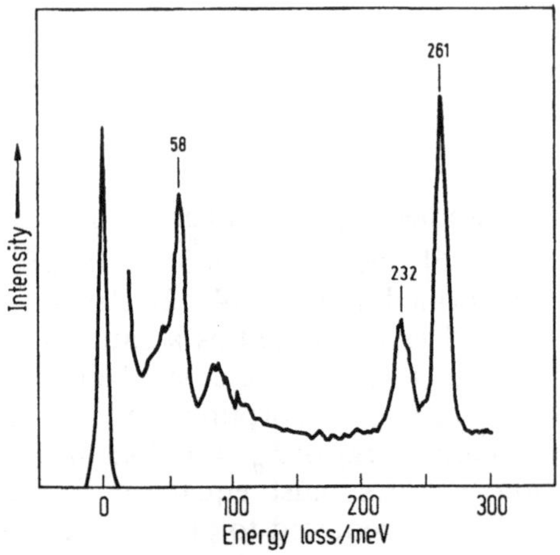

Figure 9. Electron energy loss spectra of CO adsorbed on Pt (111) at 350 K. After [79]

observed an additional, less intense, 1870 cm^{-1} band at higher coverages, as shown in Figure 8. These IR results are in excellent agreement with the EELS spectra obtained by Froitzheim, Hopster, Ibach and Lehwald [79], which are shown in Figure 9, and provide an excellent comparison of the two techniques on the same metal surface. At low coverage (<0.2 L), the two bands at 2080 cm^{-1} and 475 cm^{-1} are observed while at higher coverages two additional bands at 1870 cm^{-1} and 360 cm^{-1} are formed. Froitzheim et al. [79] also observed a levelling-out of the intensity of the 2080 cm^{-1} band at greater exposures. Figure 10 shows the CO adsorption positions available to Froitzheim et al. which provided consistency with the LEED patterns of Ertl, Neumann, and Streit [76]. Their proposal that adsorption occurs on top of Pt atoms at low coverage and additional bridged-bonding occurs between θ values of 1/4 and 1/2, represented by positions b and f in Figure 10, provided agreement between IR assignments that are normally made, the LEED results of Ertl et al. [76], and the constant IR intensity at higher coverages, thereby eliminating the need of Shegeishi and King's explanations. Their choice was also required to explain the development of the C-Pt vibrations with increasing coverage.

In a subsequent study of the Pt (111) face, Hopster and Ibach [80] reproduced both their earlier EELS results and the TPD and LEED results of Ertl and coworkers [76], and their conclusions were unchanged regarding

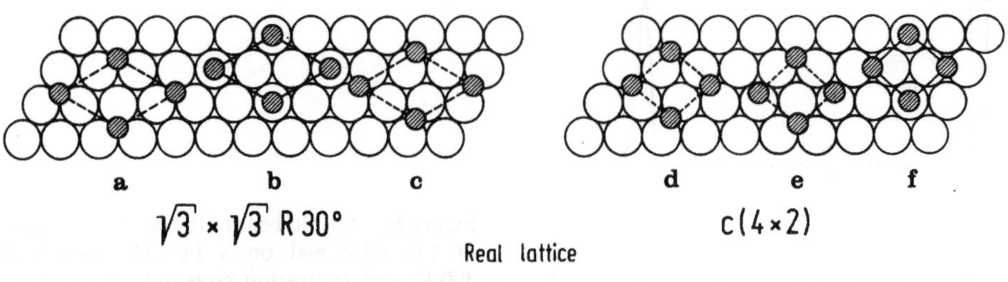

Figure 10. Different adsorption positions for CO on Pt (111) consistent with LEED patterns at low coverage (I) and higher coverage (II). After [79]

the state of CO adsorption on this plane. In addition, they examined the stepped 6 (111) × (111) surface using EELS and TPD and found an additional adsorption site which has a higher binding energy. The low CO stretching frequencies (1410 cm^{-1} and 1560 cm^{-1}) associated with these sites are similar to those observed for Ni [39], and infer a greatly weakened C–O bond which may facilitate CO dissociation on this type of Pt site. The model of Ibach and coworkers describing CO adsorption on the Pt (111) surface was supported by the study of Norton, Goodale, and Selkirk [81], who detected two adsorbed states of CO at coverages above 0.1. Measurements involving XPS, UPS, TPD, and $\Delta\varphi$ showed that the linearly adsorbed CO had a heat of adsorption about 4 kJ mol^{-1} greater than the bridged-bonded species. The former CO species had a dipole moment of 0.04 Debye, positive end out, and the latter had a value less than 0.004 D, negative end out.

McCabe and Schmidt [82] have studied CO and H_2 adsorption on the Pt (111) plane. Both their TPD spectra and their initial desorption energy agreed well with Ertl and coworkers [76]; however, their relationship, $E_d = 29.6 - 6.5\theta$, was also determined using $\nu = 10^{13}\ s^{-1}$, a value which is probably too low [76, 83]. Hydrogen desorption was more complicated and gave initial desorption energies near 80 kJ mol^{-1}. Interestingly, on the oxidized Pt surface, compared to clean Pt, the initial sticking probability declined only from 0.44 to 0.35, the desorption temperature increased, and saturation coverage remained about the same. This behavior has not yet been explained.

These same authors have summarized results describing CO adsorption on the (111), (100), (110), (211), and (210) crystal planes of platinum [84]. A comparison of these five planes is given in Table 1 and 2. Table 1 lists the binding states along with desorption peak temperatures, T_p; peak widths, $\Delta T_{1/2}$; initial desorption energies, E_{d_0}; initial sticking probabilities, s_0; and α values for the relation $E_d = E_{d_0} - \alpha\theta$. Saturation CO coverages are given in Table 2 along with Pt surface densities on the different faces.

CO adsorption on platinum films has been studied by Hoffmann and Bradshaw [56] and Stephan and Ponec [85]. The first study used reflection IR spectroscopy which resolved a single peak around 2085 cm^{-1} — results very similar to those obtained by Shegeishi and King [77] using a Pt ribbon. The second study reported a TPD spectrum showing three maxima at 110, 310, and 410 K for CO adsorbed at 295 K, then cooled to 78 K prior to the desorption run. For CO adsorbed directly at 78 K, the large 310 K peak did not occur. Using labelled CO and TPD, Bain et al. [86] have found that one-third of the CO adsorbed on Pt/γ-Al_2O_3 and Pt/SiO_2 catalysts at 293 K can be removed in a nitrogen environment. This form of CO showed a desorption peak around 330–340 K. There does not appear to be an obvious correspondence between the binding states reported in these last two studies and those listed in Table 1.

The number of different studies on the Pt (111) plane and the consistency of the results is especially rewarding. In particular, the conclusions drawn by Ibach and coworkers [79, 80] regarding the coordination of the CO molecule on close-packed surfaces is informative. Their model, which assumes

Table 1. Binding states on Pt [84]

Plane	State	T_p/K	$\Delta T_{1/2}$ Experimental	$\Delta T_{1/2}$ Calculated	E_{d_0}/kJ mol^{-1}	α/kJ mol^{-1} monolayer^{-1}	s_O/300 K
(111)	β_2	535	55	44	124	27	0.34
	β_1	400	—	—	95	—	0.64
(110)	β_2	460	50	37	109	10.5	
	β_1	~350	—	—	83	—	
(100)	β_4	550	49	45	134	<4	0.24
	β_3	525	—	—	122	—	
	β_2	450	—	—	111	—	
	β_1	~400	—	—	99	—	
(200)	β_H	625	58	50	152	12.6	0.95
	β_L	~475	—	—	114	—	
(211)	β_H	610	52	49	147	27	0.27
	β_L	~475	—	—	114	—	

Table 2. Saturation carbon monoxide densities on Pt [84]

Plane	Density relative to (111) at 300 K	CO densities/molecules cm^{-2}: Total at 300 K	Total at 195 K	High binding states	Low binding states	Site densities on (1 × 1) plane/sites cm^{-2}: Pt surface density	(111)	(110)	(100)
(111)	1.0	0.75×10^{15}	1.0×10^{15}	$\leqq 0.05 \times 10^{15}$	0.7×10^{15}	1.50×10^{15}	1.50×10^{15}	0	0
(110)	0.86 ± 0.1	0.65×10^{15}	0.72×10^{15}	$\leqq 0.05 \times 10^{15}$	0.6×10^{15}	0.92×10^{15}	0	0.92×10^{15}	0
(100)	1.33 ± 0.1	1.00×10^{15}	—	$\sim 0.5 \times 10^{15}$	$\sim 0.5 \times 10^{15}$	1.30×10^{15}	0	0	1.30×10^{15}
(210)	1.39 ± 0.1	1.05×10^{15}	—	0.64×10^{15}	0.41×10^{15}	0.58×10^{15}	0	0.58×10^{15}	0.58×10^{15}
(211)	1.33 ± 0.1	1.00×10^{15}	—	0.56×10^{15}	0.44×10^{15}	0.53×10^{15}	1.06×10^{15}	0	0.53×10^{15}

that single coordination bonding on top of Pt atoms occurs at low coverage, followed by bridged-bonding beginning at higher coverages, reconciles the IR and EELS spectroscopic results which have been reported and produces the observed LEED patterns. Whereas the initial interpretation for the $(\sqrt{3} \times \sqrt{3})$ R30° structure had invoked adsorption at sites of 3-fold coordination, the current picture that 1-fold coordination occurs instead appears correct, and it should alter current thinking regarding the relative strength of linear- *vs.* bridged-CO bonds.

H. Iridium

Fewer studies have been devoted to the CO-Ir system. Broden et al. [15] have employed synchrotron radiation in a UPS study of CO adsorption on an Ir (100) surface. The photoemission curves possessed structural features attributable to the 4σ, 1π, and 5σ energy levels of CO. The energy separation between the 1π and 4δ level is greater in the chemisorbed molecule than in the gas phase, which is expected if a stretching of the CO molecule occurs after adsorption.

Comrie and Weinberg [87] studied CO adsorption at 300 K on the Ir (111) crystal plane and again observed a $(\sqrt{3} \times \sqrt{3})$ R30° LEED pattern at low coverages. At higher coverages compression occurred, registry was lost with the surface, and a maximum value of $\theta = 7/12$ was reported. The isosteric heat of adsorption was found to be 147 kJ mol^{-1} up to $\theta = 1/3$, and at saturation coverage two binding states were measured. Zhdan et al. [88] applied UPS and XPS to this same Ir surface in a later study. An initial sticking probability near unity was calculated, and no CO dissociation occurred up to 533 K. Significant differences were associated with the occurrance of overlayer compression on the single crystal face.

Weinberg, Comrie and Lambert [89] have compared CO adsorption on single crystal surfaces of Ni, Ru, Pd, Pt, and Ir under UHV conditions. Sticking probabilities are between 0.5 and 1 on all surfaces, and an adsorption model involving a precursor state fits all surfaces but the Ir (111). Heats of adsorption for CO on these surfaces vary from 111 for Ni to 147 kJ mol^{-1} for Ir.

I. Other Metals and Alloys

Due to their higher activity in reactions involving CO, the Group VIII metals have drawn the most attention in these studies. However, several Group IB metals have been investigated because of their ability to alloy with the Group VIII metals and alter selectivity. McElhiney, Papp, and Pritchard [90] found that CO adsorption on the Ag (111) face followed a Temkin isotherm, and between 77 to 123 K the heat of adsorption fell linearly from 27 to 9 kJ mol^{-1}.

Pritchard and coworkers [91] examined Cu (110) and found that the heat of adsorption declined from an initial value of 55 kJ mol^{-1} to 19 kJ mol^{-1} at higher coverage. Two IR bands at 2088 and 2104 cm^{-1} which occurred initially later merged into a single band at higher coverages. A strong interaction between CO molecules on the Cu surface was shown by the use of isotopic mixtures of CO. Moskovits and Hulse [92] prepared 1- to 4-atom Cu carbonyl clusters by cocondensing Cu atoms and CO into a low-temperature Ar matrix. Both HF and LF IR peaks were observed; for example, Cu_2CO gave peaks at 1871 and 2116 cm^{-1} while Cu_3CO gave a single peak at 2102 cm^{-1}. The proposal was made that the LF species may involve an interaction between the oxygen atom and a Cu atom. In this system, variations in coordination number did not seem to alter the IR frequency markedly; however, this may be due to the innately weaker CO-Cu bond.

The pronounced effect of alloying on selectivity of metal catalysts has been one of the most important discoveries in heterogeneous catalysis during the past decade or two. In particular, Group VIII/Group IB alloys have exhibited changes in selectivity in hydrocarbon reforming reactions, and studies have been conducted on these systems not only to further examine the catalytic properties of these systems but also to understand more completely the nature of alloys. Certain alloys have proved to be desirable systems to determine the relative contributions of the geometric, or "ensemble", effect and the electronic, or "ligand", effect.

Sachtler and coworkers [93] conducted an intensive IR study of CO adsorbed on silica-supported Pd-Ag alloys which showed that a very pronounced increase occurred in the ratio of high frequency bands to low frequency bands as Ag was added, but the band frequencies themselves remained

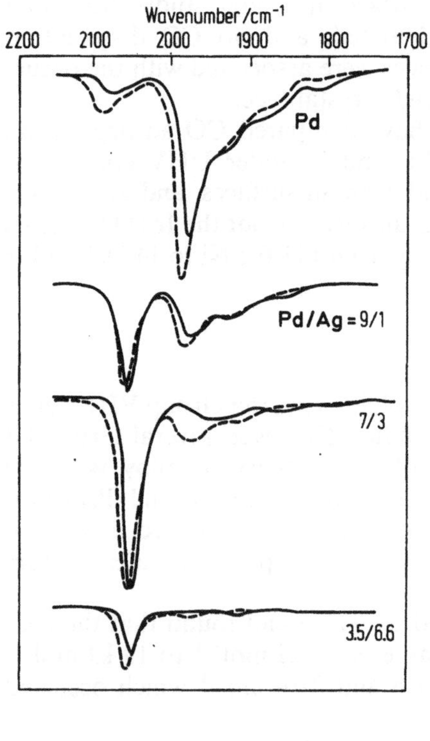

Figure 11. Transmission IR spectra of CO adsorbed on Pd-Ag alloys. (——) p_{CO} 0.01 Torr, (- - - -) p_{CO} = 0.5 Torr. After [93]

essentially unchanged. These results are shown in Figure 11. If Eischen's assignment of CO IR bands is correct, increasing the dilution of Pd with Ag would decrease the concentration of the bridged-bonded complex, which requires a pair of adjacent Pd atoms, more rapidly than the concentration of the linearly-adsorbed species, which require only single Pd atoms. However, if a change in the electronic configuration of the adsorbed molecule is the only reason for the large frequency shift, alloying Ag with Pd should cause an increasingly large shift in the IR band position as the Ag concentration increases. Therefore, these results of Sachtler et al. support Eischen's assignment of bands below 2000 cm^{-1} to multiply-bound CO species and disprove Blyholder's interpretation that all CO adsorption occurs in a linear fashion with different electronic configurations producing the different IR bands which are observed. This study also provides some of the best evidence showing the relative importance of the ensemble and the ligand effects. For CO adsorption on Pd, the former effect clearly appears to be the dominant factor and the latter is of minor importance. As further evidence that the electronic ligand factor is rather small in this system, CO adsorbed on Pd hydride and on a 30% Pd-70% Ag alloy, both of which have filled *d*-orbitals, exhibited only very small frequency shifts compared to CO adsorbed on Pd [93].

However, surface compositions of the alloy particles were not determined in these two studies, and thermodynamic consideration of surface free energies shows that surface enrichment of Ag should occur. Wise [94] plotted the results from Figure 11 versus the predicted silver surface concentration and showed that the fraction of bridged-bonded CO, relative to that on pure Pd, declined linearly with increasing Ag surface concentration. The decrease in the ratio of bridged-bonded CO molecules per surface Pd atom paralleled the decrease in probability of finding Pd-Pd pairs, and further substantiated the importance of the ensemble effect in this system.

In a subsequent study, Primet, Mathieu and Sachtler [95] used Fourier transform IR spectroscopy, and re-examined the Pd-Ag/SiO_2 catalysts described previously. The dramatic LF-to-HF shift in the major peak was again observed, but a decrease of 29 cm^{-1} was now resolved in the HF peak as the silver concentration increased. Variations in surface coverage of CO caused a frequency shift due to dipole-dipole interactions of only 9 cm^{-1}, so most of the shift in the HF band could be attributed to a small electronic effect. A 1915 cm^{-1} band at low CO coverages was also resolved in this work which indicated CO adsorbed in 3-fold coordination.

CO adsorbed on a series of Pd-Ag films was studied by Stephan, Franke, and Ponec [58] using TPD, and five different desorption states were identified. Although a linearly-adsorbed species of CO was desorbed at 300 K, it was difficult to unambiguously assign multiply-bonded CO species to the 500 K desorption peak since metal surface compositions were not known. However, the best explanation of the desorption spectra invoked surface enrichment of silver, which is consistent with previous work [94].

The influence of alloying copper with nickel on CO adsorption has also been the subject of numerous investigations although this alloy system is complicated by a bulk miscibility gap. Soma-Noto and Sachtler [96] observed IR frequency shifts for CO adsorbed on Cu-Ni/SiO_2 catalysts which were very similar to those mentioned earlier for Pd-Ag alloys [93]. Franken and Ponec [97] found that CO adsorbed on Cu-Ni films increased the work function by only 0.11 eV over a wide range of compositions. This is about 10 times lower than the $\Delta\varphi$ measured for CO-covered nickel, and the small value of $\Delta\varphi$ is attributed to a decrease in the multi-site concentration where backbonding occurs to a greater extent and withdraws more electrons from the metal. Evidence obtained by Ponec and coworkers [98] indicates that CO adsorption is not selective to nickel, and some activated adsorption occurs on surface copper atoms. However, repeated adsorption runs appeared to irreversibly pull nickel atoms from subsurface layers thereby enriching the surface in nickel. Importantly, the addition of Cu to Ni had little effect on the low temperature desorption peaks, but it facilitated CO desorption at 455 K. This high temperature (HT) desorption process was attributed either to a recombination of C and O atoms or to desorption of "horizontally-adsorbed" CO. The enhancement of this desorption step therefore has similarities to results mentioned earlier for CO desorption from stepped Ni surfaces. Spicer and coworkers [99] diffused copper to the surface of a Ni (110) crystal and determined surface compositions using Augér spectroscopy. They found four CO desorption states, as did Ponec et al. [98], but did not observe any surface enrichment as a consequence of CO adsorption. The ensemble effect was again the predominant factor as two new desorption peaks were formed over the alloy surfaces; however, the HT peak shifted downward 80 K as the Cu surface concentration increased from 0 to 65%, and this was associated with a weak ligand effect. Benndorf et al. [100] studied CO adsorption at 120 and 300 K on a CuNi (111) surface and a CuNi foil using LEED, AES, TPD, and $\Delta\varphi$ measurements. They clearly showed that the Cu and Ni sites retained most of their atomic character and produced CO heats of adsorption very close to those measured on pure Cu and Ni surfaces (49 kJ/mole on Cu, 106 kJ $mole^{-1}$ on Ni). No adsorption occurred on the Cu sites at 300 K because of the low binding energy.

The Pt-Au alloy system also adsorbs CO on surface atoms of both metals. Stephan and Ponec [101] have observed three CO desorption peaks at 110 K, 310 K, and 410 K on pure Pt whereas the last two peaks merge into a broad maximum for the alloys. Considerable CO adsorption occurred on top of a preadsorbed hydrogen monolayer at 78 K, and TPD results indicated that CO weakened the H-Pt bond. The surfaces of these Pt-Au films were formed by the Au-rich (17% Pt) phase. These authors related the HT desorption state to CO multiply-coordinated with Pt and the lower-temperature states to linearly-adsorbed CO.

Finally, Verbeek and Sachtler [102] concluded from their study of Pt-Sn alloys that CO adsorption on annealed surfaces which were initially enriched in Sn resulted in surface enrichment of Pt. Two desorption peaks around 360 and 540 K were measured, with the major effect of Sn being to decrease the intensity of the HT peak. As a consequence of these results, bridged-bonded CO was associated with the 540 K peak.

4. $CO—H_2$ Co-adsorption and Interaction on Surfaces

Much is yet to be learned about the reaction paths for the methanation and Fischer-Tropsch reactions. To enhance our understanding of these catalytic processes, attempts have been made to identify hydrogen-CO surface complexes by studying the coadsorption of these two gases, as well as different organic molecules, at various temperatures and pressures. In general, scant success has been attained regarding the observation of distinct CH_xO_y surface moities although studies have consistently revealed a mutual interaction between adsorbed CO and hydrogen on these metal surfaces.

Adsorption and decomposition of formaldehyde and methanol, as well as adsorption of CO and H_2, have recently been studied on Ni (100) and Ni (111) surfaces. In the first study, Yates, Goodman, and Madey [103] used TPD to find that H_2CO and CH_3OH adsorbed and decomposed to give CO and H_2 on Ni (100). Adsorption of CO on a saturated H layer created a new adsorbed state for each adsorbate and both states desorbed simultaneously, which provided evidence for a cooperative interaction and led these authors to suggest the possibility of a surface complex such as

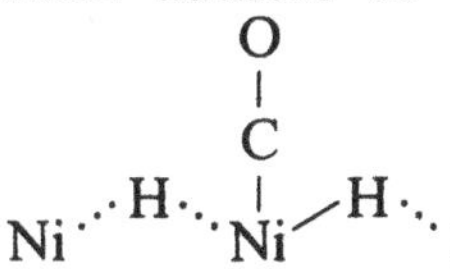

Demuth and Ibach [104] used EELS to study the adsorption and decomposition of CH_3OH on the Ni (111) plane. A pronounced shift in the O-H stretching and bending modes for chemisorbed methanol implied an interaction between the oxygen end of the molecule and the Ni surface. An intriguing result occurred at a higher temperature of 180 K where they identified a methoxy species (CH_3O) which was bonded with the O-end nearest the metal surface. This is not only one of the few examples of a surface species containing C, H, and O, but also one with an oxygen-metal bond at the surface.

From their IR study of acetone on a Ni/SiO_2 catalyst, Blyholder and Shihabi [105] also proposed coordination with the surface through the oxygen end of the molecule. At 300 K some acetone decomposed to form CO and various hydrocarbons. On Co/SiO_2, IR bands were observed which were interpreted to indicate the presence of isopropoxide,

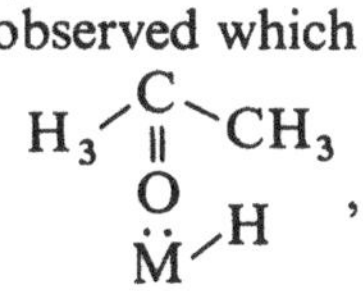

, which exhibited a metal-oxygen bond.

Two IR investigations have been conducted on the coadsorption of CO and H_2 on Ni/SiO_2 catalysts. Primet and Sheppard [106] observed IR bands at 2080, 2055, and 1950 cm^{-1} after CO adsorption on a nickel catalyst which was reduced and evacuated at 623 K. The 2080 cm^{-1} peak was removable by evacuation at 300 K. Introduction of hydrogen at 300 K to such a surface shifted the HF band to higher frequencies implying an interaction that weakened the C-Ni bond. Adsorption of CO on a hydrogen-covered surface initially gave results similar to adsorption on the evacuated Ni/SiO_2 sample; however, room temperature evacuation produced two HF bands and a stronger 1940 cm^{-1} band. As a consequence of this study, the conclusion was reached that adsorbed H withdraws electrons from the nickel. Heal, Leisegang, and Torrington [107] reported similar bands at 300 K, but after evacuation at 373 K they found only a 2074 cm^{-1} band present whereas the preceding authors observed a 2040 cm^{-1} band. When CO was adsorbed in the presence of excess H_2, the intensity of one HF band, whose frequency decreased from 2080 to 2030 cm^{-1} with increasing temperature, began a sharp decrease in intensity which occurred simultaneously with the formation of methane. Heal et al. correlated the methanation activity of this catalyst with this linearly-adsorbed CO species.

Farrauto [108] employed thermal gravimetric analysis (TGA) to study coadsorption of CO and H_2 between 373 and 673 K on a series of Ni/Al_2O_3 catalysts. The maximum weight gain varied little from 373 to 473 K, then a sharp decrease occurred up to 573 K which was concomittant with methane formation. A linear relationship was obtained between the maximum weight of the surface "complex" and the amount of H_2 chemisorbed at 300 K. Different stoichiometries of the "complex" were assumed, and Farrauto decided that the predominant surface species was $CO_{(ad)}$ or an H_2-CO complex, rather than a CH_2 or C complex, since the latter two gave surface coverages two monolayers deep compared to that determined by H_2 chemisorption.

In addition to studying CO and H_2 adsorption separately on Fe films [46, 47], Wedler and coworkers [109] also studied the coadsorption of these two gases at pressures below 10^{-1} Pa. No hydrogen adsorption occured on iron films completely covered with CO; however, at CO coverages less than unity, H_2 adsorption did occur with a higher heat of adsorption, *i.e.*, 116 *vs* 98 kJ mol^{-1} for the clean film. CO adsorption on preadsorbed hydrogen at 273 K induced a change in the adsorbed state of hydrogen, as evidenced by calorimetric, TPD, and resistivity measurements, up to a combined CO + H_2 coverage of one monolayer. Continued CO adsorption displaced hydrogen from the surface. At these low pressures, no evidence for complex formation was obtained although evidence for an interaction between CO and hydrogen was apparent. These authors stated that this interaction most likely occurs via electron transfer through the bonding electrons in the iron surface rather than the formation of a chemical complex. Kölbel and Roberg [110] have also employed microcalorimetry to examine the coadsórption of these two gases on promoted and unpromoted precipitated iron catalysts. They found that simultaneous adsorption of H_2 and CO gave higher heats of adsorption than adsorption of either gas alone although their initial value of ~ 140 kJ mol^{-1} is somewhat lower than that measured by Wedler et al. for pure CO adsorption [47]. The addition of K_2CO_3 did not alter the initial $\Delta H_{(a)}$ values appreciably, but did increase the overall heat of adsorption for CO. At the higher pressures used in this study, Kölbel and Roberg reported that coadsorption resulted in greater uptakes compared to adsorption of either gas alone, and from the relative amounts adsorbed they concluded an H_2—CO complex was formed.

Wells et al. [111] used reflection IR spectroscopy and TPD to follow the coadsorption of CO and H_2 on Rh films. The amounts of CO and hydrogen adsorbed were time dependent, with both the total gas uptake and the adsorbed H/CO ratio reaching a maximum at small exposure times. Two hydrogen desorption peaks and one CO IR band around 2075 cm^{-1} were observed, and a cooperative interaction between CO and hydrogen was proposed to explain the results. Kawasaki et al. [112] observed a single CO desorption peak near 550 K from a Rh filament which was indicative of molecular adsorption. When coadsorption was studied, three states of adsorbed hydrogen were measured which were dependent upon time of exposure at 300 K. Initially, a large desorption peak occurred near 300 K, but after an hour or so this peak had disappeared and a smaller, higher-temperature peak (β_2) formed. Finally, a large, broad peak around 600 K was produced after 20 hr or more. This complicated behavior was not completely explained, but the β_2 state was associated with a complex $H_{(ad)} + C_{(ad)}$ state while the high temperature state was attributed to the desorption of *absorbed* hydrogen from bulk rhodium. These authors have also reported that CO displaces adsorbed hydrogen on platinum at 300 K [113].

In a study of adsorbed formaldehyde and coadsorbed CO and H_2 on Ru (110), Goodman et al. [114] found that CO added to a saturated hydrogen monolayer increased the Ru—H bond strength, but CH_4 formation was was not observed at the low pressures involved. No CO dissociation occurred up to 630 K in the absence of hydrogen; however, formaldehyde decomposed to CO and hydrogen upon adsorption at 300 K. Compared to CO—H_2 coadsorption, a larger quantity of hydrogen was placed on the surface by formaldehyde decomposition, and it desorbed at a significantly higher temperature. This has been interpreted as evidence for a H_xCO complex on the surface; however, during TPD runs after H_2CO adsorption, only small quantities of CH_4, H_2O, and CO_2 were produced which infers that the complex exists only at low coverages, if at all. As suggested for iron [109], the positive CO—H_2 interaction probably results from changes induced via the metallic bonding electrons.

A recent IR study of CO adsorption at 300 K on supported Pd in the presence of hydrogen showed that hydrogen had little effect on CO adsorption when typical supports such as silica, alumina, and silica-alumina were used to disperse the Pd [115]. This behavior is consistent with earlier UHV studies [116]. However, titania, which can strongly interact with Pd [117], altered the adsorbed state of CO such that the introduction of H_2 markedly decreased the band intensities, even at 300 K. Although equilibrium H_2 and CO surface coverages are both markedly reduced by this interaction [117], this behavior infers that hydrogen is much more competitive with CO for adsorption on the remaining surface sites. This trend resulted in more active Pd methanation catalysts, as discussed later. Similar behavior also occurred for Pt [118].

5. Methanation and Fischer-Tropsch Reactions

During the decade of the 1970's, research has burgeoned on the hydrogenation of carbon monoxide over reduced metal catalysts. These efforts have been motivated by the realization that petroleum and natural gas reserves are limited and will not be available to supply all the energy requirements of the future. Coal and oil shale reserves are much greater and will provide a major portion of the energy resources in the future. However, improvements are needed for the conversion of coal into clean fuels and chemicals, to improve the overall thermodynamic efficiency and to lower the unit cost of the various energy conversion processes. Coal gasification to CO and H_2 followed by a CO hydrogenation step is one possible route to produce SNG (substitute natural gas) *via* the methanation reaction, or to manufacture longer-chain hydrocarbon fuels and chemicals via the Fischer-Tropsch synthesis reaction.

The more recent investigations have concentrated on the determination of the fundamental kinetic behavior of Group VIII metal catalysts, on the elucidation of the sequence of elementary steps in the synthesis reactions, and on the influence of the support and metal crystallite size on catalytic behavior. Obviously, the end goal is to provide synthesis catalysts which have improved properties regarding product selectivity, catalytic activity, activity maintenance, and sulfur tolerance.

Although Fischer and Tropsch [120] had studied the kinetic behavior of unsupported Group VIII metals for CO hydrogenation during the early 1920's, the *specific activities* of these metals, *i.e.*, CO reaction rates per metal surface site, were not determined until the 1970's. Vannice [121, 122] reported specific activities and other kinetic parameters in the methanation reaction over these metals supported on alumina and silica. Both studies indicated a dependence of turnover frequency (molecule $site^{-1}$ s^{-1}) on the adsorption bond strength of CO on these metal surfaces. Table 3 lists the order of specific activity for the alumina-supported metals [121], and Figure 12 shows

Table 3. Data for methanation reaction [121]

Fischer *et al.* (ranked by descending activity)[a]	Vannice [121]	
	Metal	turnover no. at 548 K × 10^3
Ru	Ru	181
Ir	Fe	57
Rh	Ni	32
Ni	Co	20
Co	Rh	13
Os	Pd	12
Pt	Pt	2.7
Fe	Ir	1.8
Pd		

[a] Fischer, F., Tropsch, H., Dilthey, P.: Brennst.-Chem. **6**, 265 (1925)

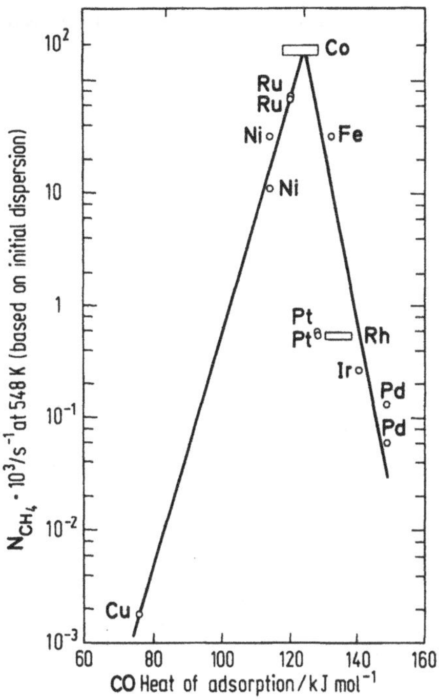

Figure 12. Turnover frequencies (N_{CH_4}) for methanation over silica-supported metals. After [122]

the dependence for the silica-supported metals, which has the shape of a "volcano" plot [122]. Subsequent studies in many different laboratories have also reported specific activities, usually in the form of a turnover frequency (TOF), which have shown remarkable consistency. These results indicate the importance of determining activities in this manner so that comparisons can be made of intrinsic activity, since variations in metal surface area are normalized out. The results of studies prior to 1977 have been analyzed and discussed in earlier reviews [10, 11] so again the emphasis in this section will be more on the current status of the work.

A. Nickel

A significant fraction of the recent CO hydrogenation studies has centered on nickel, the metal which has been accepted as the best methanation catalyst. Nickel catalysts are easily prepared and readily characterized by chemisorption techniques and, in addition, nickel is a metal easily amenable to study under UHV conditions. For years arguments have existed regarding the mechanism of the methanation and Fischer-Tropsch reactions, and the arguments for and against various pathways have been mentioned elsewhere [11, 123, 124]. The different pathways can conveniently be divided into two general groups: one which involves hydrogenation of adsorbed CO prior to the rupture of the CO bond, and the other in which CO first dissociates to produce surface carbon which is then hydrogenated. Because these surface carbon species do not have to be a metal carbide, this modification of the "carbon intermediate" mechanism, first proposed by Fischer and Tropsch themselves, removes the difficulties this model has encountered. Concurrent studies by Wentrcek, Wood and Wise [125] and by Araki and Ponec [126] have conclusively shown that surface carbon produced from dissociated CO can be readily hydrogenated to form methane. In the former study surface carbon, C_s, was deposited on a commercial Ni/Al_2O_3 catalyst via the Boudouard reaction run at 553 K in a pulse microreactor operated at 446 kPa. The quantity of C_s was determined from the amount of CO_2 produced. When hydrogen was then pulsed through the catalyst bed, the amount of CH_4 formed agreed quantitatively with the amount of C_s deposited. However, Wentrcek et al. found that this reactive (carbidic) form of surface carbon can also convert into a much less active (graphitic) form. The study by Araki and Ponec [126] involved the use of labeled CO isotopes over nickel and Ni—Cu films at low pressures (80 Pa) in a noncirculating batch reactor. When CO/H_2 mixtures were introduced to a clean, reduced Ni film, CO_2 was always produced immediately whereas CH_4 formation occurred with an induction period, as shown in Figure 13. Again it was found that C_s easily reacted with H_2 to produce methane, and the introduction of a $^{12}CO/H_2$ mixture to a $^{13}C_s$ covered surface resulted in the immediate formation of $^{13}CH_4$ with $^{12}CH_4$ and $^{12}CO_2$ being formed only after an induction period. As a consequence of these results, both groups of investigators proposed C_s as an intermediate in the methanation reaction over nickel.

Different forms of surface carbon have been found to exist, with at least one form capable of being readily gasified by hydrogen to form methane. This form of carbidic C_s is more reactive for methanation on nickel than adsorbed CO, as determined by McCarty et al. [127, 128]. In their TPR (Temperature Programmed Reaction) study of carbon deposited on a Ni/Al_2O_3 catalyst, maximum CH_4 formation from C_s occurred at 450 K while the CH_4 maximum from adsorbed CO was at 495 K. Figure 14 represents

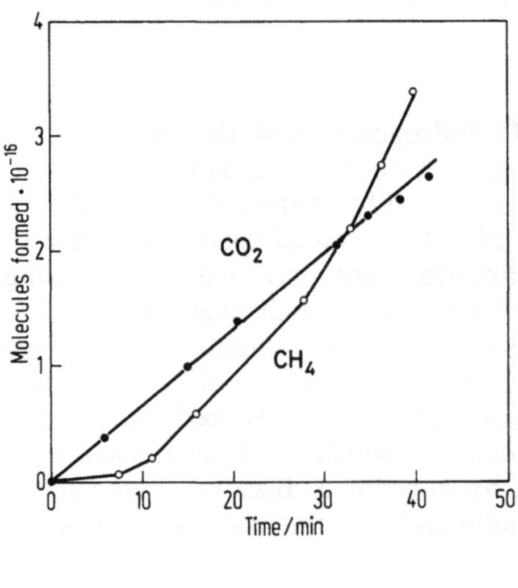

Figure 13. CO_2 and CH_4 formation from a H_2/CO mixture at 523 K over a Ni film. After [126]

these results and also shows other, less reactive, forms of C_s. The activation energy for the low-temperature C_s gasification reaction was 71 kJ mol^{-1} and for CO hydrogenation it was estimated to be $\geq$125 kJ mol^{-1}.

Since it is clear that methanation can occur *via* a reaction between amorphous surface carbon and adsorbed hydrogen, the principal question now is whether this is the *only* pathway available for CO hydrogenation on metal surfaces. Two important questions are: can molecularly-adsorbed CO react directly with hydrogen to provide another pathway for methanation, and is molecular CO an important (or necessary) reactant in the chain growth sequence?

Evidence does exist that adsorbed CO reacts differently with hydrogen than the various forms of surface carbon. As shown in Figure 14, McCarty et al. [127, 128] found that nondissociatively-adsorbed CO also appears to react directly with hydrogen although it occurs at a higher temperature, thus indicating a lower reaction rate. Zagli, Falconer, and Keenan [129] have employed both TPD and TPR techniques in a study of three supported nickel catalysts. The desorption states of CO and the formation patterns of CO_2 during TPD runs were quite different for each catalyst. They found, as others had previously [11], that CO and H_2 coadsorb at 300 K in approximately a 1:1 mole ratio, and in addition, the hydrogen remains on the surface during TPR runs to react and finally desorb as CH_4 and H_2O. During the TPR runs, CH_4 and H_2O were desorbed simultaneously which indicated that the C—O bond-breaking step was rate determining.

In an interesting study, Ho and Harriott [130] employed transient techniques in a differential reactor to study the CO disproportionation reaction, the

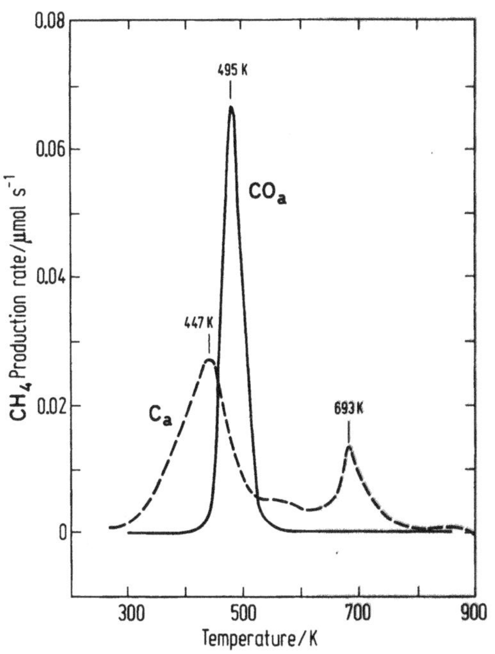

Figure 14. Temperature-Programmed Surface Reaction (TPSR) of adsorbed CO and surface carbon with hydrogen. After [128]

hydrogenation of C_s and the CO/H_2 methanation reaction over two Ni/SiO_2 catalysts. They obtained typical methanation activation energies of 84 to 105 kJ mol^{-1}, while the activation energy for CO disproportionation was 42–59 kJ mol^{-1} at monolayer coverages or less, and that for C_s gasification with hydrogen was 38–63 kJ mol^{-1}. The relative rates of the three reactions were $r_{gasif.} > r_{meth.} \gg r_{dispop.}$, as shown in Figure 15. A step increase in CO partial pressure from zero resulted in significant CO adsorption on the catalyst, no CO exiting from the catalyst bed for 3 minutes, and an immediate, approximately linear increase in CH_4 production with time up to a steady-state rate achieved in $\approx$ 3 minutes. A step change which removed CO from the H_2 + CO feed stream produced a maximum in the rate followed by a long, slow decline in the rate behavior which eliminated an Eley-Rideal mechanism. After CO removal, the H_2O/CH_4 mole ratio was 0.6 to 0.7 which showed that considerable amounts of oxygen and/or adsorbed H_2O was present in the system in addition to C_s. To obtain consistency between steady-state and transient behavior, Ho and Harriott [130] had to assume at least three types of active sites existed. They concluded that the methanation rate was very sensitive to surface concentrations of CO and hydrogen — especially the latter — as previously suggested by Polizzotti, Schwarz, and Kugler [131]. Ho and Harriott also concluded that if surface carbon is an intermediate, neither normal CO dissociation nor C_s hydrogenation could be rate controlling. Consequently, they proposed that the slow step is the reaction of adsorbed CO and hydrogen to form C_s and water, a proposal which tends to reduce the differences between the two general reaction models which have been discussed. In addition, this sequence of steps appeared consistent with the results of Zagli et al. [129].

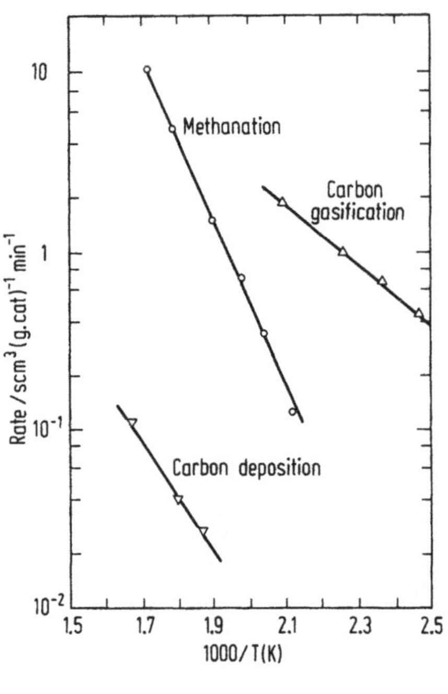

Figure 15. Rates of methanation (p_{H_2} = 0.43 atm, p_{CO} = 0.07 atm), carbon gasification (p_{H_2} = 0.2 atm), and carbon deposition (p_{CO} = 0.05 atm) over a 10% Ni/SiO_2 catalyst. (After [130])

A very nice study by Rabo et al. [132] has also shown that both C_s and $CO_{(ad)}$ apparently can react directly with hydrogen to form hydrocarbons which were predominantly methane. Using a pulse microreactor, they deposited known amounts of both surface carbon and nondissociated CO on two Ni/SiO_2 catalysts and found that C_s readily reacted at 573 K to give CH_4. However, 95% of the carbon on a nickel surface predominantly covered with adsorbed CO (64% CO, 36% C_s) reacted at 473 K to give methane. At 423 K, only 0.3% of the carbon on a surface with 98.7% CO coverage reacted to give CH_4, but significant hydrogen adsorption occurred. The higher reactivity of the C_s species was evident since a surface which was 78% covered with C_s produced small quantities of C_1—C_4 hydrocarbons at 300 K. Similar studies on Co/SiO_2 and Pd/SiO_2 also showed that both C_s and $CO_{(ad)}$ react with hydrogen. The latter catalyst was especially interesting since substantial methanation occurred although 99% of the carbon on the surface was present as adsorbed CO.

A thorough mechanistic study of CO hydrogenation has been conducted by Sachtler and coworkers [124, 133], using isotopically labelled CO. A $^{13}C_s$ species was deposited on unsupported nickel whose surface area had been measured by hydrogen chemisorption, and this $^{13}C_s$ was then reacted with hydrogen. They not only reverified the reactivity of this surface carbon, but more importantly learned that C_2—C_4 hydrocarbons contained more than one atom of ^{13}C. These results showed that CH_x surface species can participate in chain growth reactions as well as the methanation reaction. Also, they found that the fraction of $^{13}CH_4$ formed was directly proportional to the fractional surface coverage of $^{13}C_s$, the remainder being covered by $^{12}CO_{(ad)}$. This led them to propose that C_s and $CO_{(ad)}$ form a common surface intermediate whose hydrogenation is the slow step [124]. This model has been supported by the isotopic tracing studies of Happel et al. [134]. Although they acknowledged that hydrogen probably facilitates the C—O bond-breaking step, they did not consider this step to be rate determining, as opposed to the model proposed by Ho and Harriott [130]. Although differences in interpretation exist between these two studies, they provide rates of CO disproportionation in the absence of H_2 which are quite consistent. At 523 K and a CO pressure of 0.5 atm, Sachtler et al. [133] determine a TOF of $2 \times 10^{-3}\ s^{-1}$, while at 523 K and a CO pressure of 0.05 atm, a value of $\approx 1.5 \times 10^{-4}\ s^{-1}$ can be estimated from Figure 15 if the CO adsorption of 2.2 cc g^{-1} at 468 K is assumed to be a reasonable estimate of surface sites. Assuming the rate is proportional to CO pressure, the two values for CO disproportionation at 0.5 atm. are nearly identical.

Investigators at the National Bureau of Standards have conducted some particularly impressive studies of methanation on Ni (100) and (111) single crystal surfaces over a wide range of temperature and pressure [135–140]. Both turnover frequencies and activation energies for methanation are in excellent agreement with those reported by Vannice for supported Ni catalysts [141]. Analysis of the nickel surface after reaction by AES allowed the identification of both carbidic and graphitic surface carbon, and active surfaces were found to have steady-state coverages of carbidic carbon be-

Table 4. Relative activities for CH_4 formation at 478 K [141]

Catalyst	Turnover frequency for methane/$s^{-1} \times 10^3$ [a]	Rate/μmol CH_4 s^{-1} (g cat)$^{-1}$
8.8% Ni/η-Al_2O_3	2.48	0.094
5% Ni/η-Al_2O_3	1.24	0.119
16.7% Ni/SiO_2	1.10	0.240
20% Ni/Graphite	0.62	0.064
16.7% Ni/SiO_2	0.55	0.135
42% Ni/α-Al_2O_3	0.52	0.060
30% Ni/α-Al_2O_3	0.43	0.275
NiO (reduced)	0.38	0.023
42% Ni/α-A.$_2$O$_3$ (200°)	0.28	0.042

[a] Based on H_2 adsorption.

tween 0.05 and 0.25. Variations in reaction conditions which allowed this coverage to become too high initiated the formation of inactive graphitic carbon. This work indicated that a certain balance is required between carbidic carbon (CO dissociation) and adsorbed hydrogen (C hydrogenation) and that no single rate determining step may exist under certain conditions.

Several studies have been directed toward determining the influence of the support and of promoters on the catalytic properties of nickel. Vannice [141] examined nickel supported on a variety of typical support materials and observed only a small variation in specific activities. As shown in Table 4, a 5-fold variation existed in the CH_4 TOF, and a 7-fold variation occurred for total CO conversion. There was some evidence which implied that the crystallite size of Ni may influence specific activity (within an order of magnitude), and the existence of a broad methanation activity maximum for 10–20 nm Ni particles was suggested. This conclusion seems to have been verified by Mathews and coworkers [142], who claim to have observed a maximum in the turnover frequency (TOF) for methane formation for 15 nm Ni particles; however, their reported trend rested predominantly on the accuracy of one experimental point. A much more definitive verification of this relatively small (less than 10-fold) variation of TOF with crystallite size was recently reported by Bartholomew, Pannell, and Butler [143].

Huang and Richardson [144] have studied the effect of the addition of an alkali metal, in this case sodium, on the catalytic behavior of a series of Ni/silica-alumina catalysts. Sodium did not significantly poison the Ni surfaces or affect nickel reduction and crystallite size, since values of the latter determined by magnetic measurements and H_2 chemisorption were in good agreement. The turnover number on the Na-free, acidic Ni/SiO_2-Al_2O_3 catalyst was an order of magnitude lower than values reported for less acidic supports; however, adding Na initially increased the methane TOF value 6-fold before it declined at higher Na loadings. An informative kinetic analysis, using a reaction model assuming [H_xCO] intermediates, indicated a constant CO heat of adsorption of 41–45 kJ mol^{-1} under reaction conditions. This value is in good agreement with equilibrium adsorption studies and again suggests the involvement of a more weakly adsorbed form of CO. The addition of Na produced a compensation effect between the activation energy for methane formation and the pre-exponential factor. The authors concluded that small amounts of Na poison acid sites on the support and reduce carbon build-up while larger quantities result in poisoning the metal sites.

In a modified molecular beam study of Ni and Co foils, Palmer and Vroom [145] observed specific activities for methanation which were noticeably different than previously reported values [141]. At low pressures (265 Pa) and high H_2/CO ratios, they reported turnover frequencies for Ni which were 10-fold higher than most previously reported values, provided the foil was reduced at only 525–600 K in hydrogen after cleaning in O_2 at 800 to 900 K. More typical activities were obtained when the foil was reduced in H_2 at 800–900 K. Augér analysis of the used foils showed only C and O present in the 10 nm outer surface layer and led the authors to propose that subsurface oxygen may be responsible for the observed rate enhancement. They also mentioned that the rate of CO disproportionation in the absence of H_2 was comparable to the rate of methanation.

Recent investigations have found that the support utilized can markedly alter both the activity and selectivity of CO hydrogenation over nickel catalysts. Unusually high activities were first reported by Vannice and Garten [146, 147] in their study of TiO_2-supported nickel, and these results clearly show the synergism that can occur by the appropriate metal-support combination. Titania is a support which has been shown to exhibit SMSI (Strong Metal-Support Interaction) behavior with Group VIII metals [117]. Although TOF's for CO conversion on these catalysts were 1–2 orders of magnitude higher than for typical nickel catalysts, perhaps the most outstanding feature of these titania-supported nickel catalysts was the marked shift in product selectivity. Whereas nickel typically produces CH_4 as the major product, the amount of methane in the product stream was reduced to 15–20 wt % over these Ni/TiO_2 catalysts even at H_2/CO ratios of 3. These results have been reproduced by Bartholomew and coworkers [143, 148] who also found that very low loadings of highly dispersed nickel on alumina exhibited somewhat similar behavior. Nickel on garnierite (primarily a magnesia silicate) has been found to have lower TOF values [149] than typical nickel catalysts whereas Ni/ThO_2 samples prepared from oxidized ThNi intermetallics tend to have higher specific activities [150]. Titania-supported NiRu and Ru have proven to be superior methanation catalysts under commercial reactor conditions [151].

Although the exact nature of the metal-support interaction is not understood in these SMSI (Strong Metal-Support Interaction) systems, it appears that electron transfer occurs between the metal and the titania [152]. Although it was initially suggested that the Ni particles may be considered to be electron deficient, *i.e.*, partially oxidized [147], which was consistent with the results of Palmer and Vroom [145] and their conclusions regarding the presence of subsurface oxygen in the metal, recent results infer that electrons may be transferred from the partially reduced titania to the nickel [152]. Regardless, these recent results show that new improved catalysts can be developed by a judicious choice of support material; however, much remains to be learned about the chemistry and physics involved in these SMSI catalyst systems.

B. Iron

In the past, more studies were conducted on iron than any other metal, undoubtedly as a consequence of the medium-pressure synthesis process discovered by Fischer and Pichler, which produces long-chain hydrocarbons. The inexpensiveness of iron is a particularly favorable aspect of its use in CO/H_2 synthesis reactions, and promoted iron is presently used commercially by SASOL in South Africa. The difficulty in cleaning and characterizing this metal, though, has hindered surface studies of unsupported Fe conducted under UHV conditions. Only recently have such studies begun to be reported.

Dry [153] has discussed the present state of knowledge regarding commercial doubly-promoted iron catalysts, and has mentioned the limitations on product selectivity imposed by a 1-carbon atom, step-wise, chain-growth process. The effect of such a chain-growth mechanism, which follows polymerization kinetics (*i.e.* the Schulz-Flory equation), was originally proposed many years ago [154, 155] and product distributions predicted by this model are regularly observed in CO hydrogenation reactions [154–157]. Recently, however, examples of selectivity have been obtained which do not follow the chain-growth model [147, 158–162], and these results provide evidence that needed alterations in selectivity may indeed be achievable. The observed differences in product distribution are most likely introduced by modifications to the chain-growth process and the presence of additional reactions as discussed by Kibby and Kobylinski [157]. The use of shape-selective catalyst supports, such as zeolites, can also produce marked shifts in product distribution, as subsequently discussed [159–162].

Heal et al. [163] have used IR to study some 5 wt % Fe/SiO_2 catalysts as they were heated in H_2/CO mixtures. By monitoring the frequencies of gas-phase methane and methyl groups, they found their Fe samples were less active than comparable Ni and Co catalysts and methane was the predominant product. Two adsorbed CO bands were observed: a sharp band at 2170 cm^{-1} and a broad weak band around 2030–2040 cm^{-1}. Upon evacuation, one broad band below 2000 cm^{-1} remained and its intensity decreased by half as the sample was heated to 473 K. All their samples deactivated rapidly, which is almost certainly a result of carbon deposition and/or carbide formation at the Fe surface.

Dwyer and Somorjai [164, 165] have employed a unique experimental design which allows in situ high pressure reactions to be conducted on surfaces that are cleaned and characterized in an UHV environment. An iron foil and an Fe (111)-oriented single crystal were characterized by Augér and LEED in an UHV chamber which was attached to an isolation cell for kinetic studies. The necessity of a 96 hr reduction in hydrogen at 1073 K to remove bulk C and S typifies the difficulties associated with cleaning Fe samples. At reaction conditions of 6 atm. pressure, a H_2/CO ratio of 3 and

573 K, methane constituted 85% of the product. The initial methanation activation energy was 96 kJ mol^{-1}, and the initial methane TOF was 1.9 s^{-1} [164]. These results are in excellent agreement with those reported by Vannice [121], who determined an activation energy of 89 kJ mol^{-1} for a 15% Fe/Al_2O_3 catalyst and whose rate equation predicts a methane TOF of 0.9 s^{-1} under these conditions. Using AES, Dwyer and Somorjai found that multi layers of carbon at least 2 nm thick formed rapidly and decreased E_{CH_4} values to 50 kJ mol^{-1} and turnover frequencies to 0.4 s^{-1}. AES analysis did not detect the presence of oxygen in this carbon layer. Termination of the methanation run, followed by cooling, AES analysis, H_2 introduction, and reheating to 573 K produced CH_4 at the same rate as that present at the point of termination. They conducted similar experiments over an Fe foil which had been preoxidized, and observed a 10-fold increase in the initial rate. However, the oxide layer was rapidly reduced and carbon deposition again occurred. Chain growth occurred only when reduced Fe surface atoms appeared to be accessible and was not observed after C multilayers were formed, while methanation always occurred but with different kinetic parameters. Therefore, Dwyer and Somorjai concluded that at least two paths exist for methanation and that surface carbon did not participate in the chain growth process.

In a subsequent study, Dwyer and Somorjai [165] repeated this study using an Fe (111) single crystal, and measured an initial CH_4 turnover frequency of 1.35 s^{-1}. They added small quantities of ethylene and propylene to determine their role in the chain-growth process. The methanation reaction was not affected, but production of higher hydrocarbons up to C_8 was enhanced. The product distribution appeared to follow that predicted by the Schulz-Flory equation, as indicated in Figure 16. Because the total mass of the increased higher hydrocarbon production corresponded closely to the ethylene and propylene reacted (excluding hydrogenation to the corresponding paraffins), these authors concluded that olefins cannot only initiate chain growth but also be incorporated into higher hydrocarbons. Therefore, these results substantiate earlier proposals that readsorption of olefins, which have been assumed to be primary products in the F—T synthesis, may play an important role in chain-growth reactions [1].

More recent studies by Krebs and Bonzel [166, 167] have employed AES and XPS to examine the surfaces of Fe (110) and Fe foil after the synthesis

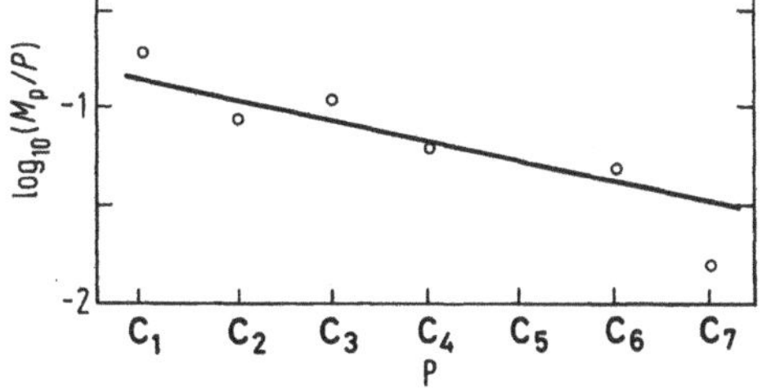

Figure 16. Schulz-Flory plot of the hydrocarbon distribution over Fe (111) at 573 K using a 3:1 H_2:CO feed containing 2.2% propylene. M_p is the mass fraction of hydrocarbons containing P carbon atoms. (After [165])

reaction had been run at 1 atm. in a chamber similar to that used by Dwyer and Somorjai. Their activation energy for methane formation was in excellent agreement with previous values [121, 164, 165] and their TOF agreed very well with that reported by Dwyer and Somorjai. Similar to results on nickel, CH_4 formation increased as carbidic carbon increased, but formation of graphitic carbon decreased activity. They identified a third carbon phase, which they considered to be CH_x species and could possibly contain oxygen. This phase and carbidic carbon could both be readily hydrogenated.

The importance of the role of olefins was shown even more dramatically by the work of Caesar et al. [159], in which they prepared a physical mixture of a commercial, promoted Fe catalyst and shape-selective ZSM-5 molecular sieve particles. The fraction of aromatics in the product stream increased enormously when this zeolite was added, and it has been shown that this zeolite converts olefins to aromatics [160]. As a consequence, the product distribution does not follow that predicted by the Schultz-Flory equation, and the highly-aromatic gasoline fraction constitutes 55–65% of the hydrocarbon product. The obvious conclusion is that gas-phase olefins are produced which are intercepted by the zeolite particles and aromaticized. These authors proposed that the major chain-growth step involves some form of CO insertion into adsorbed olefinic species.

Chang et al. [160] have recently found that Fe/ZSM-5 catalysts operated at 644 K and 3.5 MPa will convert a syngas feed to a product containing 33 wt %CH_4 and 19% C_{5+} hydrocarbons, of which 25% of the latter fraction is aromatic. Less than 3% of the aromatics had carbon numbers greater than 10 — a result indicating the molecular sieving properties of the zeolite; however, calculations indicated that diffusional effects had little effect on the measured rates. A low reaction rate appears to exist under these conditions, compared to initial rates [121, 164–167], since an estimated CO turnover frequency of only 0.5 s^{-1} is calculated at this high temperature and pressure assuming a fraction exposed of 0.1 for Fe.

Bennett and coworkers [168–171] have studied the behavior of a number

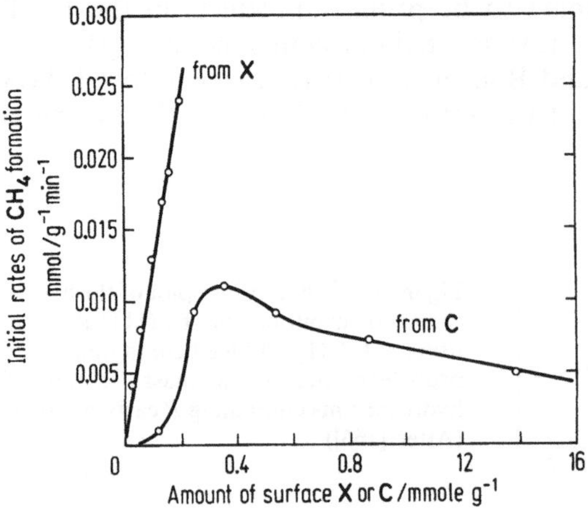

Figure 17. Initial rates of methane formation from surface carbon formed by CO disproportionation, C, and from a surface intermediate, X. (After [170])

of commercial iron catalysts and found that nitrided Fe catalysts had higher activity and better activity maintenance [168] in agreement with older work [1]. Over a fused Fe catalyst, high conversions gave detectable hydrocarbons up to C_{30}, and fit a rate equation for total CO conversion of the form $r = \frac{k[H_2]}{1 + b[H_2O]/[CO]}$ [169]. The apparent activation energy was 85 kJ mol^{-1}, in good agreement with previous values. The study of Matsumoto and Bennett [170] utilized transient methods in a reactor operated at 101 kPa to gain insight about the reaction occurring on the surface of a doubly-promoted Fe catalyst. At low, steady-state conversions (<2%) at 523 K and an H_2/CO ratio of 9, a step decrease to zero CO pressure produced a maximum in CH_4 formation analogous to the behavior discussed previously for Ni catalysts [130]. The only major product other than methane was water. Steady-state turnover frequencies extrapolated to 548 K and an H_2/CO ratio of 3 gave TOF values of 0.027 s^{-1} and 0.096 s^{-1} for methane and carbon monoxide respectively, compared to values of 0.057 s^{-1} and 0.160 s^{-1} for methane and carbon monoxide reported earlier by Vannice [121]. Surface carbon was deposited *via* exposure to CO in Ar at 523 K and then reacted with hydrogen — only CH_4 was formed, and this occurred at a much lower rate. As a consequence, the existence of a different carbon-containing intermediate was inferred. This surface species produced a higher rate of methanation, and the rate was proportional to the concentration of this species, as shown in Figure 17. In addition, this intermediate appeared to be necessary for the production of higher molecular weight hydrocarbons. They concluded that the slow step in CO hydrogenation was the reaction of adsorbed hydrogen with this intermediate, which was presumed to be $CO_{(ad)}$ or a form of C_s rather than $CHOH_{(ad)}$, because this same intermediate was

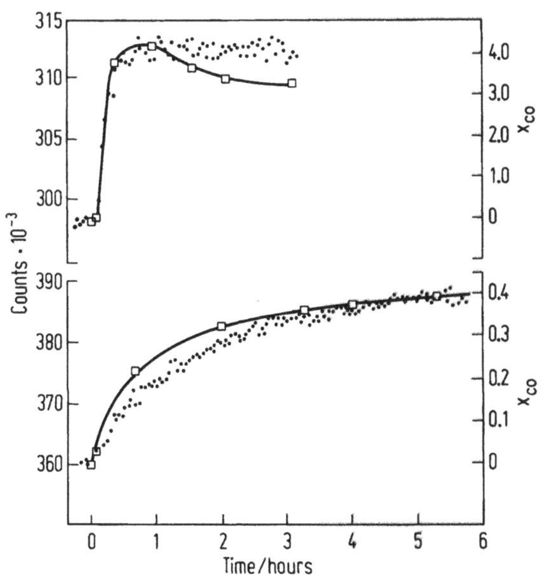

Figure 18. Percent conversion (solid lines) and extent of carbide formation (points) versus Fischer–Tropsch synthesis reaction time. The upper curve is for 10 Fe/MgO. The lower curve is for 10 Fe/SiO_2. (After [173])

also responsible for carburization of the iron catalyst. Results from the last study favored C_s as the intermediate, although oxygen was detected on the surface, and the authors presumed chain growth occurred via $CH_{x(a)}$ species [171].

The importance of surface carbon and subsequent bulk carbide formation on the catalytic behavior of iron catalysts has been clearly demonstrated by Mössbauer studies conducted by Amelse, Butt, and Schwartz [172] and by Raupp and Delgass [173, 174]. The first group of investigators determined that a 24 hr reduction in H_2 at 698 K reduced 80–90% of the iron in a 5% Fe/SiO_2 catalyst to metallic iron. After 6 hours under reaction conditions at 523 K, four carbide phases could be identified, and their relative stabilities were:

$$\varepsilon' - Fe_{2.2}C < \varepsilon - Fe_2C < \chi - Fe_2C_5 < \theta - Fe_3C .$$

Four iron carbide phases have also been identified by Niemantsverdriet et al. [175]. Kinetic studies in a separate reactor indicated that the activity of the catalyst increased with the degree of carburization. Raupp and Delgass [173] conducted *in situ* Mössbauer studies on a working 10% Fe/SiO_2 catalyst at 523 K and confirmed very nicely the increase in activity with increasing carbide formation. Figure 18 demonstrates this behavior. After complete carbidization, the activity either lined out or declined with time, the latter decrease being attributed to the formation of a carbon overlayer, as described by Dwyer and Somorjai [164], and not to sintering. An increase in chain-growth reactions also occurred concurrently with carbiding. Subsequent hydrogenation of the carbided catalyst was very slow, with only 30% of the C_s being reduced after 10 hr at 523 K, and methane was the

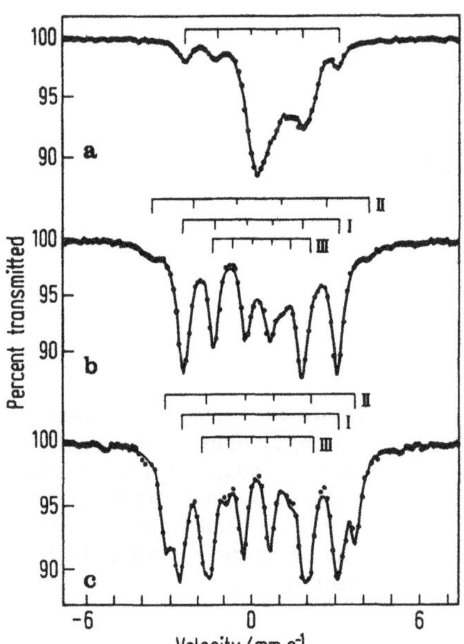

Figure 19. Room temperature Mössbauer spectra of 10 Fe/SiO_2 catalysts carbided in 3.3 H_2/CO 6 hr at 523 K, then quenched in He. The stick diagrams indicate line positions and relative intensities of the ε' carbide (a) and the I, II, and III sites in ε carbide (b) and χ carbide (c). (a) 6.1-nm particles; (b) 7.4-nm particles; (c) 10.1-nm particles. (After [174])

only product formed. In another study, Raupp and Delgass [174] found that large, silica-supported Fe particles ($\geqq$ 10 nm) formed χ (Hägg) carbide, as do bulk Fe catalysts, while smaller particles formed the less stable ε' and ε carbides. These Mössbauer results are shown in Figure 19. Both alloy formation and the nature of the support were able to affect carbidization — dispersing Fe on MgO removed carbide stabilization and gave only χ carbide, and adding Ni to Fe/SiO_2 catalysts eliminated the formation of any bulk carbides.

Benzinger and Madix [176] have used XPS and TPR to examine the adsorption of methanol, ethanol, propanol, olefins, and carboxylic acids on the Fe (100) surface. The primary alcohols formed alkoxy intermediates which were stable up to 400 K then reacted in three ways: 1) complete dissociation to $CO + H_2$; 2) rehydrogenation to alcohol; and 3) fragmentation plus hydrogenation to give hydrocarbons and organic compounds. These were 1st order reactions with activation energies of 105–110 kJ mole^{-1}. Interestingly, the isopropanol did not form a stable alkoxy intermediate and only desorbed. Stable methyl groups were put on the surface by adsorbing CH_3Cl, and these decomposed to H_2 and C_s at 475 K with an activation energy of 110 kJ mol^{-1}. Ethylene, acetylene, and cis-2-butene also formed stable species. Carboxylic acids formed carboxylate species which decomposed to H_2, CO, CO_2 and adsorbed oxygen, but these reactions had higher E_a values of 130–140 kJ mol^{-1}. These results allow the possibility that O-containing surface species may exist under appropriate F-T reaction conditions.

C. Ruthenium

This has been the third metal which has recently received considerable attention. Disadvantages such as its higher cost may well be outweighed by advantages such as its high activity, its capability to be supported in a state of high dispersion, and its unique ability to produce high molecular weight polymers at low temperatures.

Ruthenium has been more thoroughly studied under reaction conditions by IR than any other metal. Dalla Betta and Shelef [177] obtained IR spectra under reaction conditions of 523 K, a hydrogen pressure of 0.075 atm, and a CO pressure of 0.025 atm. The major CO band occurred near 2000 cm^{-1}, and other stretching frequencies were observed which appeared to represent species not involved in the synthesis reaction. An increase in the IR band at 3600 cm^{-1} was due to the formation of -OH groups on alumina, bands at 1585 and 1378 cm^{-1} were assigned to the -OCO vibration of the formate ion, and C-H stretching bands at 2928 and 2850 cm^{-1} slowly formed. From isotope substitution experiments, these authors decided that none of these species represented reaction intermediates on the metal surface; instead, they represented reaction products which were adsorbed on the alumina.

King [178] has investigated silica- and alumina-supported Ru catalysts under higher pressure conditions using IR spectroscopy. At 548 kPa (65 psig) total pressure, CO in N_2 produced a major peak at 2045 cm^{-1} which was shifted to 2020 cm^{-1} when the nitrogen was replaced with hydrogen. Heating then to 373 K and cooling to 300 K left a CO band near 2027 cm^{-1}. Under reaction conditions of 101 kPa, $H_2/CO = 1$, and 553 K, this peak had shifted below 2000 cm^{-1}. King also observed -CH_x groups, but he decided that they did exist on the Ru surface and could represent intermediates. He determined from the shifts in the IR spectra, as did Dalla Betta and Shelef [177], that an H_2—CO interaction occurs which weakens the CO bond. He concluded that the weakly-held CO is more reactive since its intensity decreased under reaction conditions.

The most recent IR study of a Ru catalyst under reaction conditions was conducted in a reactor system so that turnover frequencies could simultaneously be determined. Ekerdt and Bell [179, 180] found that the strong IR band at 2030 cm^{-1} from Ru/SiO_2 varied little with temperature, pressure, and H_2/CO ratio. They also observed bands around 2900 cm^{-1} indicative of —CH_x species but determined that they were not intermediates. Suddenly decreasing the CO pressure to zero produced a rate maximum followed by a decline, behavior similar to that reported for Ni and Fe catalysts [130, 170]. The IR absorbance remained constant until a low CO pressure was attained, and then a decline in the rate began concurrently with a decrease in the absorbance. Continued hydrocarbon formation, primarily methane, after the CO absorbance reached zero (Figure 20) indicated the presence of a carbonaceous overlayer on the surface which could be hydrogenated to products, and the rate determining step was proposed to be the reaction of —CH_x species with hydrogen. The methanation rate equation was in excellent agreement with that reported earlier [121], and turnover frequencies, which varied by a factor of four, were in satisfactory agreement with previously reported values. This study indicates the importance of surface carbon in the synthesis reaction, as C_s coverages ranged between 1 and 6 monolayers, and the authors proposed that chain-growth occurs via polymerization of —CH_x units, a scheme which coincides with that of Pichler

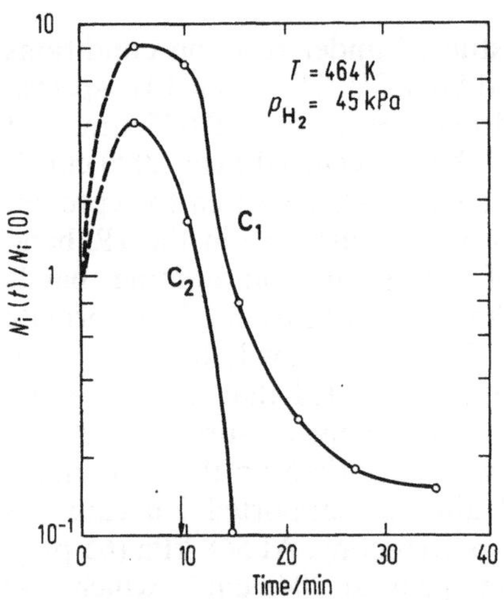

Figure 20. Relative rates of methane and ethane formation at 464 K under flowing hydrogen after steady-state reaction was ended at $t = 0$ by stopping CO flow. The arrow indicates when the CO IR absorbance goes to zero. (After [179])

[7] and Sachtler and coworkers [133], who found that surface carbon could be incorporated into higher hydrocarbons on Ru and Co as well as nickel. Ethylene and cyclohexene were added to the syngas stream in an effort to scavenge surface intermediates. The presence of alkyl and/or alkylidene species was indicated, with the cyclohexene product distribution favoring methylene groups [180].

Low and Bell [181] used TPD and TPSR to study CO adsorbed on a Ru/Al_2O_3 catalyst. Adsorption was molecular at 303 K and two peaks appeared in the TPD spectrum, with activation energies of desorption corresponding to 113 and 155 kJ mol^{-1}. Above 423 K dissociation began to occur, which formed a very reactive C_s species that could be hydrogenated at 300 K to produce methane, whereas temperatures of 400 K were required to hydrogenate adsorbed CO to produce CH_4 and C_2H_6. Although it could not be determined if the molecularly-adsorbed CO reacted directly with hydrogen or dissociated first, based on the results of their previous study [179], the authors preferred the latter path.

King [182] has studied the catalytic behavior of ruthenium on a wide variety of support materials under reaction conditions of 404 kPa, $H_2/CO = 2$, and temperatures ranging from 448 to 573 K. Methane was the principal product in all cases, and the value for the TOF for methane agreed to within a factor of two with an earlier reported value [121] when compared at comparable Ru dispersion (fraction exposed). Although the support was shown to exert some influence on the product selectivity, the most intriguing result of this study is an apparent metal crystallite size effect for ruthenium, as shown in Figure 21. Larger Ru particles possess a higher specific activity than well-dispersed Ru particles, and results from other investigators are consistent with this correlation [121, 183]. Specific activity clearly declines with increasing dispersion, and this is one of the better examples to date of a particle size effect on activity. This study provided detailed product analyses up to carbon numbers of five, while the work of Everson et al. [184] has determined product distributions up to C_{12} at pressures between 0.8 to 1.6 MPa and temperatures near 500 K.

The effect of the support on selectivity when zeolites are used has been examined by Uytterhoeven and coworkers [161]. A 5.8% Ru/NaY zeolite

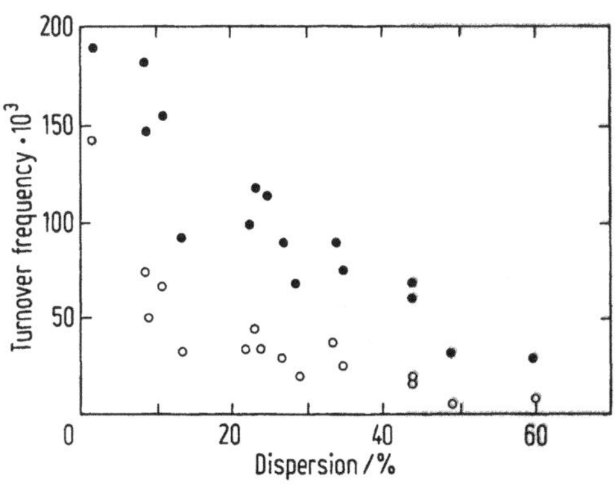

Figure 21. Catalyst turnover frequencies (per second) as a function of ruthenium dispersion at 523 K (●) CO conversion, (○) CH_4 production.

catalyst, with a fraction exposed of 1/2 based on hydrogen chemisorption, showed a sharp dropoff in hydrocarbon production above C_9 and less than 1% of the product consisted of C_{11+} hydrocarbons. A 15.6% Ru/SiO_2 sample with a fraction exposed of 0.06 showed a constant probability for chain-growth extending beyond a carbon number of 11. Under their reaction conditions of 1.42 MPa (14 atm), 525 K, and an H_2/CO ratio of 1.5, the TOF value for carbon monoxide for the zeolite was $0.014\ s^{-1}$ while that for the silica-supported catalyst was $0.043\ s^{-1}$. Although several possibilities exist to explain this difference, such as a cage effect, it is worth while noting that these activity results fit the trend found by King [182] and agree with the work of Elliott and Lunsford [185], which also showed that a Ru/zeolite catalyst possessed a lower TOF than Ru on alumina. The potential benefits of a shape-selective support are again demonstrated by this study as they were in the study by the workers at Mobil [159, 160].

An intriguing contradiction exists at this time concerning the presence of an H_2-D_2 kinetic isotope effect on Ni, Ru, and Pt catalysts. Dalla Betta and Shelef [186] have reported the absence of such an effect for supported catalysts whereas McKee [187] had previously observed a kinetic isotope effect over Ru powder. Dalla Betta and Shelef concluded that hydrogen cannot be involved in a rate determining step (RDS), and postulated that CO dissociation represents the slow step. This conclusion does not agree with most results from recent studies, such as those by Ekerdt and Bell [179], which indicate that hydrogenation of surface species is the slow step in the methanation reaction. The sequence of steps proposed by Ho and Harriott [130] involving a hydrogen-facilitated CO dissociation step may reconcile these differences, and the possibility of such a step had been proposed earlier [177]. However, the overall rate is a complicated function of kinetic and thermodynamic factors, and the absence of a kinetic isotope effect does not preclude adsorbed hydrogen from a RDS [188]. Also, in view of the variation in specific activity with particle size [143, 182], a change in the RDS over large unsupported metal particles compared to smaller supported particles could constitute another explanation.

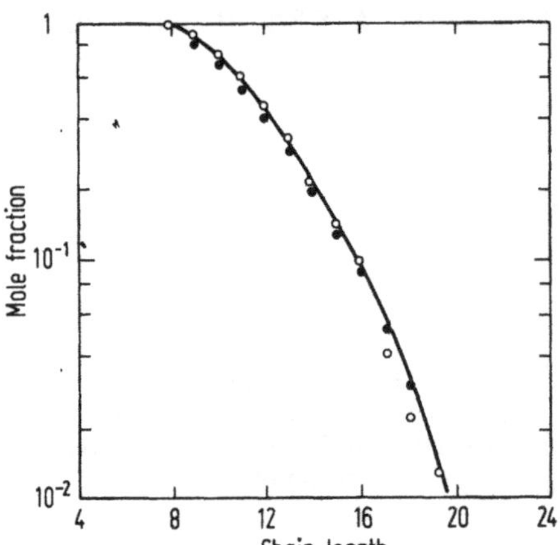

Figure 22. Liquid product distributions over a Ru/η-Al_2O_3 catalyst under transient operating conditions: ○ — Experimental, ● — Theoretical. (After [189])

A study has been conducted by Dautzenberg et al. [189] in which they employed transient reactor techniques to examine the Fischer-Tropsch chain-growth reaction over two $Ru/\gamma\text{-}Al_2O_3$ catalysts. Their intent was to determine if the relatively low activity of CO hydrogenation catalysts is due primarily to a low concentration of highly active surface sites or to a low intrinsic activity associated with most of the surface metal atoms. The catalyst was operated at 1 MPa, 483 K and an H_2/CO ratio of 1 for either 4 or 12 minutes, then the catalyst sample was heated to 623 K in pure H_2 and the desorbed products were collected. By fitting these product concentrations to a mathematical model based on a Schulz-Flory mechanism, they obtained good agreement between experiment and theory (Figure 22) and showed that the concentration of the surface intermediate is established rapidly and is time-independent. Around 70% of the Ru surface was apparently covered by growing chains at steady-state which infers that a large number of surface Ru atoms with low intrinsic activity constitutes the active sites. However, if the surface were covered with this many hydrocarbon chains, they should be detectable by IR, but as yet none have been seen. Therefore, the conclusions from this transient study must be considered questionable. Under these conditions, a relatively high carbon monoxide TOF value of $0.16\ s^{-1}$ was measured, which was in excellent agreement with a value of $0.12\ s^{-1}$ estimated at these conditions from the rate equation reported by Vannice [121]. These high TOF values may be due to the periodic cleaning in H_2 which was employed in both studies. The rate of initiation, *i.e.*, CO dissociation, was not the RDS in CO hydrogenation over this ruthenium catalyst.

D. Cobalt

Although cobalt has been extensively studied in the past and it has one of the highest specific activities for CO hydrogenation [122], it has not been the subject of many studies recently. Blanchard and Bonnet [190] compared the catalytic behavior of supported Co carbonyl clusters with precipitated Co catalysts. Adding thallium and fluorine to silica-supported Co lowered the activity; however, the effect on Co surface area was not determined. The Co/ThO_2/Kieselguhr catalyst used in this study gave a relatively flat hydrocarbon distribution from n-C_9 to n-C_{22} then dropped off sharply, again providing another apparent example of a selectivity not fitting a Schulz-Flory distribution, whereas a Co/Al_2O_3 catalyst produced a product distribution closer to that predicted by this mechanism. Blanchard et al. [191] later reported a selectivity dependence on pore diameter which was associated with a diffusional effect. However, it is possible that steady-state behavior in the catalyst bed was not attained at the time of these measurements.

Ponec et al. [192] have recently studied ^{13}CO disproportionation over Co and Ru evaporated films, and they showed that $^{13}C_s$ is incorporated into methane when a $^{13}C_s$-covered surface is exposed to a $H_2/^{12}CO$ mixture at 523 K. However, comparable amounts of $^{12}CH_4$ were also concomitantly produced and the question of whether CO dissociation is the only route to methanation over these two metals could not be answered. This behavior over Ru and Co differed significantly from that reported for nickel [126].

Sachtler et al. [133] have shown that more than one $^{13}C_s$ atom is incorporated into higher hydrocarbons (C_3H_8) indicating, as for nickel, that surface carbon appears to be able to participate in a chain-growth process over Co. Rabo et al. [132] have shown that C_s on cobalt produced *via* CO dissociation can react at 300 K, but they also found that 84% of the carbon on a cobalt surface (covered 1/3 with C_s and 2/3 with adsorbed CO) reacted at 473 K, thus providing evidence for the possible direct hydrogenation of CO to hydrocarbons.

Palmer and Vroom [145] had reported that exposure of Co and Ni foils to oxygen followed by H_2 reduction at low temperature produced a much more active catalyst than reduction at high temperature in hydrogen. This increased activity was attributed to subsurface oxygen. Ignatiev and Matsuyama [193] confirmed this with AES, XPS, and LEED measurements. The oxygen pretreatment at 900 K produced CoO at the surface, and no oxygen was detectable after a 900 K reduction whereas reduction near 550 K gave a strong oxygen signal and higher activity. After exposure to a syngas mixture, only surface carbon was detected, similar to results for Ni and Fe.

Fraenkel and Gates [162] have reported another example where the use of zeolites can significantly alter selectivity. Cobalt exchanged into an A-zeolite and reduced by Cd metal produced only propylene at 424 K, 6 atm and 1% conversion, and the TOF under these conditions was 1.1×10^{-4} s^{-1}. At higher temperatures propane was also formed. Reduction with hydrogen, rather than Cd, produced a mixture of C_1–C_5 hydrocarbons. A Cd-Co/Y-zeolite catalyst had a TOF of 2.2×10^{-3} s^{-1} at 455 K and a selectivity of 70 wt% C_4–C_7 hydrocarbons with little C_2 or C_3 production. This appears to be another example of shape selectivity although the role of the Cd was not determined.

E. Rhodium

Rhodium has attracted attention during the latter part of this decade because of its propensity to form two-carbon, oxygen-containing compounds. At relatively high pressures (up to 20 MPa) and temperatures between 473 to 573 K, Bhasin et al. [194] found that silica-supported Rh produced up to ~50 wt% of the product in the form of acetic acid, acetaldehyde, and ethanol, with the balance primarily composed of methane. Small amounts of Fe significantly altered selectivities, and favored the formation of CH_3OH and C_2H_5OH. Some evidence was obtained which indicated that methanol formation proceeded via a hydrogenation step of adsorbed formaldehyde.

Ichikawa [195–197] has also reported that supported Rh has a high selectivity to methanol and ethanol. His data indicated that basic supports such as ZnO, MgO, and CaO enhance methanol formation compared to BeO, SiO_2 and Al_2O_3 [195], while TiO_2, ZrO_2 and La_2O_3 favor ethanol production [196]. He also reported that the use of Rh carbonyl clusters favored alcohols while catalysts impregnated with $RhCl_3$ produced more hydrocarbons [197]. The relative yield of ethanol varied over La_2O_3-supported Rh clusters in the following manner:

$$Rh_2(C_2H_5)_2(CO)_3 < Rh_4(CO)_{12} > Rh_6(CO)_{16} > Rh_7(CO)_{12}[NEt_4]_3 > Rh_{13}(CO)_{23}H_{2-3}[NBu]_2 .$$

This study not only showed the capability of Rh to produce low molecular weight alcohols, but also showed the pronounced effect that the support can have on catalytic behavior.

Using their specially-designed reactor-UHV system, Sexton and Somorjai [198] characterized a Rh foil using TPD and AES and also determined its catalytic behavior at pressures near atmospheric. TPD spectra showed a single, molecularly-adsorbed CO species which desorbed at 523 K with an activation energy of 32 kcal mol^{-1}. Assuming a geometrical surface area and 10^{15} sites cm^{-2}, they calculated a methane TOF of 0.13 s^{-1} *vs* 0.034 for a previously reported value for Rh/Al_2O_3 [121] — the activation energies were nearly identical at 24 kcal mol^{-1}. AES analysis of a used foil indicated carbon, but little oxygen, was present at the surface. Treatment in oxygen increased the rate two-fold, but no oxygen Augér peak was observable after this oxygen treatment. Surface carbon was deposited during reaction, and enough C_s was present to react with H_2 at 573 K to form CH_4. The quantity of methane formed indicated the presence of 60 "monolayers" of carbon, which implied the formation of much surface carbon. The proposal was made that excess oxygen diffuses into the Rh lattice from the surface and promotes methanation, perhaps by faceting the surface. These results are quite similar to those of Palmer and Vroom [145]. When CO_2 was hydrogenated, a carbonaceous surface layer was produced which was very similar to that formed during CO hydrogenation; however, $CO_2 + H_2$ produced only methane. As a consequence of this result, it was suggested that chain growth requires CO on the surface.

In a subsequent study by Somorjai and coworkers [199] on Rh foil and the Rh (111) surface, catalytic behavior over clean and preoxidized Rh surfaces was determined at 6 atm over a range of temperature and H_2/CO ratios. The clean surfaces produced only light hydrocarbons whereas the preoxidized surface produced oxygenates and showed a more pronounced activity decline. About one monolayer of carbon was detected on the surface after reaction. Different activation energies of 24 kcal mol^{-1} on the clean surfaces and 12 kcal mol^{-1} on the preoxidized surfaces indicated that a different synthesis mechanism occurred on the two types of surfaces. A higher oxidation state for Rh appears to be required to produce oxygenates, and the CH_4 TOF was also dependent upon oxygen in the near surface region.

Kroeker, Kaska and Hansma [200] applied IETS to characterize Rh on Al_2O_3 after high pressure reactions of syngas mixtures were conducted with isotopic labeled CO. After cooling to liquid He temperatures, both Raman and IR active vibrational modes were observed on the Rh surface which indicated no oxygen-containing groups were present and only the ethylidene species existed. This is consistent with the capability of Rh to produce C_2 species, especially at high pressure [194].

F. Other Group VIII Metals

Rabo and coworkers [132, 201] have performed some very informative experiments on Pd, Pt, and Ir catalysts using a high pressure, tubular reactor and a recirculating gradientless reactor. At pressures between 1.1–110 MPa (150–16,000 psig) and temperatures of 548–623 K, they found that Pd/SiO_2 (and Pd/Al_2O_3 to a lesser extent) was a very selective catalyst for methanol formation. Methane formation became significant only outside the thermodynamically-favorable regime for CH_3OH formation. Pt and Ir also selectively produced methanol, but their activities were 10 times lower. Two IR bands at 2085 and 1970 cm^{-1} were observed on a freshly reduced Pd/SiO_2 sample, but after use in the reactor no band above 2000 cm^{-1} was visible. This behavior was attributed to a sintering process since X-ray diffraction measurements showed particle size growth. Coupled with their other study [132], which showed that Pd surfaces were predominantly covered with nondissociated CO, these authors proposed that methanol formation proceeds *via* direct hydrogenation of CO through a CH_xO species over Pd, Pt, and Ir.

A study of Pd/η-Al_2O_3 catalysts by Vannice and Garten [202] has shown that well-dispersed Pd on alumina is more active for CO conversion than silica-supported Pd. At 2.1 MPa and temperatures between 623 and 673 K, a 0.5% Pd/Al_2O_3 catalyst had a high initial activity but it declined markedly as time on-stream increased. However, at 574 K the activity stabilized within an hour and the carbon monoxide TOF under these conditions was nearly 20-fold higher than that calculated at identical conditions for a Pd/SiO_2 catalyst studied by the Union Carbide investigators [201]. A noteworthy difference between the Pd/Al_2O_3 catalysts studied by Vannice and Garten and the Pd catalysts studied by Poutsma et al. [201] is that the former produced only methane while the latter favored methanol formation. One possible explanation is that two parallel pathways exist, and the decrease in CO hydrogenation activity occurred only because of a decrease in the methanation rate, with the pathway to methanol being altered little.

Vannice and coworkers [203] have recently found that methanation activity over Pd is very dependent upon the support used, and TiO_2-supported Pd in the SMSI state is the most active catalyst. A ranking of Pd catalysts by decreasing TOF gave: Pd/TiO_2 (SMSI) > $Pd/TiO_2 \simeq Pd/Al_2O_3 \simeq Pd/SiO_2$–$Al_2O_3$ > Pd/SiO_2. An identical ranking exists for supported Pt catalysts [119, 204]. The SiO_2-supported metals are more than 100 times less active than the TiO_2-supported metals, and 10 times less active than the Al_2O_3 and SiO_2-Al_2O_3-supported metals. The lower activity of SiO_2-supported noble metals has also been observed by Fujimoto et al. [205] using TPR measurements. These higher TOF values occur for Pd/TiO_2 and Pt/TiO_2 in the SMSI state even though surface coverages of CO are markedly reduced under reaction conditions, as determined by in situ IR spectra [115, 119, 206]. These studies indicate that only a small fraction of the surface atoms

($\geq$ 1%) constitute "active sites" and that higher TOF values are due to more equal surface coverages of CO and hydrogen under reaction conditions [118, 206].

G. Alloys and Other Systems

Ponec and coworkers [126, 207] have studied Ni-Cu films and powders. The addition of Cu to Ni decreased the rate of CO conversion per unit surface area by nearly 100-fold but rates per active site, as counted by selective chemisorption of H_2 or CO, were not determined. The selectivity to C_2 and C_3 hydrocarbons increased as Cu was added. Removal of CO from the feed at steady-state conditions quickly ended production C_2 and C_3 hydrocarbons, but CH_4 formation continued. They concluded, therefore, that molecularly-adsorbed CO was required for chain growth and methane formation occurred via dissociative CO adsorption step. From IR studies, they determined that ensembles of Ni atoms provide multiple-coordination sites which adsorb CO strongly therby weakening the C–O bond and facilitating dissociation. These they proposed as initiation sites, while chain growth was assumed to occur via single Ni atoms, which adsorb CO linearly and provide molecular CO. The role of Cu was then explained — it reduces the total activity by diluting the Ni surface and reducing the number of ensembles which initiate the CO hydrogenation reactions, but increases selectivity by providing isolated Ni atoms which provide molecular CO for chain growth. More recent studies indicate, however, that CO insertion is not the most likely chain growth mechanism [124, 179].

Although Luyten et al. [208] also found that addition of Cu to silica-supported Ni (and Ru) reduced activity, the remainder of their results differed from those reported by Ponec et al. [126, 207]. They reported that Cu addition increased selectivity to methane, and by using H_2 chemisorption to count active sites, they found that carbon monoxide TOF values declined as Cu content increased. Because E_{CH_4} values were constant at 130 kJ mol^{-1} for all Cu concentrations, values noticeably higher than those reported by Araki and Ponec [126], the rate decrease was associated with a decrease in the number of ensembles on the surface. This conclusion had been reached earlier by other investigators [209]. One cause for this discrepancy could be due to the lack of characterization of the surface layer in the films and proof of uniformly alloyed particles in the supported system.

The results of Bond and Turnham [209] for silica-supported Ru-Cu catalysts also had shown a 50-fold decrease in turnover frequency and an increase in selectivity to CH_4 with increasing Cu content. They found that the activation energy remained constant; however, their value of ~88 kJ × mol^{-1} was noticeably lower than the value of 126 kJ mol^{-1} reported by Luyten et al. [208]. Bond and Turnham concluded that ensembles of surface atoms constituting an active site contain around 4 Ru atoms.

Elliott and Lunsford [185] studied the catalytic behavior of a family of thoroughly characterized, zeolite-supported Ru, Ru-Ni, Ru-Cu, and Ni catalysts. In agreement with previous studies, CO conversion rates declined as Cu content increased. However, they found that Cu addition increased the activation energy, giving values close to those reported by Luyten et al. [208], and decreased selectivity to CH_4. The addition of Ni to Ru/zeolites was found to decrease the turnover frequencies linearly with increasing Ni content and to significantly inhibit catalyst deactivation, presumably by reducing the build-up of excess carbon. It appears that conflicting results presently exist regarding effects on kinetic parameters and selectivity in Ni-Cu and Ru-Cu sytems although all studies have clearly shown that Cu addition reduces specific activity in the CO hydrogenation reaction.

Another alloy which has shown marked changes in kinetic parameters and selectivity is the Ru-Fe system. Vannice, Lam, and Garten [210] employed Mössbauer spectroscopy to prove that alloy formation did occur, since this cannot be assumed on coimpregnated silica-supported catalysts. In the range of 35–65% Ru, a noticeable decrease occurred in the turnover frequency, but a marked shift in product selectivity also occurred which reduced methane formation and enhanced the olefin/paraffin ratio. For example, a 65% Ru-35% Fe/SiO_2 catalyst run at 103 kPa, 525 K and a H_2/CO ratio of 3 gave a product containing 45 mole % C_2–C_5 olefins. Such favorable changes in selectivity have yet to be fully exploited. Subsequent studies by Ott, Fleisch, and Delgass [211, 212] on unsupported RuFe alloys using XPS and SIMS showed that surface enrichment of iron occurs and only small amounts of iron are needed to alter surface electronic properties. Addition of iron decreased methane make and increased the olefin/paraffin ratio as reported earlier [210]. No substantial carbon buildup was found on the Ru and 3% Fe-97% Ru catalysts which maintained steady activity and selectivity behavior. However, additional Fe produced a thick carbon overlayer, reduced activity, and shifted selectivity to lighter hydrocarbons. These results indicated that pure iron first forms a bulk carbide then begins to produce a carbon overlayer.

One last group of catalysts — rare earth/Group VIII metal intermetallic compounds — has been investigated by three different groups of investigators. Wallace and coworkers [213, 214] have studied $LaNi_5$, Maple and coworkers [215] examined $CeAl_2$, $CeCo_2$ and $CeNi_2$, and mischmetal-Ni alloys were investigated by Atkinson and Nicks [216]. Under reaction conditions, these systems exhibit similar behavior in that significant increases occur in the BET surface areas, due to oxidation of the rare earth component. The active catalyst can therefore be likened to a metal supported on a rare earth oxide, and turnover frequencies based on H_2 or CO chemisorption on used samples are comparable to regular supported Ni catalysts [141], although a range of values does appear to exist. All turnover frequencies in these systems are lower than those reported by Vannice and Garten for the very active Ni/TiO_2 catalyst systems [146, 147], except for the ThNi system mentioned previously [150].

Finally, Tucci and Streeter [151] have found that $RuNi/TiO_2$ and Ru/TiO_2 catalysts are superior methanation catalysts when used in a fluidized bed reactor under commercial reactor conditions (700–810 K, 6.9 MPa). Compared to alumina and SiO_2-Al_2O_3 supported catalysts, a 1% Ru, 10% Ni/TiO_2

catalyst gave the best performance as activity increased to 90% conversion over 20 days then remained constant for the remaining 15 days of the test period. These data reveal the potential of the new metal-support systems which have recently been reported.

H. Mechanism of CO Hydrogenation Reactions

The proposed reaction pathways during synthesis reactions have generally consisted of one of three types to explain carbon-carbon bond formation [11, 124]: 1) carbidic or $CH_{x(a)}$ intermediates; 2) hydrocondensation of CH_xO_y groups; or 3) CO insertion (alkyl migration). Based on the recent studies discussed here, the following statements can now be made about methanation and chain-growth reactions:

- The highest rates of hydrocarbon formation occur on those metals which dissociate CO most readily (Fe, Co, Ni, Ru).
- Carbidic surface carbon readily hydrogenates to CH_4.
- Carbidic surface carbon can participate in chain growth processes.
- No oxygen-containing surface species, other than CO, have been unambiguously identified under reaction conditions.
- On metal surfaces a number of different states of adsorbed CO can exist which can be identified by IR and EELS.
- The hydrogen surface concentration is very important and a delicate balance may exist between C_s and $H_{(a)}$ for optimum behavior.

The predominant methanation route clearly appears to be *via* dissociative CO adsorption to form a surface carbon species (carbidic) which is then hydrogenated. However, evidence still exists that adsorbed CO can react directly with hydrogen to form methane, although the pathway still most likely involves $CH_{x(a)}$ intermediates, and the rate of this reaction is lower. Either route is possible from theoretical considerations [217, 218]. The series of elementary steps involved in the overall sequence may be very dependent upon reaction conditions and a single rate determining step may not exist [140]. Because surface coverages of CO almost always appear to be much greater than hydrogen coverages, as indicated by pressure dependencies which are typically near 1st order in H_2 and zero order in CO, the rates may be as sensitive to alterations in the metal surface which enhance the surface coverage of hydrogen as those which affect the state of CO adsorption. This perspective provides an explanation for the higher rates (more than 100-fold) which have been observed in SMSI catalysts [203].

Presently, the favored model for chain-growth processes is that involving $CH_{x(a)}$ as building blocks in a step-wise polymerization reaction to normally give products which fit a Schulz-Flory distribution. Recent studies by Biloen, Sachtler and coworkers [124] and by Ekerdt and Bell [179, 180] have provided results which strongly support a model involving migration of growing-chain alkyl groups to form a C-C bond with a CH_x group. Although the value of x is not unambiguously known, and may vary as a function of metal, temperature, pressure, etc., evidence favors a carbene group [124, 179, 180, 200, 219]. Termination by desorption gives an α-olefin whereas hydrogenation gives a linear hydrocarbon. One different model which is worth noting is that of investigators at the Brookhaven National Laboratory, who propose that alkoxy surface species exist and that chain growth occurs *via* methoxy (or oxymethylene) groups [220]. IR studies have indicated such species may exist on cobalt [221], and recent UHV studies on iron also support this concept [176]. Although currently not in favor, future work is needed to clearly prove or disprove the validity of this model.

The formation of alcohols in particular, and oxygenates in general, apparently requires molecular CO adsorbed on the surface. The ability of Pd, Pt, and Ir to produce methanol is good evidence of this because CO adsorption is almost completely nondissociative on these metals [201]. Termination of chain growth by insertion of CO (or an oxymethylene group) followed by hydrogenation and desorption would produce straight-chain, terminal alcohols. Alkali metal promoters are well known for their ability to promote alcohol formation but, based upon recent studies of the behavior of potassium on iron single crystal surfaces [53], the old model involving electron donation to the iron may need revision. Instead, K atoms (or clusters) may interact directly with CO to provide special sites to enhance molecular adsorption. The role of promoters in this reaction is a long-standing, unanswered question and future study is still needed to clearly understand the effect of promoters.

6. Water Gas Shift Reaction

The water gas shift (WGS) reaction, $CO + H_2O \rightarrow H_2 + CO_2$, is catalyzed by many metals and metal oxides, and has long been used to produce hydrogen from syngas streams generated by the steam reforming of hydrocarbons. In energy conversion processes involving gasification, this reaction is important for two reasons: it is utilized to shift the H_2/CO ratio in the syngas stream to that required in the reactor for CO hydrogenation, and it can occur sequentially in the synthesis reactor to allow byproduct water to convert unreacted CO to CO_2. This alters the H_2/CO usage ratio and produces CO_2 as the oxygen-containing byproduct.

This reaction has not been the subject of intensive study, and results pertaining to the catalysis of this reaction have been discussed briefly by Happel [222] and Wagner [223] and reviewed recently by Newsome [224]. Few recent studies have been devoted to the investigation of reduced Group VIII metals as WGS catalysts. Masuda [225, 226] has studied the WGS and the reverse WGS reactions over Pt films using labeled isotopes of CO_2 and H_2. The rate equation for the reverse WGS reaction was $r = kp_{CO_2}^{0.6}p_{H_2}^{0.5}p_{CO}^{-0.5}$ with an activation energy of $103 \pm 5\ kJ\ mol^{-1}$. No kinetic isotope effect was observed when D_2 was used, and rapid equilibrium was achieved between $C^{18}O_2$ and $C^{16}O_2$. Rates of carbon atom transfer between CO and CO_2 and oxygen atom transfer among CO, CO_2 and H_2O were determined while the WGS reaction was occurring. Using the concept of stoichiometric numbers applied to a series of elementary steps describing this reaction, Masuda concluded that a [CO-OH] complex is formed on the surface, and the RDS is the decomposition of this surface intermediate to give adsorbed CO_2 and H.

Alumina-supported ruthenium as a WGS shift catalyst has been investigated by Shelef and Gandhi [227] and by Taylor et al. [228]. In the presence of 0.1% NO, Shelef and Gandhi found that the activity of three catalyst systems decreased in the following order: Ru/Al_2O_3 > $Cu\text{-}Cr\text{-}Ru/Al_2O_3$ > $Cu\text{-}Ni\text{-}Ru/Al_2O_3$. However, metal surface areas and specific activities were not determined. Taylor et al. learned that an oxidized Ru/Al_2O_3 catalyst was 6 to 70 times more active than the comparable Ru/Al_2O_3 which was reduced in hydrogen. Low loadings of Pt and Pd (0.1–0.3 wt%) on alumina showed a similar trend, but activities varied only 2-fold. The oxidized Ru catalysts did not contain Ru oxide and chemisorption measurements showed that changes in the dispersion could not account for this behavior. Again, turnover frequencies were not determined. Two possible explanations for the higher activity of the oxidized catalysts were suggested: Ru surface reconstruction, perhaps due to subsurface oxygen, or a metal-support interaction between Ru and alumina.

The possibility that a partially oxidized Ru surface could account for this activity increase is supported by the recent work of Uytterhoeven and coworkers [229], who studied zeolites which were ion-exchanged with Ru—$(NH_3)_6Cl_3$. Turnover frequencies at 393 K in the WGS reaction were higher with these catalysts than those reported for homogeneous Ru and Rh catalysts and, in addition, the activity was higher than that of a conventional copper-based catalyst. Temperature programmed H_2 reduction and IR spectroscopy indicated that a Ru complex was stabilized in the zeolite supercages with the catalytically active species alternating between Ru^{I} and Ru^{II}.

The definitive work to date on the WGS reaction catalyzed over supported Group VIII metals is that recently reported by Grenoble, Estadt, and Ollis [230]. They have measured turnover frequencies and other kinetic parameters, shown in Table 5, and have shown that the most active metal, Cu, is 50 times more active than the most active Group VIII metal, Co, and nearly 4000 times more active than Ir, the least active metal. Using bulk thermodynamic properties, they calculated that only iron should have an appreciable tendency toward bulk oxidation. Interestingly, iron had reaction orders completely different from the other metals, which may indicate that Fe is partially oxidized under reaction conditions. A significant support effect was observed, and turnover frequencies for Pt/Al_2O_3 were 10-fold higher than for Pt/SiO_2 and 100-fold higher than for Pt/carbon. A similar result was obtained for alumina- and silica-supported Rh. A detailed kinetic analysis resulted in a model proposing that the role of the support is to activate water and that a formic acid intermediate is formed. The bifunctional nature of this model led to predicted power law rate expressions in good agreement with experimentally measured equations, and it also helped explain the observed support effect because the higher Lewis acidity of Al_2O_3, compared to SiO_2 or carbon, would lead to a higher concentration of the formic acid intermediate thereby enhancing the rate. A

Table 5. Kinetic parameters of alumina-supported metals for water-gas shift reaction (Ref. [230])

Catalyst %M/Al_2O_3	Temp.[e]/K	X^b	Y^c	Activation energy/kJ mol^{-1}	Arrhenius preexponential factor/$molec^{-1}$ s^{-1} (metal site)$^{-1}$	Turnover frequency[a,d]
1% Ru	563	−0.21 ± 0.08	0.66 ± 0.08	94.3 ± 2.5	7.38×10^7	0.1929
1% Rh	603	−0.10 ± 0.02	0.44 ± 0.08	96.4 ± 5.4	5.00×10^6	0.0086
1% Pd	653	0.14 ± 0.06	0.38 ± 0.07	80.0 ± 2.9	2.61×10^5	0.0135
2% Os	593	−0.27 ± 0.08	0.63 ± 0.02	98.9 ± 11.3	6.18×10^7	0.0615
2% Ir	623	0.03 ± 0.04	0.48 ± 0.06	86.3 ± 13.8	2.31×10^5	0.0032
2% Pt	543	−0.21 ± 0.03	0.75 ± 0.04	82.1 ± 5.4	1.91×10^6	0.0635
10% Fe	623	0.58 ± 0.12	0.04 ± 0.12	80.4 ± 5.4	3.18×10^5	0.0151
5% Co	523	−0.35 ± 0.12	0.67 ± 0.12	47.3 ± 15.1	5.05×10^3	0.2472
5% Ni	523	−0.14 ± 0.05	0.62 ± 0.11	78.4 ± 1.3	1.40×10^6	0.1029
10% Re	523	−0.09 ± 0.05	0.55 ± 0.11	75.0 ± 6.7	2.58×10^6	0.3839
10% Cu	403	0.30 ± 0.05	0.38 ± 0.19	55.7 ± 3.4	1.44×10^6	12.1900
5% Au	543	0.74 ± 0.02	0.13 ± 0.10	48.6 ± 2.5	9.46×10^2	0.0356

[a] The activity of the metal-free Al_2O_3 support is one-tenth that of the least active metal, Ir
[b] Order with respect to CO
[c] Order with respect to H_2O
[d] At 1 hr on stream, 573 K, $p_{CO} = 24.3$ kPa, $p_{H_2O} = 31.4$ kPa
[e] Temperature at which orders were determined

volcano plot exists for this reaction, similar to that shown in Figure 12 for methanation, except that the optimum CO-metal bond strength seems to occur at a lower value of 84 kJ mol^{-1}. The results obtained in this thorough study represent the first reported specific activities for the WGS reaction over reduced Group VIII metals.

7. Summary

Carbon monoxide adsorption on the Group VIII metals exhibits certain patterns. Typically, as coverages exceed one-half, LEED patterns show that compression occurs in the monolayer and the molecules lose registry with the surface metal atoms. However, IR spectra at higher pressures commonly show two major CO bands, which infer that two general forms of adsorbed CO predominate on the surface. Recent studies indicate that particular sites which are associated with rough surfaces, *i.e.*, those containing steps and kinks, facilitate CO dissociation. Since one pathway to form methane has clearly been shown to involve surface carbon produced by CO dissociation, the concentration of these sites may have a significant effect on the selectivity in the CO hydrogenation reaction.

This reactive surface carbon species has been shown to participate in chain-growth reactions, and a surface polymerization sequence involving $CH_{x(a)}$ species, most likely methylene groups, is currently the model best describing the Fischer-Tropsch synthesis reaction. The precise role of molecularly adsorbed CO has not been completely defined, but formation of oxygenates, especially alcohol, clearly appears to require molecular CO. An important role of alkali metal promoters may be that of altering the surface concentration of molecular CO by direct interaction with CO.

Both metal-support interactions and metal crystallite size effects have been reported in reactions involving CO. The use of titania as a support for nickel enhances activity and shifts selectivity markedly toward higher molecular weight paraffins. The support utilized also significantly alters turnover frequencies in the WGS reaction. Ruthenium particles in a state of high dispersion appear to have lower turnover frequencies for CO hydrogenation than larger Ru crystallites. Infrared results provide evidence that the small Ru particles may be partially oxidized on the support, which alters their catalytic properties. In contrast, ruthenium in a partially oxidized state appears more active in the WGS reaction than reduced ruthenium catalysts.

Although our understanding about the adsorption, activation, and reactivity of CO on metal surfaces has increased greatly, the intriguing results summarized here show that much is yet to be learned. Indications are strong that a better knowledge of metal-support interactions combined with a more complete understanding of the surface chemistry involved in CO hydrogenation will lead to improved catalyst systems in the future.

References

1. Storch, H. H., Golumbic, H., Anderson, R. B.: in The Fischer-Tropsch and Related Synthesis, Wiley, New York, 1951
2. Anderson, R. B.: Catalysis, Vol. 4 (P. H. Emmett, ed.), New York: Reinhold 1956
3. Cohn, E. M.: Catalysis, Vol. 4 (P. H. Emmett, ed.), New York: Reinhold 1956
4. Greyson, M.: Catalysis, Vol. 4 (P. H. Emmett, ed.), New York: Reinhold 1956
5. a. Natta, G.: Catalysis, Vol. 3 (P. H. Emmett, ed.), New York Reinhold 1955
   b. Natta, G., Colombo, U., Pasquon, I.: Catalysis, Vol. 5 (P. H. Emmett, ed.), New York: Reinhold 1957
6. a. Eidus, Ya. T.: Russ. Chem. Rev. **36**, 338 (1967)
   b. Nefedov, B. K., Eidus, Ya. T.: Russ. Chem. Rev. **34**, 272 (1965)
7. a. Pichler, H.: Erdoel Kohle, Erdgas, Petrochem. Brennst.-Chem. **26**, 625 (1973)
   b. Pichler, H., Schulz, H.: Chem.-Ing.-Techn. **42**, 1162 (1970)
8. Vlasenko, V. M., Yuzefovich, G. E.: Russ. Chem. Rev. **38**, 728 (1969)
9. Mills, G. A., Steffgen, F. W.: Catal. Rev.-Sci. Eng. **8**, 159 (1973)
10. Denny, P. J., Whan, D. A.: In: Catalysis, Vol. 2, Chemical Society, London 1978, Chapt. 3
11. Vannice, M. A.: Catal. Rev.-Sci. Eng. **14**, 153 (1976)
12. Jorgensen, W. L., Salem, L.: in The Organic Chemist's Book of Orbitals, Academic Press, New York, 1973
13. Blyholder, G., Allen, M. C.: J. Am. Chem. Soc. **91**, 3158 (1969)
14. Jones, L. H.: J. Mol. Spectrosc. **9**, 130 (1962)
15. Broden, G., Rhodin, T. N., Brucker, C., Benbow, R., Hurych, Z.: Surface Sci. **59**, 593 (1976)
16. Doyen, G., Ertl, G.: Surface Sci. **43**, 197 (1974)
17. Boudart, M.: In: Proc. VI Internat. Congr. Catalysis (ed. by G. C. Bond, P. B. Wells, F. C. Tompkins), Vol. I, The Chemical Society, London 1977, p. 1–9
18. Ponec, V., Sachtler, W. M. H.: J. Catal. **24**, 250 (1972)
19. Eischens, R. P., Pliskin, W. A., Francis, S. A.: J. Phys. Chem. **60**, 194 (1965)
20. Blyholder, G.: J. Phys. Chem. **68**, 2772 (1964)
21. Hulse, J. E., Moskovits, M.: Surface Sci. **57**, 125 (1976)
22. Bradshaw, A. M., Hoffmann, F. M.: Surface Sci. **72**, 513 (1978)
23. Calabrese, J. C., Dahl, L. F., Chini, P., Longoni, G., Martinengo, S.: J. Am. Chem. Soc. **96**, 2614 (1974)
24. a. Eischens, R. P., Francis, S. A., Pliskin, W. A.: J. Chem. Phys. **22**, 1786 (1954)
    b. Eischens, R. P., Pliskin, W. A.: Adv. Catal. **10**, 1 (1958)
25. Greenler, R. G.: J. Chem. Phys. **44**, 310 (1966); Ibid. **50**, 1963 (1969)
26. Hansma, P. K.: Phys. Repts. **30**, 145 (1977)
27. a. Ibach, H.: J. Vac. Sci. Tech. **9**, 713 (1972)
    b. Ibach, H.: Phys. Rev. Lett. **27**, 253 (1971)
28. Joyner, R. W.: Surface Sci. 291 (1977)
29. Bradshaw, A. M.: Surface Sci. **80**, 215 (1979)
30. Bertolini, J. C., Dalmai-Imelik, G., Rousseau, J.: Surface Sci. **68**, 539 (1977)
31. Bertolini, J. C., Imelik, B.: Surface Sci. **80**, 586 (1979)
32. Bertolini, J. C., Tardy, B.: Surface Sci. **102**, 131 (1981)
33. Andersson, S.: Solid St. Comm. **21**, 75 (1977)
34. Andersson, S., Pendry, J. B.: Surface Sci. **71**, 75 (1978)
35. a. Passler, M., Ignatiev, A., Jona, F., Jepson, D. W., Marcus, P. M., Phys. Rev. Lett. **43**, 360 (1979)
    b. Andersson, S., Pendry, J. B.: Phys. Rev. Lett. **43**, 363 (1979)
36. Fleisch, T., Ott, G. L., Delgass, W. N., Winograd, N.: Surface Sci. **81**, 1 (1979)
37. Barber, M., Vickerman, J. C., Wolstenholme, J.: J. Chem. Soc. Farady Tr I. **72**, 40 (1976)
38. Erley, W., Wagner, H.: Surface Sci. **74**, 333 (1978)
39. Erley, W., Ibach, H., Lehwald, S., Wagner, H.: Surface Sci. **83**, 585 (1979)
40. Doering, D. L., Poppa, H., Dickinson, J. T.: J. Vac. Sci. Technol. **18**, 460 (1981)
41. Akimoto, K., Sakisaka, Y., Nishijimi, M., Onchi, M.: Surface Sci. **88**, 109 (1979)
42. Campuzano, J. C., Dus, R., Greenler, R. G.: Surface Sci. **102**, 172 (1981)

43. Van Dijk, W. L., Groenewegen, J. A., Ponec, V.: J. Catal. **45**, 277 (1976)
44. Primet, M., Dalmon, J-A., Martin, G-A.: J. Catal. **46**, 25 (1977)
45. Martin, G-A., Dalmon, J-A., Primet, M.: C. R. Acad. Sci. Paris, Series C, **284**, 163 (1977)
46. Wedler, G., Geuss, K-P., Colb, K. G., McElhiney, G.: Appl. Surface Sci. **1**, 471 (1978)
47. Wedler, G., Colb, K. G., McElhiney, G., Heinrich, W.: Appl. Surface Sci. **2**, 30 (1978)
48. Yu, K. Y., Spicer, W. E., Lindau, I., Pianetta, P., Lin, S. F.: Surface Sci. **57**, 157 (1976)
49. Broden, G., Gafner, G., Bonzel, H. P.: Appl. Phys. **13**, 333 (1977)
50. Rhodin, T. N., Brucker, C. F.: Sol. St. Comm. **23**, 275 (1977)
51. Textor, M., Gay, I. D., Mason, R.: Proc. R. Soc. London. A. **356**, 37 (1977)
52. Kroeker, R. M., Hansma, P. K., Kaska, W. C.: J. Chem. Phys. **72**, 4845 (1980)
53. Benzinger, J., Madix, R. J.: Surface Sci. **94**, 119 (1980)
54. Bridge, M. E., Comrie, C. M., Lambert, R. M.: Surface Sci. **67**, 393 (1977)
55. Prior, K. A., Schwaha, K., Lambert, R. M.: Surface Sci. **77**, 193 (1978)
56. Hoffmann, F. M., Bradshaw, A. M.: J. Catal. **44**, 328 (1976)
57. Bradshaw, A. M., Hoffmann, F. M.: Surface Sci. **72**, 513 (1978)
58. Stephan, J. J., Franke, P. L., Ponec, V.: J. Catal. **44**, 359 (1976)
59. Bader, S. D., Blakely, J. M., Brodsky, M. R., Friddle, R. J., Panosh, R. L.: Surface Sci. **74**, 405 (1978)
60. Castner, D. G., Sexton, B. A., Somorjai, G. A.: Surface Sci. **71**, 519 (1978)
61. Thiel, P. A., Williams, E. D., Yates, J. T., Jr., Weinberg, W. H.: Surface Sci. **84**, 54 (1979)
62. Yates, J. T., Jr., Williams, E. D., Weinberg, W. H.: Surface Sci. **91**, 562 (1980)
63. Dubois, L. H., Somorjai, G. A.: Surface Sci. **91** 514 (1980)
64. Yates, D. J. C., Murrell, L. L., Prestridge, E. B.: J. Catal. **57**, 41 (1979)
65. Yang, A. C., Garland, C. W.: J. Phys. Chem. **61**, 1504 (1957)
66. Yao, H. C., Rothschild, W. G.: J. Chem. Phys. **68**, 4774 (1978)
67. a. Hansma, P. K., Kaska, W. C., Laine, R. M.: J. Am. Chem. Soc. **98**, 6064 (1976)
b. Kroeker, R. M., Kaska, W. C., Hansma, P. K.: J. Catal. **57**, 72 (1979)
68. Williams, E. D., Weinberg, W. H.: Surface Sci. **82**, 93 (1979)
69. Pfnur, H., Menzel, D., Hoffmann, F. M., Ortega, A., Bradshaw, A. M.: Surface Sci. **93**, 431 (1980)
70. Madey, T. E., Surface Sci. **79**, 575 (1979)
71. Fuggle, J. C., Madey, T. E., Steinkilberg, M., Menzel, D.: Surface Sci. **52**, 521 (1975)
72. Brown, M. F., Gonzalez, R. D.: J. Phys. Chem. **80**, 1731 (1976)
73. Dalla Betta, R. A.: J. Phys. Chem. **79**, 2519 (1975)
74. McCarty, J. G., Wise, H.: Chem. Phys. Lett. **61**, 323 (1979)
75. Bossi, A., Carnisio, G., Garbossi, F., Giunchi, G., Petrini, G.: J. Catal. **65**, 16 (1980)
76. Ertl, G., Newmann, M., Streit, K. M.: Surface Sci. **64**, 393 (1977)
77. Shigeishi, R. A., King, D. A.: Surface Sci. **58**, 379 (1976)
78. Krebs, H.-J., Luth, H.: Appl. Phys. **14**, 337 (1977)
79. Froitzheim, H., Hopster, H., Ibach, H., Lehwald, S.: Appl. Phys. **13**, 147 (1977)
80. Hopster, H., Ibach, H.: Surface Sci. **77**, 109 (1978)
81. Norton, P. R., Goodale, J. W., Selkirk, E. B., Surface Sci. **83**, 189 (1979)
82. McCabe, R. W., Schmidt, L. D.: Surface Sci. **65**, 189 (1977)
83. Ibach, H., Erley, W., Wagner, H.: Surface Sci. **92**, 29 (1980)
84. McCabe, R. W., Schmidt, L. D.: Surface Sci. **66**, 101 (1977)
85. Stephan, J. J., Ponec, V.: J. Catal. **42**, 1 (1976)
86. Bain, F. T., Jackson, S. D., Thomson, S. J., Webb, G., Willocks, E.: J. Chem. Soc., Faraday Tr I, **72**, 2516 (1976)
87. Comrie, C. M., Weinberg, W. H., J. Vac. Sci. Technol. **13**, 264 (1976)
88. Zhdan, P. A., Boreskov, G. K., Borinin, A. I., Schepelin, A. P., Egelhoff, W. F., Weinberg, W. H.: Surface Sci. **71**, 267 (1978)
89. Weinberg, W. H., Comrie, C. M., Lambert, R. M.: J. Catal. **41**, 489 (1976)
90. McElhiney, G., Papp, H., Pritchard, J.: Surface Sci. **54**, 617 (1976)
91. Horn, K., Hussain, M., Pritchard, J.: Surface Sci. **63**, 244 (1977)
92. Moskovits, M., Hulse, J. E.: J. Phys. Chem. **81**, 2004 (1977); Surface Sci. **61**, 302 (1976)

93. Soma-Noto, Y., Sachtler, W. M. H.: J. Catal. **32**, 315 (1974)
94. Wise, H.: J. Catal. **43**, 373 (1976)
95. Primet, M., Matthieu, M. V., Sachtler, W. M. H.: J. Catal. **44**, 324 (1976)
96. Soma-Noto, Y., Sachtler, W. M. H.: J. Catal. **34**, 162 (1974)
97. Franken, P. E. C., Ponec, V.: J. Catal. **42**, 398 (1976)
98. Harberts, J. C. M., Bourgonje, A. F., Stephan, J. J., Ponec, V.: J. Catal. **47**, 92 (1977)
99. Yu, K. Y., Ling, D. T., Spicer, W. E.: J. Catal. **44**, 373 (1976)
100. Benndorf, C., Goetz, R., Gressmann, K-H., Kessler, J., Thieme, F.: Surface Sci. **76**, 509 (1979)
101. Stephan, J. J., Ponec, V.: J. Catal. **42**, 1 (1976)
102. Verbeek, H., Sachtler, W. M. H.: J. Catal. **42**, 257 (1976)
103. Yates, J. T., Jr., Goodman, D. W., Madey, T. E.: Proc. 7th Inter. Vac. Congr. (Vienna), 1977, p. 1
104. Demuth, J. E., Ibach, H.: Chem. Phys. Lett. **60**, 395 (1979)
105. Blyholder, G., Shihabi, D.: J. Catal. **46**, 91 (1977)
106. Primet, M., Sheppard, N.: J. Catal. **41**, 258 (1976)
107. Heal, M. J., Leisegang, E. C., Torrington, R. G.: J. Catal. **42**, 10 (1976)
108. Farrauto, R. J.: J. Catal. **41**, 482 (1976)
109. Wedler, G., Colb, K. G., Heinrich, W., McElhiney, G.: Appl. Surface Sci. **2**, 85 (1978)
110. Kolbel, H., Roberg, H.: Ber. Buns.-Gesell. physik. Chem. **81**, 634 (1977)
111. Wells, M. G., Cant, N. W., Greenler, R. G.: Surface Sci. **67**, 541 (1977)
112. Kawasaki, K., Shibata, M., Miki, H., Kioka, T.: Surface Sci. **81**, 370 (1979)
113. Kawasaki, K., Kodama, T., Miki, H., Kioka, T.: Surface Sci. **64**, 349 (1977)
114. Goodman, D. W., Madey, T. E., Ono, M., Yates, J. T., Jr.: J. Catal. **50**, 279 (1977)
115. Vannice, M. A., Wang, S-Y., Moon, S. H.: J. Catal. **71**, 152 (1981)
116. Conrad, H., Ertl, G., Latta, E. E.: J. Catal. **35**, 363 (1974)
117. Tauster, S. J., Fung, S. C., Garten, R. L.: J. Am. Chem. Soc. **100**, 170 (1978)
118. Vannice, M. A., Twu, C. C.: J. Chem. Phys. **75**, 5944 (1981)
119. Vannice, M. A., Moon, S. H., Twu, C. C.: ACS Prepr., Div. Petr. Chem. **23**, 303 (1980)
120. Fischer, F., Tropsch, H.: Brennst.-Chem. **7**, 97 (1926)
121. Vannice, M. A.: J. Catal. **37**, 449 (1975)
122. Vannice, M. A.: J. Catal. **50**, 228 (1977)
123. Ponec, V.: Catal. Rev.-Sci. Eng. **18**, 151 (1978)
124. Biloen, P., Sachtler, W. M. H.: Adv. Catal. **30**, 165 (1981)
125. Wentrcek, P. R., Wood, B. J., Wise, H.: J. Catal. **43**, 363 (1976)
126. Araki, M., Ponec, V.: J. Catal. **44**, 439 (1976)
127. McCarty, J. G., Wentrcek, P. R., Wise, H.: ACS Prepr., Div. Petr. Chem. **22**, 1315 (1977)
128. McCarty, J. G., Wise, H.: J. Catal. **57**, 406 (1979)
129. Zagli, A. E., Falconer, J. L., Keenan, C. A.: J. Catal. **56**, 453 (1979)
130. Ho, S. V., Harriott, P.: J. Catal. **64**, 272 (1980)
131. Polizzotti, R. S., Schwarz, J. A., Kugler, E. L.: ACS Prep., Div. Petr. Chem. **23**, 451 (1978)
132. Rabo, J. A., Risch, A. P., Poutsma, M. L.: J. Catal. **53**, 295 (1978)
133. Biloen, P., Helle, J. N., Sachtler, W. M. H.: J. Catal. **58**, 95 (1979)
134. Happel, J., Suzuki, I., Kokayeff, P., Fthenakis, V.: J. Catal. **65**, 59 (1980)
135. Goodman, D. W., Kelley, R. D., Madey, T. E., Yates, J. T., Jr.: ACS Preprints, Div. Petr. Chem. **23**, 446 (1978)
136. Kelley, R. D., Madey, T. E., Renesz, K., Yates, J. T., Jr.: Appl. Surface Sci. **1**, 266 (1978)
137. Goodman, D. W., Kelley, R. D., Madey, T. E., Yates, J. T., Jr.: J. Catal. **63**, 226 (1980)
138. Goodman, D. W., Kelley, R. D., Madey, T. E., White, J. M., J. Catal. **64**, 479 (1980)
139. Goodman, D. W., Kelley, R. D., Madey, T. E., White, J. M.: J. Vac. Sci. Technol. **17**, 143 (1980)
140. Kelley, R. D., Goodman, D. W.: in "The Chemical Physics of Solid Surfaces and Heterogeneous Catalysis", D. A. King, D. P. Woodruff, eds., Vol. 4, Elsevier, Amsterdam, To be published
141. Vannice, M. A.: J. Catal. **44**, 152 (1976)
142. Bhatia, S., Bakshi, N. N., Mathews, J. F.: Can. J. Chem. Eng. **56**, 575 (1978)

143. Bartholomew, C. H., Pannell, R. B., Butler, J. L.: J. Catal. **65**, 335 (1980)
144. Huang, C. P., Richardson, J. T.: J. Catal. **51**, 1 (1978)
145. Palmer, R. L., Vroom, D. A., J. Catal. **50**, 244 (1977)
146. Vannice, M. A., Garten, R. L.: J. Catal. **56**, 236 (1979)
147. Vannice, M. A., Garten, R. L.: J. Catal. **66**, 242 (1980)
148. Bartholomew, C. H., Pannell, R. B., Butler, J. L., Mustard, D. G.: IG & EC Prod. Res. Devel. **20**, 296 (1981)
149. Jacobs, P. A., Nijs, H. H., Poncelet, G.: J. Catal. **64**, 251 (1980)
150. Imamura, H., Wallace, W. E.: J. Catal. **65**, 127 (1980)
151. Tucci, E. R., Streeter, R. C.: Hydrocarbon Proc., p. 107, April, 1980
152. Tauster, S. J., Fung, S. C., Baker, R. T. K., Horsley, J. A.: Science **211**, 1121 (1981)
153. Dry, M. E.: I & EC Prod. Res. & Devel. **15**, 282 (1976)
154. Herington, E. F. G.: Chem. Ind. **65**, 347 (1946)
155. Anderson, R. B.: J. Catal. **55**, 114 (1978)
156. Henrici-Olivé, G., Olivé, S.: Adv. Polymer Sci. **15**, 1 (1974)
157. Kibby, C. L., Kobylinski, T. P.: ACS Ann. Meet., Miami Beach, Fla., September 1978
158. Madon, R. J.: J. Catal. **57**, 183 (1979)
159. Caesar, P. D., Brennan, J. A., Garwood, W. E., Ciric, J.: J. Catal. **56**, 274 (1978)
160. Chang, C. D., Lang, W. H., Silvestri, A. J.: J. Catal. **56**, 268 (1979)
161. Nijs, H. H., Jacobs, P. A., Uytterhoeven, J. B.: J. C. S. Chem. Comm., **1979**, 180
162. Fraenkel, D., Gates, B. C.: J. Am. Chem. Soc. **102**, 2478 (1980)
163. Heal, M. J., Leisegang, E. C., Torrington, R. G.: J. Catal. **51**, 314 (1978)
164. Dwyer, D. J., Somorjai, G. A.: J. Catal. **52**, 291 (1978)
165. Dwyer, D. J., Somorjai, G. A.: J. Catal. **56**, 249 (1979)
166. Krebs, H. J., Bonzel, H. P.: Surface Sci. **88**, 269 (1979)
167. Bonzel, H. P., Krebs, H. J.: Surface Sci. **91**, 499 (1980)
168. Borghard, W. G., Bennett, C. O.: I & EC Prod. Res. Devel. **18**, 18 (1979)
169. Atwood, H. E., Bennett, C. O.: I & EC Proc. Des. Dev. **18**, 163 (1979)
170. Matsumoto, H., Bennett, C. O.: J. Catal. **53**, 331 (1978)
171. Reymond, J. P., Meriaudeau, P., Pommier, B., Bennett, C. O.: J. Catal. **64**, 163 (1980)
172. Amelse, J. A., Butt, J. B., Schwartz, L. H.: J. Phys. Chem. **82**, 558 (1978)
173. Raupp, G. B., Delgass, W. N.: J. Catal. **58**, 361 (1979)
174. Raupp, G. B., Delgass, W. N.: J. Catal. **58**, 348 (1979)
175. Niemantsverdriet, J. W., van der Kraan, A. M., van Dijk, W. L., van der Baan, H. S., J. Phys. Chem. **84**, 3363 (1980)
176. Benzinger, J. B., Madix, R. J.: J. Catal. **65**, 36 (1980); Ibid. **65**, 49 (1980)
177. Dalla Betta, R. A., Shelef, M.: J. Catal. **48**, 111 (1977)
178. King, D. L.: ACS Preprints, Div. Petr. Chem. **23**, 482 (1978)
179. Ekerdt, J. G., Bell, A. T.: J. Catal. **58**, 170 (1979)
180. Ekerdt, J. G., Bell, A. T.: J. Catal. **62**, 19 (1980)
181. Low, G. G., Bell, A. T.: J. Catal. **57**, 397 (1979)
182. King, D. L.: J. Catal. **51**, 386 (1978)
183. Dalla Betta, R. A., Piken, A. G., Shelef, M.: J. Catal. **35**, 54 (1974)
184. Everson, R. C., Woodburn, E. T., Kirk, A. R. M.: J. Catal. **53**, 186 (1978)
185. Elliott, D. J., Lunsford, J. H.: J. Catal. **57**, 11 (1979)
186. Dalla Betta, R. A., Shelef, M.: J. Catal. **49**, 383 (1977)
187. McKee, D. W.: J. Catal. **8**, 240 (1967)
188. Wilson, T. P., J. Catal. **60**, 167 (1979); Shelef, M., Dalla Betta, R. A.: J. Catal. **60**, 169 (1979)
189. Dautzenberg, F. M., Helle, J. N., van Santen, R. A., Verbeek, H.: J. Catal. **50**, 8 (1977)
190. Blanchard, M., Bonnet, R.: Bull. Soc. Chim. France, **1977**, 7
191. Vanhove, D., Makambo, P., Blanchard, M.: J. C. S. Chem. Comm., p. 605 (1979)
192. Sachtler, J. W. A., Kool, J. M., Ponec, V.: J. Catal. **56**, 284 (1979)
193. Ignatiev, A., Matsuyama, T.: J. Catal. **58**, 328 (1979)
194. Bhasin, M. M., Bartley, W. J., Ellgen, P. C., Wilson, T. P.: J. Catal. **54**, 120 (1978)
195. Ichikawa, M.: Bull. Chem. Soc. Japan **51**, 2268 (1978)
196. Ichikawa, M.: J. C. S. Chem. Comm., **1978**, 566

197. Ichikawa, M.: Bull. Chem. Soc. Japan **51**, 2273 (1978)
198. Sexton, B. A., Somorjai, G. A.: J. Catal. **46**, 167 (1977)
199. Castner, D. G., Blackadar, R. L., Somorjai, G. A.: J. Catal. **66**, 257 (1980)
200. Kroeker, R. M., Kaska, W. C., Hansma, P. K.: J. Catal. **61**, 87 (1980)
201. Poutsma, M. L., Elek, L. F., Ibarbia, P. A., Risch, A. P., Rabo, J. A.: J. Catal. **52**, 157 (1978)
202. Vannice, M. A., Garten, R. L., I & EC Prod. Res. Devel. **18**, 186 (1979)
203. Wang, S-Y., Moon, S. H., Vannice, M. A.: J. Catal. **71**, 167 (1981)
204. Vannice, M. A. and Twu, C. C.: In Press
205. Fujimoto, K., Kaneyama, M., Kunugi, T.: J. Catal. **61**, 7 (1980)
206. Vannice, M. A. and Wang, S-Y., J. Phys. Chem. **85**, 2543 (1981)
207. van Barneveld, W. A. A., Ponec, V.: J. Catal. **51**, 426 (1978)
208. Luyten, L. J. M., v. Eck, M., v. Grondelle, J., v. Hoof, J. H. C.: J. Phys. Chem. **82**, 2000 (1978)
209. Bond, G. C., Turnham, B. D.: J. Catal. **45**, 128 (1976)
210. Vannice, M. A., Lam, Y. L., Garten, R. L.: ACS Prepr., Div. Petr. Chem. **23**, 495 (1978)
211. Ott, G. L., Fleisch, T., Delgass, W. N., J. Catal. **60**, 394 (1979)
212. Ott, G. L., Fleisch, T., Delgass, W. N.: J. Catal. **65**, 253 (1980)
213. Coon, V. T., Takeshita, T., Wallace, W. E., Craig, R. S.: J. Phys. Chem. **80**, 1878 (1976)
214. Elattor, A., Wallace, W. E., Craig, R. S.: ACS Prepr., Div. Petr. Chem. **23**, 464 (1978)
215. Luengo, C. A., Cabrera, A. L., MacKay, H. B., Maple, M. B.: J. Catal. **47**, 1 (1977)
216. Atkinson, G. B., Nicks, L. J.: J. Catal. **46**, 417 (1977)
217. Goddard, W. A., Walch, S. P., Rappe, A. K., T. H. Upton, J. Vac. Sci. Technol. **14**, 416 (1977)
218. Kasowski, R. V., Caruthers, E.: Phys. Rev. B. **21**, 3200 (1980)
219. Brady, R. C., Pettit, R.: J. Am. Chem. Soc. **102**, 6181 (1980); Ibid. **103**, 1287 (1981)
220. Sapienza, R., Slegeir, W., O'Hare, T., AIChE 1981 Summer National Meet., Detroit, Paper 39d
221. Blyholder, G., Shihabi, D., Wyatt, W. V., Bartlett, R.: J. Catal. **43**, 122 (1976)
222. Happel, J.: Cat. Rev. **6**, 221 (1972)
223. Wagner, C.: Adv. Catal. **21**, 323 (1970)
224. Newsome, D. S.: Catal. Rev.-Sci. Eng. **21**, 275 (1980)
225. Masuda, M., Miyahara, K.: Bull. Chem. Soc. Japan **47**, 1058 (1974)
226. Masuda, M.: J. Research Inst. Catalysis, Hokkaido Univ. **24**, 83 (1976)
227. Shelef, M., Gandhi, H. S.: I & EC Prod. Res. Devel. **13**, 80 (1974)
228. Taylor, K. C., Sinkevitch, R. M., Klimisch, R. L.: J. Catal. **35**, 34 (1974)
229. Verdonck, J. J., Jacobs, P. A., Uytterhoeven, J. B.: J. C. S. Chem. Comm. **1979**, 181
230. Grenoble, D. C., Esdadt, M. M., Ollis, D. F.: J. Catal. **67**, 90 (1981)

Chapter 4

Chemisorption on Nonmetallic Surfaces

S. R. Morrison

Energy Research Institute
Simon Fraser University
Burnaby, B.C. V5A IS6, Canada

Contents

1. Introduction

A. Bonding in Adsorption

There are many different types of active sites on the surface of the non-metallic solid and there are correspondingly many different types of bonding of an adsorbate atom or molecule to the surface. Covalent bonding, acid-base bonding, bonding based on crystal field effects, hydrogen bonding, and ionosorption are the main categories. Covalent bonding describes the case where an electron pair is shared in a hybrid bonding orbital between an adsorbing atom and the solid, the atom of the solid contributing one unpaired electron, the adsorbing atom contributing the other unpaired electron to form the pair. Acid-base bonding is the case where both electrons in the pair are contributed either by the solid or by the adsorbing atom, and this electron pair is shared in a bonding orbital betweeen the adsorbate and the surface atom of the solid. Crystal field effects are in part electrostatic effects associated with the position of the adsorbate relative to directional bonding orbitals in the solid, although again electrons are shared in bonding orbitals between the solid and the sorbate. Such crystal field effects are most important when the orbitals provided at the solid surface are *d*-orbitals of transition metal ions. Hydrogen bonding describes the sharing of a proton between oxygen on the solid and on the adsorbate. Ionosorption is the formation of an adsorbed ion by transfer of electrons between defects deep in the solid and the adsorbate. Ionosorption is unique to nonmetals, there is no parallel behavior in the case of adsorption on metals.

The concept of covalent bonding between atoms is, of course, very familiar and with a nonmetallic adsorbent the concepts are easily carried over to the adsorbate-solid bond. With metals in the bulk, metallic bonding dominates, involving a sea of electrons in the bulk of the metal. Therefore, a covalent bond between a metal atom and an adsorbate, where the electrons must be localized on a single metal atom in order to form the bond, requires a process described as "induced covalency" of the surface metal atom. Sufficient energy is released by the localization of the electron on the surface metal atom that the bond can form. With nonmetals, induced covalency is observed for semimetals on the one extreme and solids which are moderately ionic on the other. However, adsorbate/solid bonding with a highly covalent character can be expected directly for homopolar materials. Germanium and silicon, for example, automatically can form covalent bonds with many adsorbed species because the bulk material is covalent. In the bulk material one has hybrid orbitals, such that the electrons are localized between the atoms; at the surface such orbitals lead to "dangling bonds", where one of the hybridized orbitals is directed out from the surface. There is no neighboring atom available, and this orbital is readily available for covalent bonding to an adsorbing species. In the terminology of the band model, the energy level associated with the dangling bond is a "surface state". Normally the energy level of the surface state is different from the energy levels of the normal valence electrons. We will discuss this in more detail in Section 4.B.

The transition between a covalent material and an ionic solid is rather abrupt in terms of the Pauling electronegativity difference of the cation and anion, according to Kurtin and his coworkers [1]. Solids with a greater Pauling electronegativity difference than, say, gallium arsenide, act essentially as ionic materials according to their tests. There is evidence [2] that for many compound semiconductors with significantly higher ionicity than gallium arsenide, a strongly covalent adsorbate can "induce" covalency in the solid and form a strong covalent bond. But normally, a highly ionic solid is expected to form primarily acid/base bonds with an adsorbate. A strongly ionic solid will have essentially no one-electron dangling bonds available at its surface. Thus it will not tend to form covalent bonds with an adsorbing atom. However, if the occupied or unoccupied orbitals of the ionic solid are at a suitable energy level relative to the levels on the adsorbate, they can be available for sharing of electron pairs with the adsorbing species. A "suitable energy" would be a low energy unoccupied orbital or a high energy occupied orbital. CsCl would not have suitable energy levels, but Cs_2S would have high energy sulfur-like occupied orbitals and $AsCl_3$ would have low energy arsenic-like unoccupied orbitals, each available for electron pair sharing.

Consider the acid/base behavior of a series of oxides. If the cation of the oxide is very strongly electropositive, then the oxygen will have a very high negative partial charge associated with it. This highly negative charged oxygen will share its electron pair with an adsorbing species, such as a proton, that has an unoccupied orbital available. On the other hand, if the cation of the oxide has a very high electronegativity, as, for example, is the case with the cation phosphorus, the solid will have an acidic character. It will be able to accept and share an electron pair from an adsorption base such as OH^- or NH_3. As will be discussed in Section 3, there are two types of centers, one at which electron pairs are shared, the other at which protons or hydroxide groups are shared. The two are closely related. Such acid/base interaction in adsorption may be one of the more important features of nonmetals in catalysis because such interactions are not normally found on metals and adsorption of organic molecules at such sites often leads to adsorbed ions. For example, basic sites can remove an H^+ from an adsorbing organic molecule. This extraction can leave a negatively charged fragment weakly adsorbed and reactive.

Adsorption associated with crystal field or liquid field bonding is normally the adsorption of charged or polarizable ions that bond to *d*-orbitals of a transition metal ion at the surface of the solid. The features of this type of bonding are very similar to the features of bonding directly to transition metals in an elemental metallic form.

Ionosorption occurs on semiconductors that have bulk defects that either contribute electrons to a high energy level conduction band or contribute holes (unoccupied valence orbitals) to a low energy valence band. Such solid if perfect would have no such conduction band electrons to offer to an electronegative adsorbate nor would they have holes at a sufficiently low energy to accept an electron from an electropositive adsorbate. For example, ionosorption is the adsorption of a highly electronegative ion on the surface of an imperfect solid, which, were it perfect, would have no unpaired electrons at a suitable energy to share with the adsorbing electronegative species. If electrons from deep within the solid that are associated with imperfections can move to the surface and be localized at the orbital on the electronegative adsorbate, ionosorption will occur. Thus an electron from an imperfection deep in the crystal (say 1000 Å) can be captured on (reduce) gaseous oxygen to O_2^-. Correspondingly, an electropositive ion can be ionosorbed if it can give up its electron to an energy level deep in the bulk of the solid. Because an integral number of electrons are transferred in ideal ionosorption, the adsorbate has an integral number of charges.

In this introductory section we have introduced some of the ways in which atoms can be bonded to the surface of a nonmetallic solid. We have not, and will not describe the complexities of bonding in detail — for example, we have not mentioned hydrogen bonding — for a detailed discussion of bonding is beyond the scope of this Chapter. The types of bonding have been introduced in a very qualitative way, sufficient to provide a background for our discussion in the rest of Section 1 of the general processes of adsorption, including a discussion of the activation energy and heat of adsorption. In Sections B, C, and D we concentrate on adsorption with local bonding only, where the band model is not necessary to describe the process. In those three sections respectively, we will go into much more detail regarding that most important adsorbate, water, regarding the behavior of acidic and basic sites, and regarding covalent bonding processes. Then, in Section D and E, we will look at cases in the opposite extreme, where the band model dominates the adsorption and provides the only reasonable description of the adsorbed species. Thus, Section D covers cases where a description somewhere between the localized interaction model and the nonlocalized band model would be desirable to describe the system. We will attempt to show, however, that the two descriptions provide essentially the same overall picture. Finally, in the last section, we will discuss the effect of adsorption on the properties of the solid, emphasizing effects that lead to techniques for studying and understanding adsorption on nonmetals.

B. Activation Energy and Heat of Adsorption

Energy must be supplied to a molecule in general before chemisorption can occur, and this energy is termed the activation energy. As a step in the adsorption of a molecule from the gas phase, normally it becomes dissociated, and the energy of dissociation can provide a major part of the activation energy. If the molecule were required to dissociate before adsorption, this would mean a very high activation energy for the process. Even when it dissociates during the adsorption process, substantial energy input is usually necessary to induce the dissociation step and permit the adsorption to proceed. Other activated steps in addition to dissociation may also be required. Of course, this energy of activation is recovered when the strong bonds form between the fragments of the adsorbate and the solid, but the initial input of energy must be provided. Lennard-Jones [3] suggested a convenient way of viewing the energies involved in the adsorption process when dissociation is the main source of activation energy, and this model is shown in Figure 1. If we make the assumption that the only independent variable describing the energy of the system is the distance from the surface, r, then two curves as shown in Figure 1 can be used to describe the system. Curve (a) shows the energy of the molecule as it approaches the surface, and curve (b) shows the energy of the two fragments as they approach the surface. The energy difference at infinite r is the dissociation energy of the molecule. As the molecule approaches the surface, we assume a small attraction of the Van der Waals type, followed at smaller r by replusion, leading to the energy/distance curve of the form of curve (a). As the two atoms approach the surface, strong chemical bonds form, so the energy of the system decreases substantially at the equilibrium bonding distance. Closer than this distance, the repulsive forces again dominate. It is assumed in this model that the equilibrium distance of the chemisorbed species is much smaller than the equilibrium distance of the physisorbed molecule. Now, according to this model, if the molecule approaching the surface is provided the energy E_A, it can reach the distance indicated by r', where the energy of the molecule and the two fragments is equal. Then the molecule can dissociate with no further energy input. The energy E_A is termed the "activation energy", the energy Q is the "heat of chemisorption" for the molecule.

In the ideal case the heat of adsorption and the activation energy of adsorption would be independent of surface coverage. An approach to such an ideal case, for sub-monolayer coverage, is found for physical adsorption. In some rare cases, an approach to such an ideal case is found for chemisorption on metals. When Q and E_A are independent of coverage, simple adsorption isotherms, such as the Langmuir isotherm, that relates amount adsorbed, Γ, with pressure, p, are derived. In the Langmuir derivation, the rate at which molecules are adsorbed is set proportional to pressure and to the fraction of the surface that is unoccupied, $\Gamma_o - \Gamma$. (rate $= k_1 p(\Gamma_o - \Gamma)$). The rate of desorption is proportional to coverage (rate $= k_2\Gamma$), and equating the rates at steady state, we find:

$$\Gamma = \Gamma_o p/(p + b^{-1}) \tag{1}$$

where

$$b = k_1/k_2 = A \exp(-E_A/kT) [\exp(-\{E_A + Q\})/kT]$$

Here k_1 is the rate constant for adsorption, with activation energy E_A, and k_2 is the rate constant for desorption, with activation energy $(E_A + Q)$, as indicated in Figure 1. The Langmuir equation is reproduced here, where we are interested in chemisorption, as a very idealized case: for cases where E_A and Q vary with coverage Γ the expression can become extremely complex.

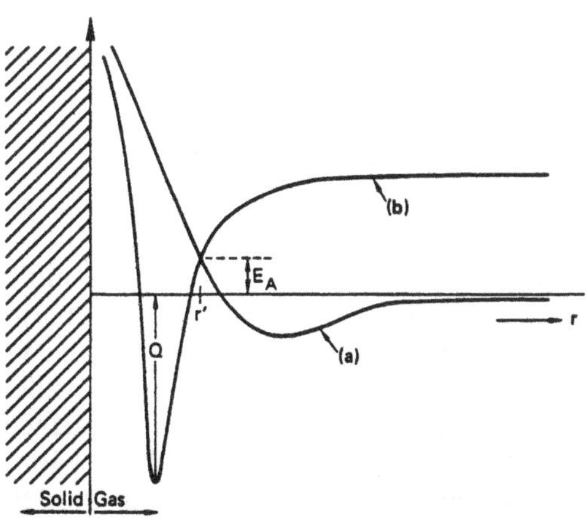

Figure 1. The energy of a system versus the distance of the adsorbate from the surface. Curve (a): Adsorption as a molecule; Curve (b): Adsorption as two atoms

On non-metallic adsorbates, the observed heat of adsorption decreases rapidly with coverage; the observed activation energy of chemisorption increases rapidly with coverage. There are two dominating reasons for these variations. The first is heterogeneity, the second is the development of surface double layers. As will be discussed in more detail in Section 1.C, heterogeneity in adsorption sites originates from many causes: surface sites associated with different crystal faces that are exposed, surface sites associated with surface steps (the edge of one or more planes that extend only partially across a crystal face), sites associated with kinks in changes in direction of surface steps, sites associated with dislocations or grain boundaries intersecting with the surface, or sites associated with impurities or adsorbed species deposited on the surface. The sites associated with such flaws will vary in their bonding energy. This Q will vary. The reason that Q decreases with coverage on a heterogeneous surface follows logically. During the adsorption process, one expects sites with the highest Q to adsorb preferentially, sites with the lowest Q to adsorb last, so the heat of adsorption Q will appear to decrease with coverage. The reason that the activation energy E_A increases with coverage on a heterogeneous surface, is not so clear. An examination of Figure 1 permits the observation to be rationalized, for in Figure 1 we observe that as Q decreases, (curve (b) is raised), E_A will increase. But Figure 1 is not based on a detailed model, so rationalization by Figure 1 is not entirely satisfying.

The other dominant reason for the variation of E_A and Q, particularly important with nonmetallic adsorbents, is the development of a double layer. In most cases of adsorption there is a partial charge located on the adsorbate resulting in a plane of charge, countered by the opposite charge on the surface plane of the adsorbent. Although the approximation is poor because the lateral distance between charges compares to, and many even be larger than, the distance between the charged planes, it is instructive to consider the configuration as a parallel plate capacitor, with the voltage V between the planes given by:

$$V = \delta q N_t / C = \delta q N x_o / \varepsilon\varepsilon_o \tag{3}$$

with q the electronic charge, N_t the density of adsorbate atoms, δ the partial charge on the adsorbate, and C the capacity per unit area. Here, as indicated by the second equality of equation (3), C varies with the dielectric constant ε, the permittitivity of free space ε_o, and varies inversely with the distance between the planes, x_o. Now for local bonding, if we assume a partial charge of, say, 0.3, a dielectric constant of unity, an adsorbate density of 5×10^{18} m^{-2}, and a value for x_o of 3×10^{-10} m, we find from equation (4) that a huge double layer voltage drop, namely 8 volts, is predicted. This is clearly much too high — it would represent, as discussed below, a decrease in Q of 8 eV. There are two dominant errors, first when we idealize the system to a parallel plate capacity and second when we ignore the fact that movement of surface species will lead to an effective distance x_o of less than 3×10^{-10} m. More realistic models, with x_o much less and some improvement in the model, would, if they could be made quantitative, combine to bring the estimated 8 eV down to a reasonable value.

However, at least we can indicate the direction that Q and E change with coverage, even without knowing the exact amount of the change. If the adsorption of the first atom involves an energy Q_o, the simple model indicates that as the coverage increases and V in equation (3) increases, the movement of charge to the adsorbate, being repelled by the field, will require an energy $V\delta$ eV. Then the heat of adsorption will decrease with coverage. In the idealized case, it will become:

$$Q = Q_o - V\delta \qquad \text{(in electron volts)} \tag{4}$$

Thus Q will decrease rapidly with coverage.
On the other hand, the activation energy will increase because the movement of the charge requires this extra energy $V\delta$ before the chemical bonding takes place:

$$E_A = E_A^o + V\delta \qquad \text{(in electron volts)} \tag{5}$$

Thus the double layer mechanism does offer one case where Q decreases and E_A increases with coverage.

The case of ionosorption, wherein charge is transferred from ionized centers deep in the crystal to the adsorbate at the surface, is a case where double layer effects are particularly dominant. This case will be discussed in detail in Section 5. It will be shown that because of double layer effects the surface coverage can only reach a value of the order of 10^{16} m^{-2} before Q goes to zero and adsorption must stop. And adsorption may cease with much less than this equilibrium coverage, especially at low temperature, because E_A becomes too high.

C. Surface Heterogeneity

There are many site configurations on the surface of the solid, and the heat of adsorption with local bonding will depend on the site of the adsorption. As described in the last section one expects the most attractive sites to be

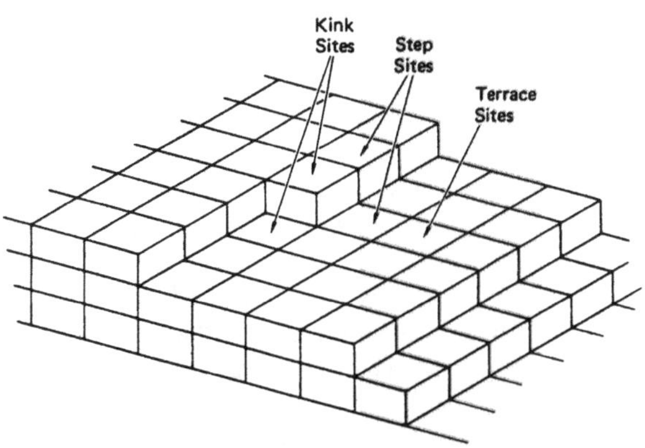

Figure 2. Adsorption sites of various activity on a heterogeneous surface.

occupied by adsorbates first, the least attractive sites to be occupied last, and so the heat of adsorption will decrease with coverage due to the heterogeneity of the sites available. In the present section we will describe the general classifications of surface sites that lead to heterogeneity, to provide an understanding of the spectrum of sites available in adsorption. In subsequent sections, however, we will not emphasize the surface heterogeneity, for although heterogeneity clearly is important in adsorption, it is difficult to develop quantitative theories.

One of the most important types of active site on the otherwise uniform surface is the surface step. When a plane of atoms covers only part of the surface, the edge of this plane is the step, the flat region is a "terrace". A step is shown in Figure 2. Usually, when a crystal is growing, the new atoms are deposited at a step and the step moves across the surface. When a crystal is being etched, usually the atoms are removed from a step, and the step moves across the surface as the extra half plane disappears. Steps are generated at dislocations, or grain boundaries, as discussed below, or are generated at a corner where two surface planes meet. Well grown crystals may have few surface steps, but a badly deformed, multi-phase catalyst may have many. An important class of experiment has been the preparation of crystals with many steps (and few other defects), permitting the study of the role of steps in the adsorption and catalytic properties of the solid. The preparation of a stepped crystal [4] involves cutting the crystal a few degrees off a major (low index) plane, and annealing while observing the results by LEED (low energy electron diffraction). By a certain LEED pattern, the development of regular surface steps is observed. Then, the enhanced adsorption associated with these steps can be (and has been) [5] observed. Most of the work to date has been done with metal adsorbents, but undoubtedly many of the features apply to nonmetal adsorbents. Steps provide two forms of attractive adsorption sites, first sites where an adsorbate can coordinate to many substrate atoms, and second sites where a host atom is very poorly coordinated and so bonds readily with the adsorbate.

A turn in the direction of the surface step is called a kink, and kink sites are especially active adsorption sites. An inside corner on a step results in a location of particularly high coordination for the adsorbing species; an outside corner results in an atom of the host solid that is extremely poorly coordinated and will often provide a high heat of adsorption, interacting strongly with an adsorbing molecule.

Another form of surface imperfections is the intersection of the surface with a dislocation pipe. A dislocation pipe is a line at the edge of an extra half plane inserted into the bulk of the crystal. An atom where the dislocation pipe meets the surface is poorly coordinated. However, there are expected to be few dislocations intersecting the surface (between circa $10^8\ m^{-2}$ for a well-grown crystal and circa $10^{15}\ m^{-2}$ for a badly damaged crystal). So on the average, these may provide 10^{-4} monolayers of adsorption sites. Their major importance in surface properties is in etching, crystal growth, and surface step generation, not adsorption.

Other forms of flaws that lead to heterogeneity are a multiplicity of crystal faces exposed, impurities segregated or deposited on the surface, and bulk defects (*e.g.*, non-stoichiometry) that can concentrate at the surface. The importance of bulk defects that are mobile at the ambient temperature will be noted when discussing ionosorption in Section 5.

Heterogeneity not only results in a varying heat of adsorption, but affects the activation energy of adsorption. One of the most striking examples of this effect is the observation that molecules can adsorb, with a low heat of adsorption, at a defect, and then diffuse over the surface to sites where otherwise the activation energy would be too high for adsorption. Two examples are common. One is the surface that is heterogeneous because impurities, such as platinum, have been deposited. There has been considerable interest [6, 7] in "hydrogen spillover" or "oxygen spillower" where the adsorbed molecule is dissociated and adsorbed, on, say, platinum, and then spills over onto the nonmetal surface. Another example, so far noted primarily with experiments on metals, is adsorption on a step, followed by diffusion onto a terrace, leaving the active step site available for further adsorption.

The geometry, the spacing of atoms on the surface during adsorption is of interest and is related to heterogeneity. In many cases on an adsorbate-free surface the crystal spacing does not carry through to the surface atoms — the surface atoms are displaced, either parallel or perpendicular to the surface. An extreme example of this is the case of polar surfaces, where a layer of charge would be exposed if there were no relocation. In this case, on the clean surface, the atoms almost always are found (by LEED) to be displaced. Displacement normal to the surface is termed "relaxation", displacement in an ordered manner to form a periodic structure in the surface plane (detectable by LEED) is called "reconstruction", and a general term that does not specify the direction of the displacement is "relocation". Relocation often occurs even when the surface of a nonmetal is adsorbate-free, and that pattern relocates again (sometimes back to the original crystal spacing) upon adsorption of a species. The adsorbate itself may show a regular structure — for example, it may occupy every second site until the surface is half covered, and only then, with a lower heat of adsorption, it may begin to fill in the rest of the monolayer. This, of course, is primarily to be expected when there is some repulsive force between the adsorbate atoms.

Relocation of the surface atoms of the host crystal during adsorption is not always toward the regular bulk atom positions. Clearly major relocation can occur representing the formation of an incipient new phase. This case becomes a grey area between adsorption and a gas/solid reaction.

Another form of heterogeneity in adsorption is adsorption in "islands", where the attraction between the adsorbate atoms in high compared to their attraction to the substrate. Such island formation means that the most active sites for further adsorption are at the circumference of the islands. Island formation is more common in physisorption than in the chemisorption under discussion here.

D. Desorption

As indicated in Figure 1, the activation energy for desorption, E_D, is related by:

$$E_D = E_A + Q \tag{6}$$

Such a relation holds by the Lennard-Jones model, Figure 1, and, as will be derived later, holds for the simplest electrostatic model of ionosorption. The relation cannot be proved for cases where Q and E_A vary because of surface heterogeneity, but it seems from experimental observation to be a common rule that the activation energies and the heat of adsorption are connected by a relation similar to that of equation (6).

The origin of the relationship is, as noted in Section 1.B, that E_A increases and Q decreases as the surface coverage of the adsorbate increases, and thus, to a first approximation, E_D is independent of coverage. And experimentally, for many cases of adsorption, E_D does seem to be independent of coverage.

A technique termed Temperature Programmed Desorption, TPD, takes advantage of the observed independence of the activation energy of desorption on the coverage. Specifically, after a gas is adsorbed on the surface, the temperatury is slowly raised and the desorption of the gas from the surface is noted as a function of time (or temperature if the temperature is raised linearly with time). If the gas desorbs as well defined peaks (in the pressure/time, or equivalently, the pressure/temperature plot), well defined energies of desorption are indicated.

The success of the TPD experiments not only provides an indication that equation (6) is valid, but provides an extremely valuable tool for the study of the various bonding modes of the adsorbed species.

The peaks of the TPD measurements can be related to various bonding mechanisms and various surface reactions, especially when one combines the technique with other methods of detecting the desorbed particle and the properties of the adsorbate that are lost as the peak is passed.

The various surface reactions that can occur before an adsorbate is desorbed and are noted especially in TPD experiments, include dissociation, dimerization, reaction with another adsorbate, and reaction with the solid. The first three are equally common with metal substrates, the last is seldom observed with metal substrates. Such sorbate/solid reactions where the desorbing species includes an atom or more of the host solid is particularly important in catalysis. This, for example, consider the case where an organic molecule is adsorbed on an oxide that is an oxidation catalyst. When the organic molecule desorbs it may be oxidized; a lattice oxygen atom is attached to the molecule. Then, in a later step in the catalytic reaction, this oxygen atom that was lost from the solid is restored by adsorption of gaseous oxygen. Such an exchange of atoms, as well as of electrons, during the adsorption desorption processes of catalysis is rarely, if at all, observed with metal adsorbates, but is very common with very active nonmetal catalysts. This case is discussed further in Section 5.B.

2. The Adsorption of Water

On ionic nonmetals, such as oxides, the adsorption of water is a dominant factor in the surface characteristics, both with respect to adsorption of other species and to catalysis. As will be described in the next subsection, the adsorption of water can change the acid or base sites on the surface from Lewis to Brønsted sites. The adsorption of water can passivate the surface with respect to the adsorption of many other species, and, indeed, many of the catalyst activation processes required in catalytic chemistry are needed, to a large extent, to remove a certain fraction of the adsorbed water from the surface. The adsorption of water also has an effect on the electronic properties of semiconducting nonmetals, usually acting as a donor.

A hydroxylated surface is formed on an oxide by the chemisorption of a monolayer of water, the surface appearing "hydroxylated" because the OH^- ions tend to associate with the cations of the solid, the H^+ ions tend to associate with the lattice oxygen ions, leading to a hydroxylated configuration as illustrated by Figure 3. The surface appears to be covered by OH^- ions. Evidence that such a simple model is realistic arises in the work of Boehm [8], for example, who exchanged the adsorbed protons for sodium ions on hydroxylated TiO_2, and found two distinct bonding energies for the protons. One bonding energy could be associated with the protons bonded to the lattice oxygen, the other bonding energy could be associated with the proton associated with the adsorbed OH^-.

Hydroxylation as illustrated in Figure 3 is, in a sense, an intermediate stage in the interaction of water with the oxide. It is intermediate between hydration of the surface and physical adsorption of water. Long exposure can lead to hydration of the surface layer. With silica, for example, at room temperature, exposure to water vapor sufficient to hydroxylate the surface for the order of 100 years will lead [9] to hydration to the depth of the order of 0.5 μm. The hydration of silica follows a parabolic law, and only a few

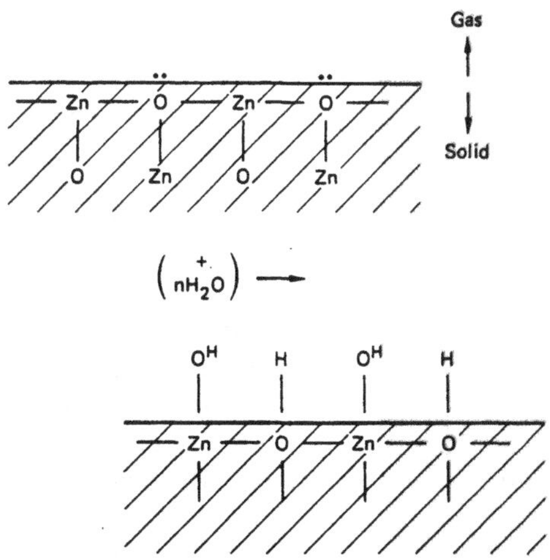

Figure 3. Idealized hydroxylation of a ZnO surface. On the water-free surface the surface oxygen formally has an unshared electron pair, the surface zinc formally has an unoccupied orbital available to share an electron pair. These sites are occupied by H_2O with strong acid/base bond formation

years is required to hydrate the surface to 0.1 μm. On the other hand, physical adsorption over the hydroxylated layer will occur immediately upon exposure of the surface to moderate pressure of water vapor. In general, at room temperature the physisorption of water is reversible, the stronger interactions irreversible.

Hydration of silica is conveniently observed by measurements on the thermally grown silica layer on silicon. In measurements of capacity of a silica layer, made by depositing a metal layer over the silica, one can detect easily the introduction of positive ions (*e.g.*, protons) into the silica. Singh et al. [10] observed that heating the SiO_2 in a 95% humidity bath for 1 hour at 313 K was sufficient to provide a source of protons for diffusion into the silica. If 150 Å was etched from the silica, however, the source of protons was removed. Thus, the hydrous oxide layer appears to be less than 150 Å thick following such a treatment. Gammage et al. [11], studying thorium oxide, observed for this case, the slow conversion of a physically adsorbed layer of water (adsorbed over the hydroxylated surface) into a hydrated surface. In general, however, hydration is difficult to study and is not well documented.

The chemisorption of water leading to hydroxylation, on the other hand, has been studied on many oxides, primarily using the techniques of infrared absorption and temperature programmed desorption (TPD), as described in Section 1.D. Using these techniques, various workers have found two dominant forms of chemisorbed water, one with the oxygen ion coordinatively bonded to the cation, and a second with the water held to the surface by hydrogen bonding to an oxygen ion. Physisorbed water, with its weak bonding, is also detected. The bonding for the three cases is detected by the observation of infrared bands, and the three dominant modes of bonding are identified further by TPD measurements.

For example, the adsorbed water an TiO_2 held by coordination bonds is found [12, 13] to desorb at 623–673 K; at about 473 K hydrogen bonded water comes off; physisorbed water desorbs below 373 K. The physisorbed water has a heat of desorption of about 35 kJ mol^{-1}, the hydrogen bonded water about 60 kJ mol^{-1}, and the coordinatively bonded water about 105 kJ mol^{-1}. Above 773 K there are still hydroxyl groups remaining on the TiO_2, however. In fact, Marimoto and Naono [14] show gross amounts of water still on the TiO_2 up to 873 K.

The protons in hydroxyl ions can be displaced by ion exchange in solution. If the hydroxylated surface is ion-exchanged with fluoride ions, in other words, fluoride ions replace many of the OH^- groups on the surface, the behavior changes. Such fluoride treatment, as will be discussed below, makes the surface more acidic. With a fluoride treatment to the TiO_2 surface, the hydroxyl desorption occurs at about 723 K [15], a much lower tυmperature than the 773–823 K required to complete the hydroxyl desorption for the non-fluorided surface. Thus, above 723 K on the fluorided surface, only Lewis acid centers remain (see discussion of Lewis acids and Brønsted acids below).

Other oxides have analogous behavior patterns. For example, silica retains residual hydroxyl groups [14] to temperature exceeding 1270 K. An interesting feature observed with silica is the transition from reversible adsorption to irreversible adsorption. If the silica is heated above about 723 K, the sites for water adsorption are lost, presumably because reconstruction of the surface. Movement of the lattice ions to new positions at the high temperature is such as to bring oxygen ligands to the active coordination sites of the cations. The surface structure is assumed to move toward a siloxane structure. Prolonged exposure of the silica to liquid water will lead to reactivation of the surface for water adsorption, changing the siloxane structure to a silanol structure (hydroxylated surface).

An interesting feature of a partially hydroxylated surface, observed in particular with silica [16] is that the hydroxyls tend to be present in pairs. After drying the SiO_2 at 873 K, more than 85% of the protons were still present as pairs. The presence of pairs was detected by Peri and Hensley [16] by the interaction of the partially hydroxylated surface with $AlCl_3$. With two adjacent protons, two molecules of HCl would be evolved, so by measuring the number of HCl molecules generated per $AlCl_3$ adsorbed, the number of pairs on the surface, where the H atoms are close enough to react with the same $AlCl_3$ molecule, could be determined. Unpaired hydroxyls and paired hydroxyls can also be distinguished by infrared absorption. Primet et al. [17] concluded the stretch frequency of isolated OH^- is 3685 cm^{-1}, that for paired is 3410 cm^{-1}.

On ZnO one obtains TPD peaks of water at 493 K and 543 K. Morishige et al. [18] conclude from infrared measurements that these represent hydroxyl desorption with the high energy peak due to pairs. Again, the surface is still substantially hydroxylated at temperatures beyond the peaks, and indeed, up to about 773 K.

There are, in addition, several special modes of bonding on a hydroxylated surface, variations on the coordination bond and the hydrogen bond described above. Water will normally "dissociatively" chemisorb on adjacent cation-anion pairs as observed in Figure 1, but as the surface becomes covered, more and more lone cationic or anionic sites remain, and water molecules can chemisorb non-dissociatively, coordinatively bonding to a lone cation, or hydrogen-bonding to a lone oxygen ion. The dissociatively adsorbed hydroxyl ions behave differently if there is another hydroxyl ion as a neighbor, for then hydrogen bonding to the neighbor will increase the bonding energy and increase the bond energy. In addition, such pairs of hydroxyl groups provide more activity; hydrolysis of silanes, for example is facilitated by the presence of a hydroxyl network on a silica catalyst. Finally, Griffiths and Rochester [19] conclude the OH^- coordinatively bonded to cations can be isolated or bridged, bonding to two cations.

In some cases, the presence of hydroxyl groups on an oxide catalyst accelerates catalytic reactions, at other times, it retards reactions. On ZnO Nagao and Morimoto [20] find gaseous bases (*e.g.*, NH_3) adsorb on dehydroxylated sites, whereas on MgO and SiO_2 the surface hydroxyls act as the adsorption sites. The surface hydroxyls on the silica, the silanol groups, are acidic, but on ZnO, the bare surface is more acidic.

Electronically, water adsorption leads to electron injection on many semiconductors. However, it is difficult to believe that on most ionic semiconductors OH radicals could form by electron injection from OH^- on the surface. It has been suggested [21] that the polar water molecule interacts with charged acceptor surface states on the surface, raising the energy level of the acceptor surface states until they inject electrons. A particularly interesting possibility would be acceptor states that are the same sites as strong Lewis acid sites [22]. With low hydroxyl coverage, these sites could accept electrons from the semicoductor, but when water is adsorbed, bonding to the acid sites, the electron is reinjected.

Ionosorption, adsorption with charge transfer to the bulk of the semiconductor as described in the introduction, does, in some cases, depend on the presence of water. For example, Boonstra and Mutsaers [23] observed that the ionosorption of oxygen on TiO_2 increases proportionally to the concentration of hydroxyls on the surface of the semiconductor.

Thus, with respect to ionosorption, water has a strong but poorly understood effect; namely its action as a donor, and another poorly explored effect; namely its effect on the ionosorption of other gases,

3. Adsorption on Acidic and Basic Sites

As discussed in the Introduction, acid/base activity relates to the tendency of a nonmetal to share an electron pair or to share a proton or hydroxide ion with an adsorbate. There are two forms of acid/base sites, Lewis and Brønsted. A Lewis acid sites is a site on the solid where an unoccupied orbital is present at a low energy, such that it is energetically favorable for it to share an electron pair offered by an adsorbing gaseous base such as ammonia. A Lewis base site is a site where a high energy orbital occupied by two unshared electrons is present and able to share these electrons with an adsorbing gaseous acid such as phenol. A Brønsted acid site is a site with a weakly held proton attached that either can share this proton with an adsorbing gaseous base or can pair this proton with an H^- ion from an adsorbing organic molecule. A Brønsted base site is a site with a weakly held OH^- group attached that can share this group with an adsorbing acid. Lewis sites and Brønsted site are often, but not always, interchanged by the adsorption or desorption of water:

$$L_s^+ + H_2O = L_s : OH + H^+ \tag{7}$$

Here an acidic Lewis site L_s^+ interacts strongly with the OH^- group of the water, leaving the H^+, weakly bonded, elsewhere on the surface acting as a Brønsted acid. Similarly, a surface with strong Lewis base sites can interact with the proton from a water molecule, leaving the hydroxyl group, loosely adsorbed, to provide Brønsted base activity.

Depending on the strength of the Lewis acid and base sites, either the OH^-

group on the lattice cation or the H^+ ion attached to the lattice anion will be bonded more strongly. If the H^+ ion is bonded more energetically (to strong Lewis base sites) the surface will show Brønsted basicity because the residual OH^- groups can be given up more easily than the protons. If the OH^- group is bonded more energetically, the surface will show Brønsted acidity. Thus, a Lewis acid is transformed to a Brønsted acid, a Lewis base to a Brønsted base.

With this general relation between Lewis and Brønsted activity in mind, we can easily reconcile the general behavior of acidity and basicity with respect to the adsorption of water. The sites are most active usually after outgassing at an intermediate temperature. If a sample is fully hydroxylated and has a layer of hydrogen-bonded, or even physisorbed water on top of the hydroxylated surface, then the Brønsted activity will be weak. The active sites will be occupied by adsorbed undissociated water and not be available to adsorb other species. As the sample is heated to remove the undissociated water, the Brønsted sites will become active. The Brønsted activity will go through a maximum with increasing outgassing temperature and decline as the surface approaches complete dehydroxylation. The Lewis activity will begin as the sample begins to be dehydroxylated, and is expected to increase as the percentage of dehydroxylated sites increases. An interesting example of this behavior, the nickel silicate adsorbent, is described by Sohn and Ozaki [24]. At very high temperature, even the Lewis activity may go through a maximum. If the sample is thoroughly outgassed of water, Lewis sites are the only possible sites. However, if, as in the case of silica, the dehydroxylated surface atoms relocate at the high temperature to a lower energy configuration, the surface becomes passive due to the relocation. Then the Lewis acid and Lewis base sites will be weaker (lower energy acid or base strength and lower bonding energy for the adsorbate) than Lewis sites that are produced by partial dehydration of the surface, where the temperature is too low for relocation.

Adsorption with only acid/base interaction often shows a negligible activation energy. This is one of the few cases, other than physisorption, where E_A is low and independent of coverage. This characteristic arises in many cases, presumably because the adsorbing molecule does not have to dissociate to form the strong bond. On the other hand, the heat of adsorption (the "acid strength" or "base strength" of the sites) can be high, so desorption can require a very high temperature.

The acid strength or the base strength of sites is measured in general in two ways. One is by "Hammett indicators", which are essentially normal acid/base indicators. Indicators are chosen that are uncolored or lightly colored when in solution, but become colored when adsorbed on the strong Lewis or Brønsted site. By studying indicators of various pK_a, and noting the pK_a required for adsorption on the sites, the acid or base strength of the sites can be estimated. The other dominant technique is by TPD, temperature programmed desorption. When examining acid sites, a gaseous base (*e.g.*, NH_3) is adsorbed, usually at room temperature, and then the temperature is increased with time, and, as discussed in Section 1.D, the evolution of the gas monitored as a function of temperature. Similarly, for the characterization of basic sites a gaseous acid, *e.g.*, phenol, is monitored in a TPD measurement.

Almost inevitably, there is a wide spread in the acid of basic strength of the sites on a solid. The measurement of TPD was described earlier as a technique to obtain the heat of desorption by observing peaks in the pressure as the temperature was raised. The observation of such a peak permits the measurement of the heat of desorption. With TPD measurements of adsorption on acid or basic sites, usually there is negligible interpretable structure in the desorption/temperature curve. In other words, there is no single heat of desorption, but a continuous spectrum of heats of desorption. Such a spread in the heats of desorption implies a very heterogeneous surface with respect to the energy of the bonding sites. Such heterogeneity is understandable when one considers the probably origin of the sites, especially on solids that are of interest for their acidic or basic character (and thus are most likely to be studied). Solids that are studied for their acidic or basic properties are prepared in such a way they will be imperfect.

Much more active sites will appear on a surface with surface steps, kinks in surface steps, or rough surfaces such that many surface planes are exposed than will appear on terraces. At the corners that are numerous on such materials, an O^{2-} or an M^{n+} ion (M is the cation), may be very poorly coordinated indeed. Instead of having perhaps only one oppositely charged ion missing from its coordination shell, as will often be the case on the uniform surface, it may have two or three such ions missing. The lack of neighboring counterions has a strong effect on the potential at the site. The effect can be estimated quantitatively by the Madelung approach discussed in a later section, but it is clear qualitatively that if we remove many positive ions from around a surface O^{2-} (by placing it at a kink, say) the potential at the O^{2-} will become very negative. Then the O^{2-} will react very strongly indeed with an adsorbing gaseous acid — it will be a strong basic site. Now many different geometries can be visualized for poorly coordinated surface ions. Thus, it is not unexpected that a wide range of desorption energies will be found on solids that owe their strong acid or basic nature to this type of heterogeneity. Naturally, because these solids are the most active (compared to solids with few defects) these are the ones are usually studied.

Active acidic or basic adsorption sites can also be induced by deposited impurities. For example, the ion exchange of fluoride ions for hydroxyl ions on an oxide surface leads to very strong acid sites. This behaviour has been explained by an induction mode. The fluoride ion is much more electronegative than the OH^-, and so the adjacent cations are much more positively charged than they would be with the original adsorbed OH. Thus, the adjacent cations become much stronger Lewis acid centers (more attractive to electron pairs from an adsorbing donor molecule), or by the mechanism discussed in connection with equation (7), the surface shows a stronger Brønsted activity.

More quantitative quantum mechanical (cluster model) analysis of the effect of such ions on the acid or basic character of the sites have also been made [25].

As a final mechanism for strong acid centers, and the one most used, we have mixed oxides. It is particularly effective in obtaining strong sites to prepare a mixed oxide where two cations M_1 and M_2 have different coordination numbers and perhaps different formal valences. The cation M_1 can find itself in a surface site where M_2 should be, and be very poorly coordinated compared to an M_1 on the surface of a pure M_1O oxide. Thus, a highly positive charge can develop at such a site, lowering the energy level of the electronic orbitals. Then one will have a strong acid center, attractive to electrons. Silica/alumina acid catalysts are the classic case. Tanabe and his coworkers [26] have developed a theory for the expected acidity of mixed oxides of various coordination number and charge for the cation.

4. Adsorption by Local Bonding: Surface States

A. Comparison with Adsorption on Metals

As described previously the covalent adsorption of gases on metals can be discussed primarly under two headings (see Section 1.A): first an induced covalent bond, and second, crystal field effects, the latter most common with transition metals. The induced covalent bond on metal is similar to the incipient formation of another phase — the adsorption of oxygen on a metal surface, for example, localizes an electron to form a metal-oxygen bond. In the case of the transition metals, there are dangling orbitals that do not need to be "induced", so covalent bonding with ligand field effects will occur directly. Such bonding can be fairly complicated. For example, bonds are formed with electrons from the metal shared in unoccupied orbitals of the adsorbate, and simultaneously electrons from the adsorbate shared by unoccupied *d*-orbitals of the surface metal atom.

Such bonding mechanisms can also occur with nonmetallic solids, with suitable modification in concept. For example, transition metal oxides will still be able to bond adsorbates to the localized *d*-orbitals in a way similar to the transition metals. In this chapter we will not emphasize covalent bonding that is so similar to metal-adsorbate bonding, relying for this on discussions in chapters dealing with adsorption on metals. We will, rather, emphasize deviations from the metal-like covalent bonding that arise because of the presence of the bandgap in the nonmetals, and arise because of the unique characteristics available with mixed valency compounds (*e.g.*, a mixture of V_3O_4 and V_2O_5).

B. Surface Orbital (Surface State) Model

The coordinatively unsaturated surface atom on a nonmetallic solid will often have strong acid or basic sites, if the solid is highly ionic, or will probably have a localized "dangling bond" if the solid is highly covalent, simply because the atom is at the surface. We have discussed the acidic and basic sites that offer or accept electron pairs for sharing in Section 3. Some further features of acid and base sites will be described in this section, but we will emphasize the sites on the covalent solid (including reasonably ionic solids) with unpaired electrons available to make covalent bonds, sites that can accept or donate single electrons. The energy levels associated with these sites are termed "surface states".

Surface states are localized energy levels as opposed to the nonlocalized states in the bands of the bulk solid. First, let us discuss the origin of the bands of the solid, then the origin of the surface states. Quantum analysis of the bulk energy levels of the nonmetallic solid shows that the valence electrons in the solid occupy a band of levels, termed the valence band; the next excited state, normally unoccupied, is also a band of levels, termed the conduction band. There is a gap in energy, a forbidden region, in between. The bands and bandgap are indicated in Figure 4. The bands form from the atomic levels because of orbital overlap when the atoms are close together in the solid. The greater the overlap of the atomic orbitals, the broader the bands. If the overlap is slight, for example, when the orbitals of interest are the orbitals of *d*-electrons, then the bands are narrow. Narrow bands thus correspond to electrons quite localized near their atoms. Broad bands arising, for example, by overlap of *s* or *p*-orbitals, correspond to quasi-free electrons with essentially delocalized wave functions.

Electrons can move freely in the conduction band, holes (unoccupied valence orbitals) can move freely in the valence band. Thus, either of these species can conduct electricity, and are termed "carriers". Equally important, either of these species can move to the surface and be captured by an adsorbate, where clearly an oxidizing agent will capture electrons, a reducing agent captures holes.

Quantum analysis of the surface (or other imperfections in the solid) shows the possibility of localized orbitals. In particular, the surface atoms can

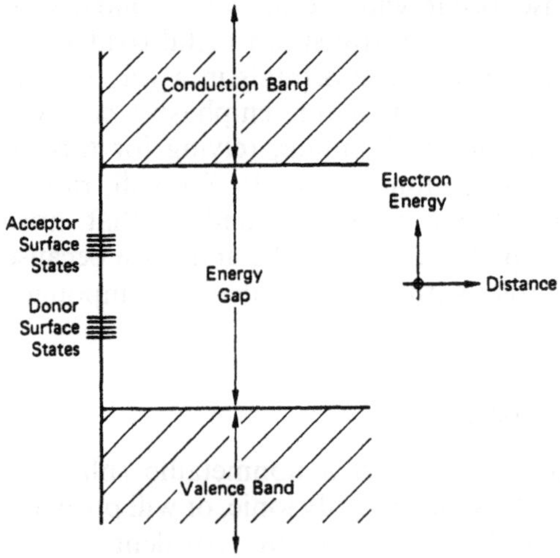

Figure 4. The band model of the solid. The energy levels of interest in adsorption are the energy levels occupied by the valence electrons, the valence band, the energy levels of the next excited state (an essentially unoccupied band), the conduction band, and localized energy levels at the surface, called surface states, which are the energy of "dangling bonds" associated with surface atoms

posses localized levels in the forbidden energy region, the gap, termed "localized" because the wave function of the electrons decays exponentially with distance from the surface. The energy levels of these localized orbitals are the intrinsic surface states. Such states are also indicated in Figure 4.

If the crystal is highly covalent, the surface states are called Shockley surface states and can be thought of as a dangling bond, the hybridized orbital extending out from the surface atom. The valence orbitals in the covalent solid are hydridized to yield the highest electron density between the bonded atoms, and at the surface this orbital persists as a "dangling bond" even when there is no nearest neighbor to pair the election. Two states are present (in the absence of an adsorbed atom), an acceptor state and a donor state. The acceptor state is the energy level at which a second electron is accepted from the solid to pair the electron in the dangling bond. The donor state is the level occupied by the unpaired electron, and this unpaired electron can be donated to a band in the crystal. One measures the energy level of the latter, a nonbonding orbital, by many techniques of semiconductor surface physics such as changes in the conductance of the solid as electrons are accepted by or donated by the surface states. Naturally, if a bond is formed between this surface state and an adsorbing radical, new energy levels will form, corresponding to the bonding and antibonding orbitals of the adsorbate/surface atom complex.

If the crystal is ionic, so the occupied orbitals have an ionic (as opposed to hybrid) character and the unoccupied (conduction band) orbitals have cationic character, then the surface states are called Tamm surface states. The levels can, in principle, be analyzed using the Madelung potential approach, based on simple electrostatic arguments [27]. Because it is sur-

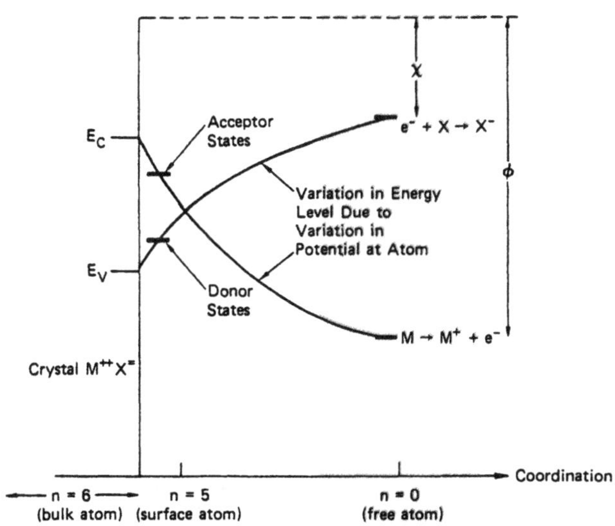

Figure 5. Madelung model showing the origin of Tamm surface states. M is the electropositive, X the electronegative species making the solid MX. *n* is the number of nearest neighbors of opposite charge

rounded by positively charged cations, an anion in the bulk of the solid is a very attractive site for electrons, so the energy level of the electrons that occupy the orbitals of the bulk anion is very low. However, a surface anion, with an incomplete complement of surrounding positive charges, has a lower potential and hence is less attractive to its electrons. Figure 5 illustrates the case. With no ions nearby ($n = 0$) the electron affinity of the anion, χ, is low, the ionization potential of the cation, φ, is high. In the bulk of the solid, when coordinated to 6 negatively charged anions, the energy level on the cation is high, simply because the potential at the cation is very negative. With 6 positively charged cations as nearest neighbors, the energy level on the anion is low. At the surface, where only 5 oppositely charged ions are neighbors, an intermediate potential (and thus energy level) obtains. As the energy of the electrons on the surface anions will be higher than the valence band (in this simple non-quantum-mechanical electrostatic picture), the energy of the elctrons on the surface anions will be higher than the valence band by the difference in potential for the two cases. Thus, the surface state will have an energy above the valence band. The potential at a particular ion is termed the Madelung potential and is influenced in principle by every charged ion in the crystal. Methods of estimating the potential depending on the partial charge on the ions and the lattice spacing, etc., have been worked out. With this simple model, the Tamm states are expected to consist of an acceptor level somewhat below the conduction band, where an electron can be captured by a surface cation, and a donor state somewhat above the valence band, where an electron can be given up from a surface anion.

Now the energy levels of the Tamm states may be far from their idealized locations indicated in Figure 6 from the Madelung model, due to quantum effects and due to the chemical nature of the ions. The Madelung potential arguments ignore the quantum mechanical effects leading to the broadening of the energy levels of the solid (the band formation). The bulk levels are expected to be broadened much more than the surface levels, so the surface states may well end up in the band region rather than the gap region, and thus be ineffective as donors or acceptors of single electrons. The energy of the levels by the electrostatic analysis also ignores the chemical behavior of the ion itself. Ions that can undergo one-electron reactions will be more likely to exhibit Tamm states in the bandgap region. Thus, in both our examples of Tamm states discussed below, lattice oxygen as a donor state and lattice chromic ions as acceptor states, the ions commonly exhibit one-electron reactions ($O^{2-} \rightarrow O^-$ and $Cr^{3+} \rightarrow Cr^{4+}$). Thus, the Madelung approach to Tamm surface states provides at best only a qualitative understanding of the behaviour of ionic crystals, but for quantitative estimates, one must look [21] to quantum calculation and also consider the general chemical behavior of the ions.

Two examples of adsorption sites that are important in catalysis and that can be considered Tamm states will be described. The first is that of surface oxygen sites on p-type semiconductors, the second is that of chromic ions at the surface of chromium oxide.

Surface states associated with the donor states of surfaces oxide atoms are very often active in adsorption. It is well known, for example, that p-type semiconductors, semiconductors that have unfilled levels in the valence-band (holes), are particularly active catalysts from oxidation reactions. Examples are NiO, Cu_2O, Co_3O_4. It is also well known that n-type semiconductors, irradiated to form holes, become photocatalytically active. Now the mechanism for this activity is related to the capture of holes by the surface states — the injection of electrons from the surface states into the unoccupied valence band level that is the hole. The surface state of interest can be the OH^- state of the hydroxylated surface — in the case of TiO_2 it has been shown that in paints, the capture of holes by the OH^- groups, resulting in a radical [28], leads to a chain reaction in the resin of the paint and causes chalking of the paint. Or the hole can be captured on a lattice oxide ion, leading to oxygen in the —1 oxidation state, which has been shown [29] to be very active in oxidation reactions. For example, it has been suggested [21] that such sites can adsorb hydrogen atoms strongly, so that OH^- groups can form by hydrogen atom extraction from hydrocarbons, leading to active hydrocarbon radicals on the surface.

$$O_L^- + R \cdot H \rightarrow O_L \cdot H^- + R \tag{8}$$

The requirement for this active adsorption site, the O^-, is the presence of holes and a donor surface state associated with the surface oxide ion. And, of course, one of the reasons for the stability of the hole on the surface oxide ion is the favorable Madelung potential at that site.

In the case of chromium oxide, the valence band (the highest energy occupied orbitals) is presumed to be the *d*-orbitals of chromium, specifically Cr^{3+}. In this case the surface states are associated with the surface chromium atoms, and the donor Tamm states are associated with the electrons on the Cr^{3+} which can give up an electron to form Cr^{4+}. These donor states can offer their electrons to the valence band, (hole capture by the Tamm state), or can offer their electrons to an adsorbate such as O_2 or O.

$$Cr^{3+} + O_2 \rightarrow Cr^{+4}{-}O_2^-$$

At moderate temperature, where the activation energy for the dissociation of the oxygen molecule is possible, O atoms become adsorbed, providing a very active species for oxidation catalysis.

$$2\,Cr^{+3} + O_2 \rightarrow 2\,Cr^{+4}{-}O^-$$

By electrical measurements it has been shown [30] that the donor surface states are about 0.3 eV above the valence band (where the valence band, as mentioned above consists of the same orbitals, but in the bulk of the solid).

C. Mixed Valency Oxides

By a mixed valency oxide, we refer to highly non-stoichiometric binary oxides. These are often narrow bandwidth oxides such as transition metal oxides. Most oxides with wide bands, (*e.g.*, bands based on the overlap of *s* or *p*-orbitals) do not deviate far from stoichiometry. The slight non-stoichiometriy simply provides donors or acceptors in the oxide. Thus, for example, ZnO or TiO_2 normally will be oxygen deficient, and the resulting interstitial cations or oxygen ion vacancies will act as donors and provide electrons to the conduction band. Or NiO will have a slight oxygen excess — the bulk cation vacancies will be acceptors and make the sample p-type, emitting holes into the dominantly oxygen 2*p* valence band. Adsorption on such wide bandwidth oxides was covered partially in the preceding section (on surface states), and will be covered in more detail in the next section (ionosorption).

Here we wish to emphasize the catalytically important case of binary semiconductors where the cation has two valences differing by unity. An example would be a mixture of V_2O_4 and V_2O_5.

In cases of semiconductors dominated by a *d*-band, where the cation is clearly of two valencies, the band of interest is often narrow, as discussed in Section 4.B. However, the *d*-orbitals have enough overlap to form a band, although the overlap is low so the carrier mobility is low, carriers (electrons or holes) can move through the crystal. In particular, a mixed valency oxide will present a partially filled *d*-band and act very similarly to a metal. Electrons can be readily exchanged either way between the solid and an adsorbate, permitting a high surface coverage without causing large electrostatic double layers.

There are several ways to describe this capability, depending on the model one wishes to use to describe the solid. One can describe it as induced covalency of the surface atoms, thus relating the behavior to metallic behavior. One can describe it in terms of ionosorption concepts, considering the carriers transfered to the adsorbate from the band, and that the carriers lost to the adsorbate can be replenished from the bulk of the crystal. Or, if we view the behavior in terms of localized orbitals, a large fraction of the surface atoms can accept an electron, the rest can donate an electron to an adsorbate, so a coverage approaching a monolayer can be reached with the double layer only a few angstroms thick.

Although the band is narrow, it is often wide enough (*e.g.*, 1 eV) that an electron from the top of the band can be given to an adsorbate of interest where electrons from the bottom of the band cannot. If the rate limiting step in the oxidation of a species is the reduction of oxygen, the band can be made almost completely occupied to provide high energy electrons for the reduction process. If the rate limiting step in the oxidation of a species is removal of an electron from some adsorbate or intermediate, the band can be made almost completely unoccupied to provide low energy empty levels for this process. Thus, for example, in the oxidation of 1-butene on the vanadia catalyst ($V_2O_5/V_2O_4 + P_2O_5$) it is found [31] that the catalyst is most active with a stoichiometry approaching that of V_2O_4, *i.e.*, almost a completely occupied *d*-band, in order to accelerate the rate limiting step, the adsorption of oxygen. On the other hand, with bismuth molybdate, where the adsorption of the organic is rate limiting, the Mo^{6+}/Mo^{5+} ratio must bekept as high as possible [32] (low Fermi energy [33]) to encourage electron or hydrogen extraction from the adsorbing organic.

As indicated such solids are of special interest because they can be viewed either from an adsorption site model or from the band model, for they represent a transition between the two. With a solid such as V_2O_{4+x}, the V^{4+} atoms at the surface can be viewed as electron donor sites for the adsorption of oxygen, the V^{5+} atoms at the surface can be viewed as electron acceptor sites for the adsorption of an electron donor. The band picture does not disagree with this model, for, as discussed in Section 4.B, the wave functions of the electrons on these *d*-orbitals, while forming a narrow band, can be considered quite localized, so the surface atoms can be considered to act independently and such a representation will be quite satisfactory. On the other hand, one can describe the adsorption in terms of the band model, considering the source and sink of the electrons as the *d*-band and the actual bonding orbitals as surface states, where the surface states are the energy levels of the adsorbate-solid bonding orbitals. To accelerate the adsorption of oxygen, the localized site model suggests a high ratio of V^{4+} to V^{5+} in the oxide. To accelerate the adsorption of oxygen, the band model suggests a high Fermi energy. The two models are interchangeable in this case.

5. Ionosorption

Ionosorption is the adsorption of species by electron transfer, where an electron or a hole from an energy band in a semiconductor becomes localized on an orbital of the adsorbate. As discussed in Section 1.A, for electron transfer to an adsorbate the electron must move to a lower surface energy level, for hole transfer to an adsorbate, the hole must move to a higher surface energy level. The origin of the electron or the hole is asually an imperfection deep in the crystal, so there is no direct bonding between the source of the electron or hole and the adsorbate. For example, an interstitial zinc atom in zinc oxide may donate an electron to the conduction band of the zinc oxide. Oxygen becomes adsorbed as O_2^-:

$$e^- + O_2 = O_2^-$$

where the electrons are exhausted from the conduction band of the zinc oxide near the surface (*e.g.*, in a layer 2000 Å thick at the surface) and become located on the adsorbed oxygen. The exhaustion or the depletion layer can be up to several thousand angstroms thick. Effectively the interstitial zinc in the exhaustion layer have contributed their electrons to the oxygen, but there is clearly no local bonding between the interstitial zinc ions and the adsorbed oxygen. Now a local bond can be formed between the ionosorbed species and a surface atom, but it is a bond between the ion (formed by electron transfer) and the surface, not between the atom and the surface. Thus, the term ionosorption seems appropriate. Figure 6(b) shows how the band model (Figure 4) is drawn at the surface to describe ionosorption.

This case is one of the few that provide a simple quantitative model for the observed large changes of the heat of adsorption and the activation energy as a function of surface coverage. The effects of the double layer are so great

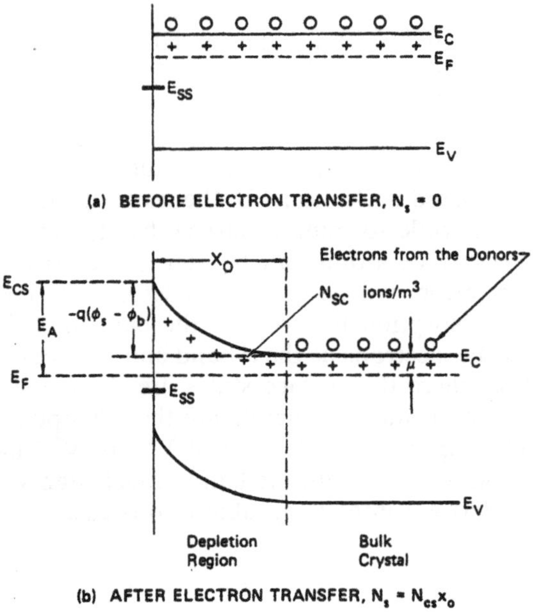

Figure 6. Model for ionosorption. Electrons, contributed to the conduction band E_c by donor impurities in the solid indicated by "+", are captured by oxygen molecules (or some other acceptor surface state) at the surface. A space charge region due to the immobile donor ions develops near the surface. The double layer voltage between the positively charged donor ions and the negatively charged adsorbed oxygen results in band bending as indicated. Figure 6a shows the model before electron transfer, 6b after. The symbols are used in the analysis of Section 5.B

that even surface heterogeneity seems often to be negligible. The development of this double layer and its role in determining the activation energy and heat of adsorption will be described in Section 5.A, before going on to specifics of models for n- and p-type material.

The adsorption of oxygen provides an example of ionosorption under the various conditions described above. On an n-type semiconductor, "simple" adsorption of oxygen as O_2^- or O^- is observed, with the conductance of the semiconductor decreasing due to the loss of bulk electrons to the surface. For example, Chon and Pajares [34] monitored the number of electrons lost from ZnO as a function of the number of oxygen molecules adsorbed. They showed that oxygen adsorbs as O_2^- below about 500 K, and as O^- above about 500 K. Oxygen also can adsorb as O_3^-, often formed by the bonding of O_2 to adsorbed O^-. An extreme case of "pure" ionosorption is the ionosorption of oxygen on insulators, where the electrons that are captured by the oxygen must be formed by some type of radiation [21], or by the donation of an electron to a surface state by a gasous reducing agent [35] because there are no free carries otherwise available. On the other extreme, there are many cases of local adsorption of oxygen on transition metal oxides where local bonding concepts are best used for the model, either using surface states, narrow bands, or non-stoichiometry as the basic model for the bonding, but noting usually that the partial charge on the oxygen is high.

A. Model

As described in the above introduction, and illustrated in Figure 6, carriers are extracted from a moderately thick "exhaustion" or "depletion" layer at the surface of the semiconductor and are transferred to the adsorbate. Figure 6 shows election capture by an oxidizing agent on n-type material; the picture is symmetric for hole capture (electron injection) by a reducing agent on p-type material. This capture of carriers leaves the immobile impurity ions, those which provided the holes or the electrons to the band, as the charged species in this depletion layer. The charge density in the depletion layer is given by qN_{sc}, where N_{sc} is the density of impurity ions. Thus, in this layer, Poisson's equation becomes:

$$d^2\varphi/dx^2 = -qN_{sc}/\varepsilon\varepsilon_o \tag{9}$$

where we assume the potential is not a function of y and z, the coordinates parallel to the plane of the surface, but only changes with the distance from the surface x. We let x_o be the thickness of the depletion layer. If we adopt the boundary conditions that $\varphi = \varphi_b$ and $d\varphi/dx = 0$ in the bulk of the crystal for all values of $x \geqq x_o$, when equation (9) is integrated twice we find:

$$\varphi - \varphi_b = -(qN_{sc}/2\varepsilon\varepsilon_o)(x - x_o)^2 \tag{10}$$

valid in the depletion region, between $x = 0$ and $x = x_o$. Thus, the band edge in Figure 6(b) varies parabolically with distance. The potential at the surface (at $x = 0$) is thus given by:

$$\varphi_s - \varphi_b = -(qN_{sc}/2\varepsilon\varepsilon_o)\, x_o^2 \tag{11}$$

Now the surface density of charges N_s is given by $N_{sc}x_o$, for this is the number of carriers per unit area that were transferred to the surface. Substituting N_s into equation 11 and eliminating x_o, we have:

$$\varphi_s - \varphi_b = -qN_s^2/2N_{sc}\varepsilon\varepsilon_o \tag{12}$$

which gives the potential barrier, $\varphi_s - \varphi_b$, that the next carrier must overcome in order to reach the surface, in terms of the density of adsorbed species.

With the help of Figure 6, we can discuss the heat and activation energy of ionosorption. Figure 6(a) shows the model before adsorption. With an n-type semiconductor described by Figure 6(a), the electrons can flow freely from the conduction band (energetically they come from the Fermi energy) to the electron accepting adsorbate. The activation energy will thus be μ, which is 0 to 0.2 eV, usually. The heat of adsorption is given by $E_F - E_{ss}$, where E_{ss} is the energy level on the adsorbate. As adsorption proceeds, the potential at the surface changes in accordance with equation (12) ro resist further transfer of carriers to the surface, and the band diagram appears as shown in Figure 6(b). The important consequence is that the contribution heat of adsorption of the electron transfer, varying as $E_F - E_{ss}$, decreases as the adsorption proceeds. If the bands were to bend sufficiently to bring E_{ss} up to E_F, the heat of adsorption contributed by the electron transfer process would be zero.

While the heat of adsorption is decreasing to zero with coverage in a way calculable from the properties of the semiconductor, the activation energy is increasing. The activation energy for the electron transfer is at least the energy

$$E^{o}_{A} = \mu - (\varphi_{s} - \varphi_{b})\, q \tag{13}$$

This is the energy necessary to bring an electron from the bulk to the surface over the potential barrier shown in Figure 6(b). There may be other contributions to the activation energy, associated perhaps with the dissociation of the adsorbate molecule, which will add to this electronic activation energy. But aside from this possibly constant contribution, the activation energy increases with coverage in accordance with Equations 13 and 12.

Other properties of the solid change during the adsorption. For example, the conductance changes because carriers are immobilized at the surface, and the work function, the difference between the free electron energy and the Fermi energy, changes as indicated in the figure. Such changes will be discussed further in Section 6.

If we assume that when the adsorption starts there is a non-negligible surface barrier, $-(\varphi_{s} - \varphi_{b})_{o}$, due to electron capture on surface states or on some other adsorbate, and the adsorption of interest induces an extra potential $\Delta\varphi_{s}$, then the rate of adsorption is given by:

$$dN_{s}/dt = A \exp\,[-(\varphi_{s} - \varphi_{b})_{o} + \Delta\varphi_{s}]/kT = d\Delta N_{s}/dt \tag{14}$$

and using the first term of the expansion of equation (12), *viz.* $\Delta\varphi_{s} = -b\Delta N_{s}$, then the adsorption rate is given by:

$$d\Delta N_{s}/dt = B \exp\,(-b\,\Delta N_{s}) \tag{15}$$

Equation (15) is termed the Elovich equation, and is a commonly obeserved relationship for adsorption on nonmetals. However, it must be pointed out that with the two arbitrary constants and the exponential dependence, much adsorption data can be fitted to the "Elovich equation" when there is no theoretical reason to do so. In fact, many cases, where the increase in activation energy undoubtedly arises because of heterogeneity of adsorption sites rather than because of the above double layer model, the Elovich equation can be fitted quite satisfactorily. Thus, a reasonable fit of an adsorption/time curve to equation (15) should not be considered a proof of the double layer model without supplementary evidence such as corresponding work function changes.

B. Catalytic Activity of Ionosorbed Species

In this section we indicate that adsorbates ionosorbed as above with no complications are, in general, ineffective in catalytic reactions primarily because of a low density, and that other features must be present to activate solids adsorbing by ionosorption.

The maximum adsorption density in ionosorption is very limited. Inverting equation (12), we obtain:

$$N_{s} = (2N_{sc}\varepsilon\varepsilon_{o})^{1/2}\,(\varphi_{s} - \varphi_{b})^{1/2}/q^{1/2} \tag{16}$$

Now, as mentioned above, the heat of adsorption becomes zero and so adsorption will cease when the energy level on the adsorbate crosses the Fermi energy. The energy level of the most reactive species are, at most, only the order of 1 eV above or below the Fermi energy of a typical solid. Thus, when $\varphi_s - \varphi_b = \pm 1$ eV or so, N_s has reached its saturation value $N_{s,max}$. If we assume a dielectric constant ε of the order of 5, then $N_{s,max}$ becomes:

$$N_{s,max} = 2.3 \times 10^4 N_{sc}^{1/2} \text{ m}^{-3} \tag{17}$$

and even with N_{sc} fairly high, the order of 10^{24} m^{-3}, $N_{s,max}$ is only 2.3×10^{16} m^{-2} or about 10^{-3} monolayers. Because $N_{s,max}$ varies as the square root of the doping level, its value does not ever approach a monolayer, while the solid still retains the properties of a semiconductor. Thus, straightforward ionosorption as described above is a low coverage process.

The low coverage associated with ionosorption usually means poor catalytic activity. Additionally, the high activation energy, as in equation (13), will lead to slow reactions. And indeed, semiconductors adsorbing by an ionosorption mechanism with no special features usually are found to be inferior catalysts. Many n-type semiconductors fall into this group.

Other features must be present to permit ionosorbed species to become catalytically active. These features lead to "ionosorption" with substantial local bonding, so it becomes a matter of choice whether the ionosorption description or one of the local bonding descriptions is preferable. One modification of the simple ionosorption model that leads to catalytically active semiconducting substrates in the case of oxides is the rapid exchange of lattice oxygen with adsorbing species (leading to oxidation catalysis), the other is the rapid diffusion of defects within the solid. We will discuss the influence of defect diffusion on the adsorption first, and the role of lattice oxygen second.

Mobile defects in the solid allow a high density of ionosorbed species. In discussing simple ionosorption in Section E.1, it was assumed that the bulk defects were immobile. However, if mobile, the defects will move to the surface, attracted by the surface charge, and the effective defect concentration N_{sc} becomes very high. An example would be cuprous oxide, where cation vacancies are the acceptor defects which are highly mobile and can move to the surface, providing holes for the adsorption of reducing agents. Another possible example, is bismuth molybdate, where oxygen ion vacancies are highly mobile at temperatures only slightly above room temperature, and these vacancies, acting as donors, can move to the surface to provide electrons for adsorbing oxygen.

The mobility of the defects is particularly important when they are created or destroyed by the adsorption, permitting oxygen exchange, the removal of lattice oxygen by the adsorbate. Both the copper oxide and bismuth molybdate cases also provide good examples of such processes. In these cases, ionosorbing oxygen will interact with defects, introducing them or removing them. The adsorption of oxygen on bismuth molybdate will be increased as oxygen ion vacancies, the donors, move to the surface. This will further lead to absorption of the oxygen when an oxygen ion vacancy reaches the surface allowing the adsorbed oxygen to be incorporated into the crystal. In the cuprous oxide case also oxygen is incorporated into the lattice, in this case with the generation of defects. The incorporation of oxygen into cuprous oxide will induce copper ion vacancies, and the absorption can continue because these vacancies are mobile and will diffuse into the crystal. Thus, in both cases, defect mobility allows a substantial accumulation or loss of a reactant that would be impossible with simple ionosorption. The reverse but less well defined process on these oxides permits the extraction of lattice oxygen by an adsorbing organic molecule such as propene. In this extraction process the lattice oxygen becomes added to the organic molecule or to hydrogen atoms. In the cuprous oxide case, the lattice oxygen loses electrons to the valence band and in the bismuth molybdate case the lattice oxygen loses electrons to oxygen ion vacancies. The resulting "adsorbed" oxygen species (perhaps the O species as discussed in Section 4.B) is the adsorption site for, say, a hydrogen atom from an organic molecule or the organic molecule itself. When the OH^- group or the organic desorbs eventually as water or as an oxidized organic molecule, the oxygen ion vacancy will be created (or with cuprous oxide a copper ion vacancy is destroyed). Again, the mobility of such defects is very important, allowing the defects to move away and the process to continue.

As a general class p-type semiconductors, such as cuprous oxide, are not found to follow the simplest theories of ionosorption, primarily because there are surface states available to capture the majority carrier, the hole, and provide local dangling bonds of a covalent nature. When a hole occupies a bonding orbital, that bond is broken, and that is the key to the high adsorption and catalytic activity of p-type semiconductors. If sufficient holes collect at the bonding orbitals of a particular surface atom, the atom is no longer bonded to the solid — the surface atom becomes highly reactive. With p-type oxides, this leads to easy lattice oxygen incorporation onto an adsorbate, because when the hole is captured by the surface oxygen ion (or hydroxyl ion), its bonding to the surface is weakened and it can bond strongly to an adsorbate. The cuprous oxide case discussed above provides an example of this behavior. At the solid/liquid interface such hole capture can cause the solid to dissolve into the solution. At the solid/oxygen interface, an elemental semiconductor or any compound semiconductor except oxides can develop an adsorbed layer or even a new surface phase by oxidation of the anion — *e.g.*, sulfides are oxidized to sulfates. Or, if the temperature is high enough that the products are volatile. the solid can decompose; *e.g.*, copper oxides decomposing to copper. The reason that such reactions are so common on p-type semiconductors and not on n-type semiconductors is that the "decomposition potential" associated with holes is usually in the bandgap region, whereas the decomposition potential associated with electrons is usually not [36, 37]. The term "decomposition potential" arises from electrochemical studies wherein it is calculated quantitatively whether bonds between the substrate and the surface atom can be broken (and the surface atom rendered active or removable) by capture of carriers from the solid. And such quantitative studies show, indeed, that for most compound semiconductors, hole capture rendering the surface atom active is to be expected, but electron capture with the same result is rare. Thus, p-type semiconductors can be expected to be more active as adsorbents and catalysts than n-type.

6. The Effect of Adsorption on the Properties of the Solid

The two electrical properties of the semiconducting solid that are most affected by adsorption are its work function and its electrical transport properties (conductivity, Hall effect, etc.). In this section measurement of these and many other properties will be discussed that provides information regarding adsorption isotherms and regarding the bonding in adsorption. We will not emphasize the effect of adsorption on the chemical properties. As has been discussed and implied in earlier sections, the main chemical effect of adsorption is usually surface passivation.

The work function is affected during adsorption simply because a double layer forms. During ionosorption, as discussed in the last section, high voltage double layers can form with a very low coverage of adsorbate. Very high voltages do not form, because the voltage of the double layer (in eV) is subtracted from the heat of adsorption, and typically after less than 0.5 volts or so of double layer, the heat of adsorption becomes zero and the adsorption stops. The same effect can occur with bonding to local sites as was discussed in relation to equation (3). Even with the local adsorbate bonding, coverage of less than a monolayer can lead to double layer voltages approaching a volt, and this double layer voltage is directly added to the work function of the solid. Thus, the work function is a sensitive measurement of the adsorption, although the direct measurement of the work function can be susceptible to error because of the growth of new phases rather than, or in addition to, an adsorbed layer.

The measurement of the conductance is also a sensitive measure of adsorption, especially on a semiconductor during ionosorption, because ionosorption means carrier exchange with the bands of the bulk solid. Local bonding in principle does not affect the conductivity of the semiconductor, because such injection or capture of carriers is not implied. However, local bonding of an adsorbate to a surface state may indirectly affect the conductance because the surface state may exchange electronic carriers with the bands of the semiconductor. But ionosorption directly affects the conductance. The number of carriers lost or gained during ionosorption can be a significant fraction of the total number in the solid, especially if the solid is a thin film. For example, if the adsorbate extracts carriers from a layer extending 2000 Å from the surface, and the film is only 1 μm thick, the density of carriers is lowered by 20%.

In a powdered semiconductor, as are often used in catalytic studies, a distinctive effect is observed that can lead to many orders of magnitude change in conductance during adsorption. Figure 7 illustrates the phenomenon. As indicated in Figure 6, the double layer V_s builds during adsorption because of localization of change at the surface. But in a powder, the carriers must reach the surface to cross to the next particle, or there is no conductance. Thus, the surface barrier becomes an activation energy for conductance in the present case in the same way it was the activation energy for adsorption in equation (13). As indicated in Figure 7, the carriers must gain the energy $qV_s = -q(\varphi_s - \varphi_b)$, the energy of the surface barrier, in order to cross, so the conductance varies as

$$G = G \exp \{\mu - (\varphi_s - \varphi_b)\}/kT \tag{19}$$

As the surface barrier varies with the amount adsorbed in accordance with equation (12), it is clear that the conductance will vary by orders of magnitude during adsorption on a pressed powder pellet. This becomes a very sensitive measure of adsorption, although without a well-characterized solid the quantitative interpretation of the measurement is difficult.

Many other properties of the solid change upon adsorption and have been used to monitor adsorption. In measurements of infrared absorption of a powder, for example, absorption bands develop characteristics of the adsorbate as affected by its bonding to the solid. The electron spin resonance (ESR) spectrum of a powder may change during adsorption or during reaction of the adsorbed species when the ESR spectrum depends on the properties of the adsorbate. The ESR spectrum also may change during ionosorption in cases where the ESR spectrum depends on bulk carrier density. An example of the former, direct observation of the adsorbate, is that of some cases of photoreaction: *e.g.*, on TiO_2 an ESR peak due to OH · radicals is observed [28] when the TiO_2 is irradiated with uv capable of producing holes. An example of the latter is found upon ionosorption on ZnO or TiO_2, where a characteristic peak (at g = 1.96 for ZnO) is found related to the conduction electrons. When these conduction electrons are removed, for example, of changes in bulk properties due to adsorption is in fluorescence. For many nonmetallic wide bandgap semiconductors, the fluorescence depends on the production of minority carriers. However, surface adsorbates can scavenge the minority carriers in a photocatalytic reaction, lowering the intensity.

In conclusion, we should mention a few of the many surface properties of solids as measured in high vacuum that are affected by adsorption and

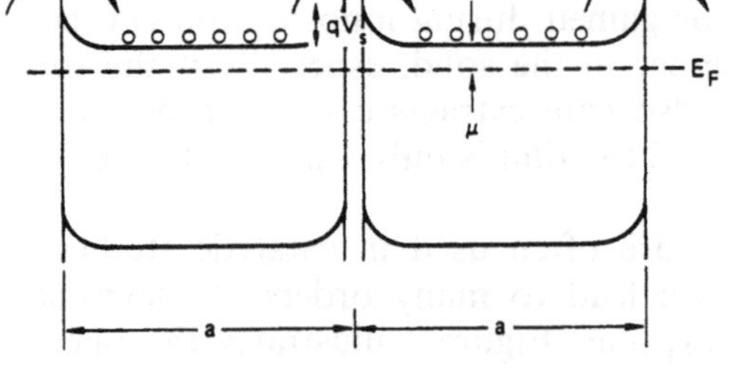

Figure 7. Band model of an n-type powdered semiconductor, showing two grains of diameter *a* in contact. Adsorption causes the band bending V_s, and for electrons to move through the pellet, they must be activated over this surface barrier

are being used [21] to study adsorption. For example, there is LEED, low energy electron diffraction, wherein the lattice spacing at the surface is observed to depend on adsorption. Then there is UPS, ultraviolet photoelectron spectroscopy, where the energy of photoelectrons emitted, with the solid used as a photocathode, shows peaks characteristic of the bonding orbitals between the adsorbates and the solid. Similar measurements, such as the measurement of characteristic values of energy lost by scattered electrons, energy loss spectroscopy, (ELS). are found to depend on adsorption. And naturally, adsorption affects the measurements that determine the chemical composition of the surface, such as AES, Auger electron spectroscopy, and SIMS, secondary ion mass spectrometry.

References

1. Kurtin, A., McGill, T. C., Mead, C. A.: Phys. Rev. Lett. **22**, 1433 (1969)
2. van Laar, J., Scheer, J. J.: Surface Sci. **8**, 342 (1967)
3. Lennard-Jones, J. E.: Proc. Roy. **A 106**, 463 (1924)
4. Ellis, W. P., and Schwoebel, R. L.: Surface Sci. **11**, 82 (1968)
5. Baron, K., Blakely, D. W., Somorjai, G. A.: Surface Sci. **41**, 45 (1974).
6. Bianchi, D., Lacroix, M., Pajouk, G., Teichner, S. J.: Catal. **59**, 467 (1979)
7. Sancier, K. M.: J. Catal. **20**, 106 (1971)
8. Boehm, H. P.: Disc. Faraday Soc. **52**, 264 (1971)
9. Lanford, W. A.: Science **196**, 975 (1977)
10. Singh, B. R., Tyagi, B. D., Marathe, B. R.: Phys. Stat. Sol **A 19**, K 143 (1973)
11. Gammage, R. B., Brey, W. S. Jr., Davis, B. H.: J. Coll. Int. Sci. **32**, 256 (1970)
12. Egashira, M., Kawasumi, S., Kagawe, S., Seizama, T.: Bull. Chem. Soc. Jap. **51**, 3144 (1978)
13. Munuera, G., Stone, F. S.: Disc. Taraday Soc. **52**, 205 (1971)
14. Morimoto, T., Naono, H.: Bull. Chem. Soc. Jap. **46**, 2000 (1973)
15. Chuckin, G. D., Khrustaleva, S. V.: Russ. J. Phys. Chem. **47**, 1155 (1973)
16. Peri, J. B., Hensley, A. L., Jr.: J. Phys. Chem. **72**, 2926 (1968)
17. Primet, M., Pichat, P., Mathieu, M. V.: J. Phys. Chem. **75**, 1216 (1971)
18. Morishige, K., Kittaka, S., Moriyasu, T.: J. C. S. Faraday I **76**, 728 (1980)
19. Griffiths, D. M., Rochester, C. H.: J. C. S. Faraday I **10**, 1510 (1977)
20. Nagao, M., Morimoto, T.: Bull. Chem. Soc. Jap. **49**, 2977 (1976)
21. Morrison, S. R.: The Chemical Physics of Surfaces, Plenum Press, New York, 1977
22. Morrison, S. R.: Surface Sci. **34**, 462 (1975)
23. Boonstra, A. H., Mutsaers, C. A. H. A.: J. Phys. Chem. **79**, 1964 (1975)
24. Sohn, J. R., Ozaki, A.: J. Catal. **61**, 29 (1980)
25. Grabowski, W., Misono, M., Yoneda, Y.: J. Catal. **61**, 103 (1980)
26. Tanabe, K., Sumijoshi, T., Shibata, K., Kirjoura, T., Kitagawa, J.: Bul. Chem. Soc. Jap. **47**, 1064 (1974)
27. Levine, J. D., Mack, P.: Phys. Rev. **144**, 751 (1966)
28. Völz, H. G., Kämf, G., Fitzky, H. G.: Farbe u. Lack **78**, 1037 (1972)
29. for example see Gravelli, P. C., Teichner, S. J.: J. Catal. **20**, 168 (1970)
30. Morrison, S. R.: J. Catal. **47**, 69 (1977)
31. Nakamura, M., Kawai, K., Jujiwara, Y.: J. Catal. **34**, 345 (1974)
32. Sancier, K. M., Aoshima, A., and Wise, H.: J. Catal. **34**, 257 (1974)
33. Morrison, S. R.: Russian J. Phys. Chem. **12**, 1734 (1978)
34. Chon, H., Pajares, J.: J. Catal. **14**, 257 (1969)
35. Iizuka, T., Tanabe, K.: Bull. Chem. Soc. Jap. **48**, 2527 (1975)
36. Gerischer, H.: J. Vac. Sci. Tech. **15**, 1422 (1978)
37. Morrison, S. R.: Electrochemistry at Semiconducting and Oxidized Metal Electrodes, New York: Plenum Press, 1980

Chapter 5

Chemisorption of Dihydrogen

Z. Knor

Czechoslovak Academy of Sciences
The J. Heyrovský Institute of
Physical Chemistry and Electrochemistry
121 38 Prague 2, Máchova 7, ČSSR

Contents

1. Introduction

In the catalytic activation of dihydrogen, chemisorption is the process of central importance. This has been a subject of investigation since the very beginning of catalytic research. This is understandable in the light of its importance in catalytic technology, and because dihydrogen chemisorption has appeared to be a particularly simple system which permitted rigorous theoretical and experimental treatments. Indeed, this system has been extensively used as an archetype for the formulation of theories of surface bonding in chemisorption.

Because the chemisorption of dihydrogen plays this very important role in catalytic research, this chapter is devoted exclusively to it, with particular emphasis on the nature of the chemisorption bond, and the relation between theory and experiment. The problem has here been approached at essentially a descriptive level. The reasons for this are twofold. In a presentation designed for use by practicing catalytic chemists a treatment involving detailed quantum mechanics would not be particularly appropriate. Furthermore, as will become apparent from the following discussion, *ab initio* calculations of dihydrogen chemisorption parameters do not yet appear to be generally very satisfactory when judged against experiment. Thus at the present, theory seems best suited to providing an essentially qualitative framework within which the experimentalist may be able to consider the significance of his results.

This chapter is not intended to provide a comprehensive account of the catalytic reaction chemistry of catalytically activated dihydrogen. This is a very broad and diverse subject and it will be dealt with in detail in several chapters in subsequent volumes.

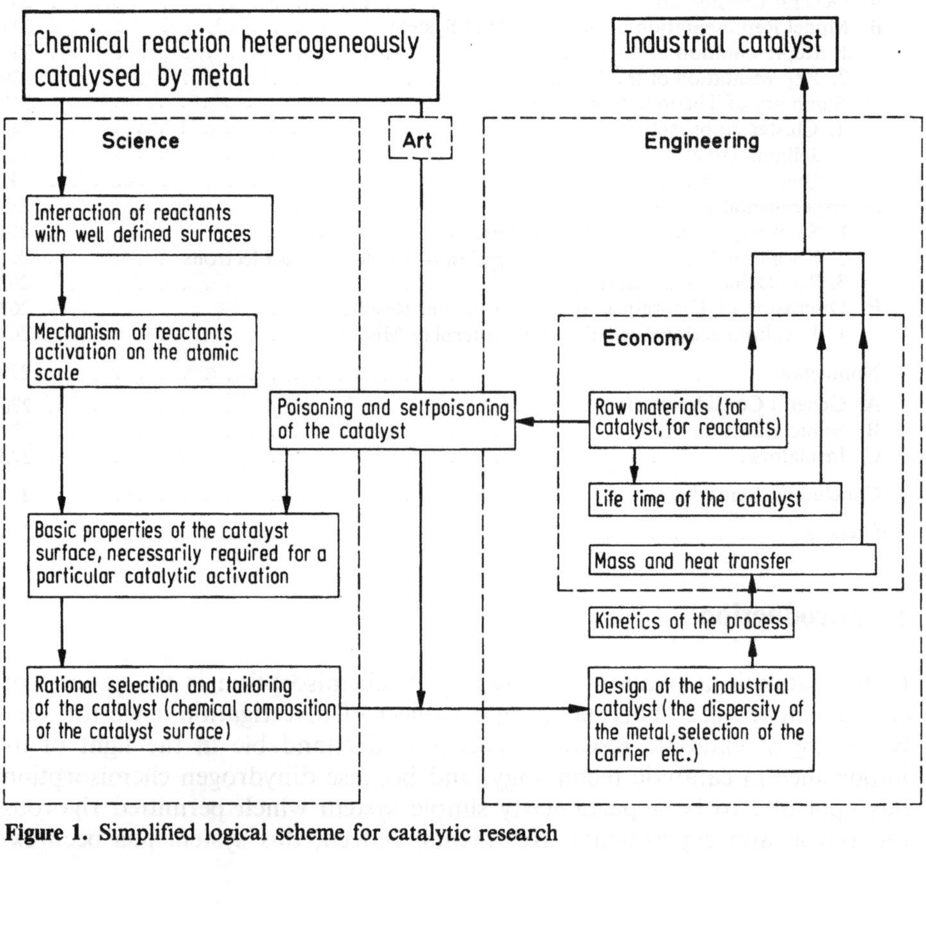

Figure 1. Simplified logical scheme for catalytic research

The amount of work done on this subject during the past decades is so vast that it is almost untractable in one article. Therefore it is necessary to limit oneself to a narrow view of the full problem. Its complexity çan be seen from the accompanying scheme of catalytic research (Figure 1). From this scheme there immediately appears the inevitability of engineering on the one hand, and the position of the science of catalysis as a kind of luxury on the other hand, at least at the present state of development. However, as can be found in the literature, several examples already exist when surface science has helped directly [1] or indirectly [1–3] to develop a new catalyst for a particular reaction. Despite the complexity of the above mentioned scheme, and the great number of factors which influence the attainment of the final goal — the preparation of a commercial catalyst — the individual materials (catalysts or groups of catalysts) exhibit considerable qualitative differences in their activity as well as in the selectivity patterns.

The activation of a gas molecule for a particular reaction is due to the perturbation of its electronic structure by the catalyst, that is due to their mutual chemical interaction. Thus we can assume that the essential difference between various catalysts is certainly connected with their different chemical reactivity (depending on the electronic and geometric structure of their surfaces) towards both reactants and poisons (or selfpoisons). Here one is facing a general problem of the theory of surface effects, namely whether the understanding of these effects necessitates as a first step an exact knowledge of the electronic and geometric structure of the isolated participants of the interaction, that is, of the reactant molecules and of the solid surface (in other words, if it is possible to understand surface-adsorbate bonding by relating the energies and symmetry of the molecular orbitals of the resulting adsorption complex to those of the isolated systems). If one is interested in the final ("equilibrium") state (*e.g.* the geometric structure of the surface layer, the vibration energy of a particular surface bond) then in developing the theory one obviously has to consider both partners in their mutual interaction in order to account for the possible reconstruction of the the virgin surface, while the properties of the isolated reactants and of the solid would be less important [4]. However, the problem of catalytic activation is in principle a dynamic problem [5] and, consequently, the knowledge of the properties of the isolated participants represents a good starting point for the theory [4], forming to some extent the basis for its predictive power. In the present article the chemical reactivity of a catalyst surface is assumed to be the central problem of the science of catalysis and we shall concentrate our attention on it.

From the theoretical point of view, all the surface processes can be divided into two classes: emission processes and chemical processes. Any particle — atom, molecule, ion, electron, photon or phonon — can be a primary agent giving rise to a particular process. However, the two types of events differ in the kind of secondary particle involved. In emission processes an electron or photon results from the interaction of the primary particle with the surface; in chemical processes a new particle (atom, molecule, ion resulting from the chemical reaction on the surface) appears on the surface and/or leaves it. In the former group of processes one obviously cannot neglect many-body interactions of the secondary particle in its initial state, whereas in the latter group one deals with "chemical" forces (acting at short distances), resulting thus predominantly in a localized interaction with the nearest neighbourhood only[1]. From this it logically follows that the key point of the microscopic description of a surface chemical interaction must be detailed knowledge of the nearest neighbourhood of the interacting reactant molecule, that is, the characterization of the "active" site. We might ask then what kind of abilities the active site should possess in order to make the desired reaction path feasible, in another words what kind of geometric and electronic structure of an active site is decisive for a particular reaction route. Besides theoretical speculation on the perturbation of individual bonds due to the interaction of a reactant molecule with a simplified model of the catalyst, one can try to characterize the active site in a particular reaction experimentally. In this case an additional assumption has to be made, namely that the active site is identical with the "adsorption" site (the trapping center with its nearest neighbourhood). The characterization of an adsorption site in classical experiments (volumetric, calorimetric and work function measurements, infrared spectra, etc.) is obtained in the form of information averaged over the whole surface. Obviously, without further assumptions these results cannot be directly compared with the theoretical ones obtained with oversimplified models. We could ask then what changes have resulted in this field due to the introduction of new experimental techniques (LEED, AES, UPS, XPS etc.). With these methods the information obtained is averaged over the cross-section of the beam of primary electrons or photons (*ca.* 1 mm^2 and 25 mm^2, respectively). The situation is further complicated by the presence of monoatomic steps (islands) on the surface of the macroscopic crystals which, even when carefully prepared, are endowed with this type of defects [6]. These steps might qualitatively change the chemical properties of the surface [6–8]. If their size is less than the coherence region of LEED experiments (*ca.* 10 nm), they are undetectable. Within an area equal to the cross-section of the primary beam there might be about 10^{11}–10^{12} such islands and consequently the average information becomes still less straightforward. In conclusion of this consideration one can state that in the field of surface science, theory approaches experiment from two sides:

Theory		*Experiment*		*Theory*
Quantum Chemical Approach				Solid State Physics Approach
Cluster of 1–10^2 atoms	→	10^{13}–10^{14} adsorption sites	←	Semi-infinite Crystal

[1] Naturally, if there were available a sufficiently powerful nonempirical theoretical method, the above mentioned classification of surface phenomena would be meaningless.

From the variety of surface sensitive experimental techniques, only the field ion microscope (FIM) gives us information directly on the atomic scale. However, it suffers from another drawback, that the surface events might be perturbed by the inevitable presence of an extremely high electric field (10^{10} V m^{-1}).

A necessary prerequisite for understanding and predictability in the field of heterogeneous catalysis would be the development of sufficiently powerful theoretical methods. However, the contemporary theory of surfaces and surface processes is not able to account for all the factors of influence. Instead of a real system one has to use a physical model, where only the most important features of the real system, relevant to the catalytic action, are being preserved. Because of this, any simplified model used as an empirical information input into the theory has to be considered always as an optimum model for a particular class of related systems, and it cannot be used universally [9]. Furthermore, even in sophisticated theoretical methods the empirical contribution is always included, though in a less obvious way which is usually not understandable in terms of the initial model [9]. Typical examples are: the empirical estimation of the diagonal and off-diagonal matrix elements in the EMHO and CNDO approaches [10–13], the semiempirical estimation of exchange and correlation terms [14, 15], the empirical limits for the pseudopotential in the density functional approach (the values for r_c) [14, 16, 17], the semiempirical estimation of the average α values [18–20], and the problem of the overlapping and non-overlapping spheres in the SCF Xα SW method [21, 22]. From the preceding considerations one can conclude that besides the models, the various theoretical approaches (empirical, semiempirical and "nonempirical") which differ only in the amount of the empirical admixture, are merely optimum approaches (methods) for the solution of a class of closely related systems and consequently they are not universally applicable [9]. Because of all the above mentioned limitations, the theoretical results can, in the most favourable cases only, be applied to the ideal experimental situation, that is, to the interaction of a gas molecule with an atomically smooth single crystallographic plane. Moreover, when comparing the theoretical and experimental results both quantities compared must have the same physical meaning. Unfortunately, this obvious condition is not always fulfilled (*e.g.* the binding energy of an adatom is compared with the experimental activation energy of desorption, the theoretically calculated value of the work function is compared with experimental value obtained on polycrystalline samples; see references in ref. [9]).

In spite of the intensive work during the past decade in this field, the experimental results still do not represent a complete and reliable basis for a general theory of surface processes. This is due to the fact that the contemporary sophisticated experimental methods are not self-sufficient. Their use necessitates the exploitation of contemporary theory for obtaining the desired information from the measured quantities (*e.g.* the structure of the surface from the diffraction pattern of electrons, the density of states distribution from the photoelectron spectrum).

All the experimental results can be divided into two groups: those obtained on well defined surfaces (being to some extent accessible to theoretical analysis), and the others obtained on more or less technical catalysts. Recently one can observe a tendency towards bridging the gap between these two groups of experiments in order to justify the transferability of conclusions between them. On the one hand the experiments are performed on well defined surfaces at conditions approaching those of industrial catalytic reactions (high pressures, intentional poisoning etc.) and on the other hand industrial catalysts are studied by modern analytical techniques to obtain information about the surface as completely as possible (ESCA, AES etc.). Unfortunately, the latter systems are usually so complicated that a full understanding of the elementary steps of the catalytic process is hardly achievable. One is therefore forced to choose as simple a systems as possible. Consequently, the present article will be limited mostly to results obtained on well defined surfaces. The applicability of the conclusions from such model systems to real catalysis has been recently proven by the results of kinetic measurements. It has been shown for several catalytic reactions that identical equations can be used for a successful description of processes on well defined surfaces as well as on technical catalysts [8, 24–26]. Thus the fundamental importance of kinetics for bridging the gap between model systems and technical catalytic processes can be demonstrated, in spite of its inability to solve unambiguously the molecular mechanism of a particular catalytic process [23].

A last comment concerns the significance of studies of surface processes on single crystal planes by those experimental methods which give us mostly information about strongly bound species (long living surface complexes) only. It is well known that in the course of a catalytic reaction the reactants considerably modify the surface properties [27], obviously due to strongly bound species. Beside this, the strongly interacting particles can sometimes play an active role in the catalytic reaction itself (*e.g.* in the methanation reaction [25, 28, 29]).

Metals are typical catalysts for the heterogeneous activation of dihydrogen. In technological practice metal catalysts are advantageously used in a dispersed form, supported by various non-metallic carriers. This is important for attaining a large surface-to-volume ratio, and thereby the mass of the metal (which is often an expensive one — *e.g.* Pt, Pd, Rh) can be economically well exploited. In such cases the transfer of the activated hydrogen onto the bare carrier surface (Al_2O_3, SiO_2, carbon) — the spillover effect — might change the activity and/or selectivity of this complex catalyst. Some special non-metallic catalysts for dihydrogen activation are also used in chemical technology (*e.g.* MoS_2, WC, ZnO). However, experimental data on hydrogen interaction with well defined surfaces of oxides, sulphides, carbides etc. are very scarce and the same holds for the theoretical treatment of catalytic reactions of dihydrogen on these surfaces. Consequently the main attention in the present article will be paid to metals.

Having in mind all the limitations we shall try to draw some general conclusions from the available theoretical and experimental material.

2. Metals

A. General Considerations

Starting with the above mentioned statement that catalytic activation is closely related to the ordinary chemical interaction of the reactant molecules with the catalyst surface, we shall proceed along the traditional line of quantum chemistry. Accordingly, the chemical reactivity is determined by the outermost (frontier) orbitals of both partners in the interaction. Quantum chemical experience has led to the formulation of two essential conditions for attaining a low activation energy of a chemical reaction: i) the correspondence of the orbital symmetry of the reactant and product molecules; ii) the correspondence of the relevant energy levels (these factors concern the frontier orbitals, both occupied and unoccupied ones). The main problem which we are facing here is the electronic structure of the metal surface, because for the reactant molecules — at least the simple ones — the needed data are available in the literature (*e.g.* ref. [30]). The transition metals represent the most interesting group from the view-point of heterogeneous catalysis and consequently we shall concentrate our discussion mainly on them.

B. Model Representation of the Gas-Metal System

As already mentioned, the chemical interaction of gas molecules with the surface results in the formation of a "surface compound". The wave functions of the adsorbate and substrate overlap significantly over the first coordination shell of atoms surrounding the bonding region and also the major change in potential acting on the electrons is primarily localized here [82]. Obviously the most simplified model for this case would be the interaction of two individual atoms: one representing the metal and the other one representing the gas particle. This model can to some extent correctly reflect on a qualitative level the localized character of "equilibrium" chemisorption; however, the influence both of the rest of the crystal and of the rest of the reactant molecule — which both might play a decisive role for a particular reaction route — are completely neglected in this model.

1. Representation of Gas Molecules

The representation of a gas molecule by an individual atom suffers from a fundamental drawback. It is well known both from general chemistry and from experimental evidence in surface science itself that atoms possess much higher reactivity than the corresponding molecules (high reactivity of systems with unpaired spins). On the other hand, the atomic representation of gas molecules can be used sucessfully if one's aim is a mere description of the final state of the chemisorbed layer (*e.g.* its structure). The conclusions based on these considerations must not be, however, applied to the case of the initial interaction of gas molecules with the metal surface. Even in the case of the description of the final state, there remains a problem connected with the comparison of theoretical and experimental results. When the depth of the potential well for chemisorption of an atom is theoretically calculated, then this quantity should be identified with the binding energy of the adsorbate atom with the surface. However, the binding energy is not accessible to direct measurement and usually it is roughly calculated from the heat of adsorption. For hydrogen the following equation is frequently used [13, 16, 31, 32]

$$E_{H-M} = 0.5(D_{H-H} + Q_{H_2}) \tag{1}$$

where E_{H-M} is the binding energy between a hydrogen atom and the metal surface, D_{H-H} is the dissociation energy of the hydrogen molecule and Q_{H_2} is the heat of hydrogen chemisorption. Alternatively, the activation energy of desorption can be used instead of Q_{H_2} for systems having zero activation energy of chemisorption; this is the case of hydrogen chemisorption on transition metals.

Equation (1) is only a rough approximation of the overall energy balance of the chemisorption process:

$$Q_{H_2} = 2E_{H-M} - D_{H-H} - \alpha E^s_{M-M} \tag{2}$$

where α is the number of metal-metal bonds which have to be broken for the chemisorption of one adatom H, and E^s_{M-M} is the metal-metal bond energy in the surface layer. The term αE^s_{M-M} is usually neglected without any physical reasoning [13, 16, 31, 32]. Unfortunately, there are no experimental values of E^s_{M-M} accessible. Only a magnitude estimate of surface bond energies can be found in the literature based either on the value of the surface tension or on the sublimation energy and the number of the nearest neighbours. These E^s_{M-M} values range from about 60 to 200 kJ mol^{-1} [33]. Thus α should approach zero if equation (1) is to be valid.

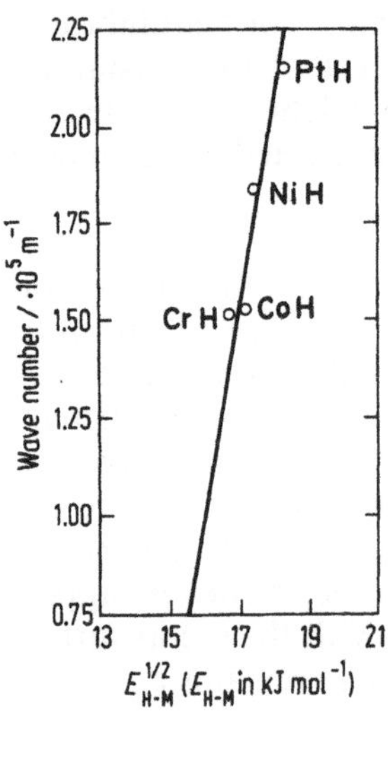

Figure 2. Correlation between the wave number of vibration and the square root of bond energy $E^{1/2}_{H-M}$ for diatomic molecules. (After [36])

The value of the term αE^s_{M-M} can be judged from equation (2) by using vibrational spectroscopy data. The vibration of a strongly bound adatom perpendicular to the surface can be considered as a close analogue of the vibration of the relevant diatomic molecule (*e.g.* NiH, PtH) [34, 35]. The vibration energy is, in the harmonic oscilator approximation, proportional to the square root of the oscilator dissociation energy [37, 45] Figure 2. This type of correlation has been verified also for the surface complexes of undissociated adsorbate [37]. One can estimate from this correlation the desired bond energies E_{H-M} in the surface complexes by using the experimental values of the vibration wave number obtained in electron-energy-loss spectroscopic (EELS) and infrared studies (Table 1).

The individual terms in equation (2) for the metals listed in Table 1 lead to the following average values: $E_{H-M} = 250\text{–}280$ kJ mol^{-1}, $D_{H-H} = 435$ kJ $\times$ mol^{-1} and $Q_{H_2} = 70\text{–}140$ kJ mol^{-1} (see Table 3, p. 252). The term αE^s_{M-M} approaches zero within the limits of the experimental errors. Thus one is tempted to assume that the model with free valences ($\alpha \to 0$) may be applied for the description of hydrogen chemisorption by transition metals (see later the experimental evidence for the exceptional role of localized atomic-like orbitals in trapping the gas molecules).

2. *Representation of the Metal*

The model representation of the transition metal surface in terms of an individual atom or a small cluster of atoms is based on the assumption that the surface atoms retain much of their atomic character, the adsorption properties being strongly influenced by local site effects and band structure effects being of secondary importance [46]. The localized character of chemisorption obviously results from the short screening length in metals, due

Table 1. Rough estimation of bond energies E_{H-M} based on Figure 2

Metal	Vibration Wave Number of the Surface complex 10^5 m^{-1}	Bond Energy E_{H-M}/ kJ mol^{-1}	Ref.
Fe, Co, Ni, Rh, Pd on Al_2O_3	1.85–1.94	300–310	[38]
Ni (Raney)	1.13	264	[38]
Ni (111)	1.135	264	[39]
Ni (100)	0.595	238	[35]
Pd	0.880	250	[38]
Pt on Al_2O_3	2.12	310	[38]
Pt (111)	1.23	280	[40]
Ir om Al_2O_3	2.12	310	[38]
W (111)	1.291	284	[41], [45]
W (110)	1.267	283	[41]
W (100)	1.046	259	[42]
	1.275	283	[43]
	1.251	281	[44]

to the high density of delocalized electrons. However, if the high density of delocalized electrons is needed for this successful model representation, one should use either large clusters or embedded cluster models [4, 47]. The surface chemical interactions thus implicitly contain both local (chemical) and delocalized (solid state) aspects [48–50]. Accordingly, it has proven useful already in previous theories to divide the metal valence electrons into two groups [51]: partly localized *d*-type electrons and delocalized *s*- (*sp*-) electrons. Then one can consider the role of these two types of electrons in the catalytic activation of gas molecules. This can be followed either on separate models (*e.g.* refs. [52, 53]), or simultaneously on one model (*e.g.* refs. [47, 54]).

C. Summary of Theoretical Results

In this paragraph, the results of various theoretical methods are collected (Table 2). Considering these results one has to bear in mind that to some extent they might possess a different "physical contents" due to the differences in the technique of calculation (different theoretical methods; different kind and number of empirical parameters) and in the input data (semiinfinitite crystal; type and number of slabs; the number of atoms in a cluster; the type of cluster — linear, planar, bulk — and its geometry; the position of the adsorbate — over an atom or over a hole, bridge position — and its distances from the surface plane; the number and type of the orbitals at the individual atoms) [9]. This is a very serious problem, because each of these factors can qualitatively change the results. Thus unrealistic results can be obtained with an inappropriate choice of orbitals within the frame of a particular theoretical method: if in the EHMO approach only *s*- orbitals are used for the representation of the noble metal atoms, the Fermi level of large clusters shifts towards positive values [55]. Neverthless, the final results of any theoretical calculation (the heat of adsorption, the charge on the adsorbate, etc.) have to be comparable with the relevant experimental data and hence comparable also with each other.

The detailed analysis of the theoretical approaches is beyond the scope of this article and it can be found in specialized literature [32, 77–84]. However, several general comments will be presented here which might help in a judgement of the physical basis of various approaches, and thus facilitate the attainment of some conclusions useful for heterogeneous catalysis.

As can be seen from Table 2, two basic types of methods are used in this field: quantum chemical methods and methods of solid state physics. The cluster models are typical for the former and the jellium type models for the latter.

1. Cluster Approach

The representation of a metal by a single atom or by a cluster of several atoms seems to correspond well with the generally accepted localized character of surface chemical interactions (see *e.g.* refs. [4, 61, 64, 69, 77, 80, 82, 85]). Unfortunately, the quantitative results of these calculations are not very encouraging (see Table 2): unrealistic negative values of heats

Table 2. Summary of the Theoretical Results

Metal	Gas	Type of site	E_{H-M}/ kJ mol^{-1}	Q_{H_2}/ kJ mol^{-1}	q/electron charge	D/ 10^{-30} C · m	E_{vib}/ meV	d_{H-M}/ Å	Note	Ref.
Cr J	H		208	− 18 c	*ca.* −0.1	2.6	146	0.64	DFF (d_{H-M} measured from the edge of positive charge distrib.)	[56]
Cr_1	H	A	94	−247 c				2.2	EHMO	[57]
Cr_1	H	A	385	334 c				1.3	EHMO	[58]
Fe J	H		208	− 18 c	*ca.* −0.1	2.6	146	0.64	DFF (see note at CrJ)	[56]
Fe_1	H	A	53	−329 c				1.95	EHMO	[57]
Fe_1	H	A	230	25 c				1.1	EHMO	[58]
Fe_1	H		193	− 49 c	−0.36			1.6	HeF	[59]
Fe_2	H		193	113	−0.42			1.6		
				− 49 c						
Fe_4 (100)	H		220	5 c	−0.40			1.6		
Co J	H		211	− 13 c	*ca.* −0.1	2.6	146	0.64	DFF (see note at Cr J)	[56]
Co_1	H	A	53	−329 c				1.81	EHMO	[57]
Co_1	H	A	167	−101 c				1.0	EHMO	[58]
Co_9	H_2			12					EHMO	[60]
Ni J	H		211	− 13 c	*ca.* −0.1	2.6	144	0.64	DFF (see note at Cr J)	[56]
Ni_1	H	A	713	990 c	−0.64			1.47	EHMO (N_1)	[12]
Ni_1	H	A	366	297 c				1.5	EHMO (A_{max})	[61]
Ni_1	H	A	55	−325 c				1.75	EHMO	[57]
Ni_1	H	A	63	−309 c				0.5	EHMO	[58]
Ni_1	H	A	−20	−475 c					CNDO	[13]
Ni_1	H	A		−134					IB	[62]
Ni_1	H	A	304	154	−0.32				WH	[63]
				173 c						
Ni_2	H	B		− 21					IB	[62]
Ni_9	H_2			− 1					EHMO	[60]

Table 2. (Continued)

Metal	Gas	Type of site	E_{H-M}/ kJ mol^{-1}	Q_{H_2}/ kJ mol^{-1}	q/electron charge	D/ 10^{-30} C · m	E_{vib}/ meV	d_{H-M}/ Å	Note	Ref.
Ni_4 L	H_2			203				1.0	EHMO	[64]
Ni_5 L	H_2			221				1.0	EHMO	[64]
Ni_7 edge	H	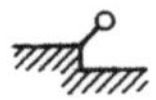 A	270	106 c				1.5	EHMO (A_{max})	[61]
Ni_{11} notch	H	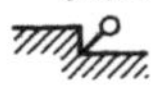 A	239	43 c				1.5		
Ni_{20}	H	A	154	−126 c			275	1.50	HF	[65]
		B	270	104 c			170	1.59		
		3 CN	308	181 c			150	1.63		
		4 CN	289	143 c			73	1.78		
Ni_{32} PC	H_2			50				2.0	EHMO (calculated for constant d_{H-H} = 1.2 Å)	[53]
Ni_{14}	H_2			57				2.0		
Ni_4 (100)	H	C		59					IB	[62]
Ni_5 (100)	H	O	326	163	−0.30				WH	[63]
Ni_5 (100)	H	A	30	−375 c	−0.23			1.6	CNDO	[13]
Ni_{10} (100)	H	A	100	−235 c	−0.19			1.6		
Ni_5 (100)	H	O in surface plane	299	163 c	+0.17			1.77		
Ni_{10} (100)	H	O in surface plane	199	− 37 c	+0.16			1.77		
Ni_9 (100)	H	A	572	707 c					EHMO	[53]
		O	433	431 c						
Ni_9 (100)	H	A	256	78 c				1.54	EHMO (A_{max})	[61]
Ni_{13} (100)	H	A	520	605 c	−0.67			1.47	EHMO	[12]
Ni_{14} (100)	H_2			(−58)–(−193)					EHMO (A_{max})	[61]

Ni_{250} (100) S_5	H layer	A	250	66 c	—0.36	1.5	EHMO (A_{max}) (N_2)	[66]
Ni_{250} (100) S_6	H layer	A	240	45 c		1.5	EHMO (A_{max}) (N_2)	[66]
Ni (100) PC	H		336	238 c		0.3	SMD (B)	[67]
Ni (100)		A	216	— 3 c	—0.41	1.27	TB BC	[68]
infinite	H	B	255	76 c	—0.34	1.38		
S		C	270	105 c	—0.36	1.8		
Ni_2 (110)	H	B	202	— 70	—0.68		WH	[63]
				— 30 c				
Ni_4 (110)	H	O	237	1, 39 c	—0.53			
Ni_7 (110)	H	A	551	668 c			EHMO	[53]
		O	369	304 c				
Ni_7 (110)	H	O	254	73	—0.01	1.65	CNDO	[13]
Ni_8 (110)	H	A	267	99 c		1.54	EHMO (A_{max})	[61]
Ni_{13} (110)	H	A	534	634 c	—0.64	1.47	EHMO (N_1)	[12]
Ni_{250} (110) S_5	H layer	A	270	105 c	—0.41	1.5	EHMO (A_{max})	[66]
Ni_{250} (110) S_6	H layer	A	271	106 c		1.5		
Ni (110) PC	H		334	234 c		0.3	SMD (B)	[67]
Ni_3 (111)	H	O	218	— 37, 1 c	—0.60		WH	[63]
Ni_5 (111)	H	A	20	—395 c	—0.20	1.6	CNDO	[13]
Ni_{10} (111)	H	A	110	—215 c	—0.23	1.6		
Ni_5 (111)	H	O	280	125 c	—0.03	1.75		
Ni_{10} (111)	H	O	290	145 c	—0.05	1.75		
Ni_{10} (111)	H	A	242	49 c		1.54	EHMO (A_{max})	[61]
Ni_{10} (111)	H	A	524	614 c			EHMO	[53]
		O	558	682 c				
Ni_{13} (111)	H	A	491	548 c	—0.66	1.47	EHMO (N_1)	[12]
Ni_{250} (111) S_5	H layer	A	237	39 c	—0.33	1.5	EHMO (A_{max}) (N_2)	[66]

Table 2. (Continued)

Metal	Gas	Type of site	E_{H-M}/ kJ mol^{-1}	Q_{H_2}/ kJ mol^{-1}	q/electron charge	D/ 10^{-30} C · m	E_{vib}/ meV	d_{H-M}/ Å	Note	Ref.
Ni_{250} (111) S_6	H layer	A	229	24 c				1.5	EHMO (A_{max}) (N_2)	[66]
Ni (111) PC	H		322	209 c				0.3	SMD (B)	[67]
Mo J	H		219	3 c	*ca.* −0.1	3.1	146	0.69	DFF (see note at Cr J)	[56]
Mo_3	H	A	170 289	− 95 c 143 c					HM (results for two values of hopping terms are shown)	[69]
Ru_3	H	A	220 301	5 c 167 c					HM (see note at Mo_3)	[69]
Rh_3	H	A	208 278	− 19 c 121 c					HM (see note at Mo_3)	[69]
Pd J	H		211	− 12 c	*ca.* −0.1	2.6	144	0.64	DFF (see note at Cr J)	[56]
Pd_1	H	A		−117					IB	[62]
Pd_2	H	B		8						
Pd_2	H_2 2 H			99 418				1.68	EHMO	[60]
Pd_3	H	A	170 231	− 95 c 27 c					HM (see note at Mo_3)	[69]
Pd_9 P 3 × 3	H_2			104					EHMO	[60]
Pd_{13} P fcc	H_2 2 H			66 433						
Pd PC semiinfinite									SMD (B)	[67]

(100)	H		301	167 c						
(110)	H		313	192 c						
(111)	H		284	134 c						
Pd_4 (100)	H	C		84					IB	[62]
W J	H^+		67	−300 c		small	200		DFF	[16]
W J	H		219	3 c	*ca.* −0.1	3.1	148	0.69	DFF (see note at Cr J)	[56]
W_1	H	A	434	433 c				1.6	EHMO	[70]
W_3	H	A	170	− 95 c					HM (see note at Mo_3)	[69]
			289	143 c						
W_9 (100)	H	A	289	143 c	−0.36			1.65	EHMO	[70]
		5 CN	166	−103 c	+0.04			1.61		
W_{12} (100)	H	B	264	94 c	−0.26			1.96	EHMO	[70]
W (100)	H	A	328	221 c	−0.57			1.7	TB BC	[71]
S	H	B	366	298 c	−0.48			1.75		
infinite	H	C	347	259 c	−0.38			1.75		
W (100)	H		216	− 3 c				0.53	HF GF (E_{H-M} value for the overlap integral s = 0.17 is shown here)	[72]
Re J	H		224	13 c	*ca.* −0.1	3.3	145	0.69	DFF (see note at Cr J)	[56]
Os_3	H	A	220	5 c					HM (see note at Mo_3)	[69]
			301	167 c						
Ir J	H		230	26 c	*ca.* −0.1	3.6	86	0.74	DFF (see note at Cr J)	[56]
Ir_3	H	A	208	− 19 c					HM (see note at Mo_3)	[69]
			278	121 c						
Pt J	H		200	36 c	*ca.* −0.1	2.5	92	0.64	DFF (see note at Cr J)	[56]
Pt_1	H	A		−125					IB	[62]
Pt_2	H	B		17						
Pt_3	H	A	170	− 95 c					HM (see note at Mo_3)	[69]
			231	27 c						

Table 2. (Continued)

Metal	Gas	Type of site	E_{H-M}/ kJ mol^{-1}	Q_{H_2}/ kJ mol^{-1}	*q*/electron charge	*D*/ 10^{-30} C · m	E_{vib}/ meV	d_{H-M}/ Å	Note	Ref.
Pt_4 (100)	H	C		121					IB	[62]
Pt (100)	H	A	267	99 c	—0.59			1.7	TB BC	[68]
infinite	H	B	274	113 c	—0.51			1.8		
S	H	C	283	132 c	—0.49			1.96		
Pt PC										
unit	H_2	ideal								
cell		plane								
6 atoms		∞							EHMO (H 1*s* 12 eV)	[73]
				125				2.0	d_{H-H} = 1.2 Å	
		notch		122				2.0	d_{H-H} = 1.2 Å	
		plane		85				2.5	d_{H-H} = 1.0 Å	
		notch		89				2.5	d_{H-H} = 1.0 Å	
		plane		18				3.0	d_{H-H} = 0.74 Å	
		notch		58				3.0	d_{H-H} = 0.74 Å	
Al J	H_2	∞		3.6				2–2.5	DFF (see note at Cr J)	[74]
		8		2.4						
Al J	H		144	—146 c		1.7		0.58	DFF (see note at Cr J)	[75]
Al J	H	A	164	—107 c				0.66	DFF (see note at Cr J)	[76]
structured	H	B	183	— 69 c				0.26		
	H	C	96	—242 c				0.38		
Al J	H		202	— 30 c	*ca.* —0.1	2.3	165	0.64	DFF (see note at Cr J)	[56]
Li J	H		173	— 88 c	*caj* —0.1	0.4	92	0.21	DFF (see note at Cr J)	[56]
K J	H		156	—122 c	*ca.* —0.1	—0.3		0	DFF (see note at Cr J)	[56]

Cu J	H		202	− 30 c	*ca.* −0.1	2.0	92	0.55	DFF (see note at Cr J)	[56]
Cu_1	H	A		−104					IB	[62]
Cu_1	H	A	25	−385 c				>2.4	EHMO	[57]
Cu_4 (100)	H	C		− 33 c					IB	[62]
Cu_9	H_2			− 70					EHMO	[60]
Ag J	H		193	− 49 c	*ca.* −0.1	1.4	100	0.42	DFF (see note at Cr J)	[56]
Ag_1	H	A		− 84					IB	[62]
Ag_4 (100)	H	C		−150					IB	[62]
Ag_9	H_2		− 59						EHMO	[60]

E_{H-M} — the binding energy of hydrogen atoms; Q_{H_2} — the heat of chemisorption of hydrogen molecules; q — the charge on the chemisorbed hydrogen atom (fraction of an electron charge); D — dipole moment; E_{vib} — the vibration energy of a particular surface bond; d_{H-M} — hydrogen-metal distance; J — jellium model; S_x — the slab, consisting of x atomic layers; PC — the periodic crystal; L — linear chain; P — plannar cluster; M_x(hkl) — the cluster of x metal atoms (of the kind M), simulating the crystallographic plane (hkl); A, B, C and O — the types of adsorption sites; top, bridge, center and over a hole; *n*CN — *n* coordinated adsorption site; c — the heats of chemisorption, estimated from equation (1) for $D_{H-M} = 435$ kJ mol^{-1} (negative values representing physically unrealistic results); (N_1) — no minimum on the potential energy curve has been observed, E_{H-M} being estimated for the listed value of d_{H-M}; (N_2) — input parameters: H 1*s* 10 eV, H layers on both sides of the slab; (A_{max}) — only A sites are listed here, because they exhibit highest values of E_{H-M}; (B) — in this case the binding energy profiles were estimated for the whole unit cell; DFF — density functional formalism; EHMO — extended Hückel molecular orbital method; CNDO — complete neglect of differential overlap method; IB — interactiong bond method; WH — Wolfsberg-Helmhotz method; SMD — semiempirical method of Doyen; TB BC — tight binding method of Bullett and Cohen; HM — Hubbard model; HF GF — Hartree-Fock Green function approach; HeF — Hellmann-Feynman approach.

of chemisorption Q_{H_2} have been obtained in many cases. This is not very surprising, because even for the most simple systems, represented by diatomic molecules, these methods do not always give reliable values of binding energies; for instance: Ni-H (experimental values for this diatomic molecule; 299 kJ mol^{-1} [53], 251 kJ mol^{-1} [33a, 86], compared with theoretical values; 742 kJ mol^{-1} [53], 713 kJ mol^{-1} [12], 366 kJ mol^{-1} [61], 304 kJ mol^{-1} [63], 55 kJ mol^{-1} [57], 63 kJ mol^{-1} [58]): H—H (experimental value for this diatomic molecule; 435 kJ mol^{-1}, compared with theoretical values 626 kJ mol^{-1} [71], 728 kJ mol^{-1} [59], 928 kJ mol^{-1} [59].

Moreover, unrealistic input parameters have been used in some cases. For example, the ionization energy of a hydrogen atom was taken as 10 eV [66], 12 eV [73], 16.1 eV [59], instead of a correct value of 13.6 eV. As a matter of fact, this quantity has been used as a fitting parameter for obtaining better agreement with the experimental values of the adsorption heat [66], of the hydrogen dissociation energy [59] or of the ionization potential of hydrogen molecule [73]. The last example throws some doubt on the physical basis of the application of the mentioned methods to such problems. This can be explicitly seen in the case of an extremely small "equilibrium" distance d_{M-H} (corresponding to the minimum of the energy of the system under study), which is sometimes even smaller than the sum of the atomic radii. Consequently, in such cases the resulting equations should be considered as interpolation formulae only.

2. *Jellium Model*

The application of the jellium model to the theoretical description of surface phenomena on transition metals suffers from some fundamental difficulties. Jellium is in principle an isotropic model, whereas all the surface properties exhibit more or less anisotropic character. This problem is in most cases "overcome" by "structuring" of the jellium (*e.g.* in the form of a corrugated surface) which is introduced as a perturbation [14, 87–89]. However, it seems to be inappropriate to introduce as a perturbation a quantity, the anisotropy of which is of the same order of magnitude as is the range of values of this quantity for all the transition metals [9]. Moreover, if the fundamental equation of the density functional formalism is to be correctly applied, two basic conditions have to be fulfilled. This means that either the electron density distribution should change slowly (regardless of the magnitude of the change), or the overall change of this quantity should be small, no matter how rapid this variation is [14, 87–89]. Neither of these conditions is fulfilled near the metal surface. Neverthless, some authors still believe in successful universal applicability of this approach to surface problems (see ref. [1], pp. 326, 568). Additionally, when the fundamental equation of the density functional formalism is applied, only the first term in the expansion is usually used, the other ones being omitted. However, the inclusion of the second term (which should give a better approximation) worsened the results [88]. Nor are the quantitative results of this approach very encouraging (see Table 2). For example, the heats of hydrogen chemisorption Q_{H_2} are in many cases negative which would mean that metals like Fe, Co, Ni and Mo do not chemisorb hydrogen molecules from the gas phase at all. This, of course, contradicts the general experience [5, 32, 79].

3. *Qualitative Considerations*

The examples in the preceding paragraphs were selected to illustrate the view that the absolute values of many computed quantities may be neither very interesting, useful [90, 91], nor particularly credible [91]. However, the theory should be in principle capable of providing qualitative models that could be of great assistance in correlating, assigning and interpreting experimental results [61, 90, 91]. The qualitative trends resulting from changes of geometry and/or chemical constitution are particularly important [61, 90]. In this sense, for instance, the empirical quantum chemical schemes might be useful for treating surface reactivity problems [92, 93]. These general trends will be discussed later, after summarizing also the experimental data.

What kind of general qualitative conclusions can be drawn from the theory of surface processes? Because of the lack of a complete theory of these processes we must limit ourselves to partial problems only. The problem of trapping a gas molecule and its further activation in the surface region will be considered here as a central catalytic problem. The importance of this point is reflected in the recently increased theoretical interest in the sticking probability problem [94–97]. It is concluded that the crucial point is the dissipation of the energy of the arriving particle by localized electron excitation [94, 95], by collective excitations of electrons [94–96], and/or by the lattice [98]. For light gas particles (simple gases) the first of these modes of energy dissipation seems to be the most efficient one [95]. Obviously, this problem is closely related to the yet generally unsolved question of the role of various types of orbitals in the surface interactions.

In searching the theoretical literature, one might sometimes meet with explicit discrepancies. For example, a major contribution of *d*-electrons in the hydrogen—nickel system is advocated by refs. [61, 66, 85, 99, 100], a minor or even negligible contribution is proposed in refs. [13, 65, 67, 101], and finally the remaining assumption of an intermediate contribution is taken in refs. [77, 102]. In similar cases one has to bear in mind two important points. Firstly, the individual theoretical models and methods might preferentially stress the role of certain orbitals (*e.g.* UPS studies on Ni show a peak below the *d*-band induced by hydrogen chemisorption which can be described in Anderson-model theories as a 1*s*-*d* bonding state, whereas in the density functional approach this state is considered to be largely a 1*s* resonance [100]). Secondly, the role of *d*-orbitals in trapping the hydrogen molecule (explicitly expressed in the major role of *d*-orbitals or their hybrids in trapping [60, 61, 85, 92, 95, 102, 103]) is probably different from their participation in the "equilibrium" bond of a hydrogen atom on the surface (explicitly expressed in the minor role of *d*-electrons in the chemisorption bond of hydrogen atoms on Ni surface [13, 65, 67, 101]). This might explain at least some of the discrepancies. The essential role of *d*-orbitals for trapping hydrogen molecules manifests itself also in the anisotropy of adsorption properties of individual crystallographic planes, because the localized metal *d*-electrons reach further out through the conduction electron cloud on high index planes than on close-packed ones [95]. The directionality of the surface *d*-orbitals can be judged directly from the theoretical paper on the asphericity of *d*-electron clouds near the transition metal surfaces [104]. Interestingly, the "lobes" of the *d*-electron density distribution depicted in this paper [104] correspond to the directions towards nearest missing neighbours (in agreement with the speculative assumption used for the interpretation of the anisotropy of the FIM images [105]). A single exception is the (0001) basal plane of the hcp lattice, where surprisingly only two lobes were reported, instead of the expected three which would correspond to the trigonal symmetry of this plane.

Furthermore, it has been theoretically shown for several *d*-metals [106, 107] that the electrons in the upper part of the *d*-band (near the Fermi level) have a localized (atomic-like) character, whereas those near the bottom of the *d*-band appear to be delocalized (in an interesting analogy to the intuitive Pauling model of transition metals). Since the frontier orbitals play a decisive role in chemical interactions, this effect might be of some importance in discussing the chemical reactivity of transition metals.

Besides the perturbation of the trapped molecule due to the localized bonds with the metal surface, the high density of the delocalized (*s*, *sp*) electrons might play a nonnegligible role in surface interactions. This effect has been studied separately from the localized interactions as a response of the electron gas (spilled out from the metal surface) to the arriving hydrogen molecule or atom (the latter case was already mentioned above). The interaction of the hydrogen molecule with the *s* (*sp*) electrons can be described either in terms of quantum chemistry as the filling of the antibonding molecular orbitals (similarly as in the case of the *d*-electron interaction) [102], or in terms of screening [108–112]. In all cases where the hydrogen molecule in the configuration parallel to the surface enters the "spilled out" region of the metallic electrons, the binding within the molecule is weakened [108 to 110]. The only exception [111] makes an unrealistic assumption that the hydrogen molecule can be treated as an inert gas. The weakening of the bonds due to the screening by delocalized electrons directly follows from the Berlin's formulation of the Hellmann-Feynman theorem [113], namely that a bond between two atoms is automatically weakened if there arises a nonzero concentration of electrons in the antibonding region outside the two atoms [113]. This is exactly what happens when a gas molecule is "immersed" into the spilled out region of the delocalized electrons in the metal surface.

D. Experimental Results

It has been already mentioned in the preceding paragraphs that the exploitation of sophisticated contemporary experimental methods suffers from some fundamental difficulties connected with the necessity to use theory in order to extract the desired information from the measured quantities. Apart from that and from the problem of the surface smoothness on the atomic scale, additional problems are encountered in this field: the influence of the probing particles (electrons, photons, ions) onto the state of the adsorbate layer; the role of trace contaminants in the "clean" metal surfaces (*e.g.* the segregation of carbon, sulphur, alkali metals etc. into the surface region from the bulk of the crystal, which is sometimes initiated or accelerated by the surface chemical process itself [114, 115]). These effects can change the rate and/or the route of the surface reaction and might be even responsible for some discrepancies in experimental surface studies. The influence of some additives has been clarified by studying artificially poisoned and promoted surfaces in terms of either the geometric (*e.g.* refs. [116, 117]) or the electronic (*e.g.* refs. [118–120]) structures. These effects, however, are certainly interrelated. Similar are the problems with carbon and sulphur contamination coming from the reactant mixture [27, 121].

In addition to the above mentioned difficulties, there are several yet generally unsolved problems connected directly with the application of individual experimental methods. As an illustration, some unclarified topics can be listed. A great deal of difficulty is connected with the reference levels in surface photoelectron emission studies [9, 122–124]. There is also a problem of the interpretation of the work function in terms of the charge residing on the adsorbed particle and needed for comparison with theoretical results; the work function strongly depends on the generally unknown position of the adparticle with respect to the surface (the work function depends on the dipole moment of the surface layer). The latter factor is important particularly in the case of such small particles like hydrogen atoms or protons capable of penetration into the metal lattice in the subsurface region. This might result even in reversing the sign of the dipole moment without changing the real charge of the adparticle [9]. Finally, there is the problem of the "average" work function and its changes in the case of stepped surfaces [6, 9, 125, 126].

1. Summary of the Experimental Results

Some of the above mentioned difficulties can be eliminated by exploiting simultaneously several reasonably chosen experimental techniques *in situ*. This has been practicised frequently in the past decade when an increasing number of combined experimental techniques have become commercially available. The quantitative results of the experimental studies on well defined surfaces are summarized in Table 3 (this field has been recently reviewed also in ref. [214]). The existence of "multiple adsorbed states" which is an interesting feature exhibited by many metals, is apparent in this table. Among the various "types" of adparticles, the weakly bound species and particularly their role in catalytic reactions have attracted some interest [207–213]. However, the experimental studies of this problem on well defined surfaces are scarce and therefore in this special case also the results obtained on more or less industrial type catalysts will be briefly commented in the present article.

Table 3. Summary of the experimental results

Metal (crystallographic plane)	N_s	Type of site	Structure of the adsorbed layer	E_D/ kJ mol^{-1}	Q_{H_2}/ kJ mol^{-1}	$\Delta\Phi_{max}$/ eV	E_{vib}/ meV	d_{H-M}/ Å	$s(T)$	Note	Ref.
Fe (100)	2		N	100, 75a		0.07					[127]
	2			87, 92a 69a					0.03		[128]
(110)	2		(1 × 1) hc	108, 77a		—0.09			0.16 (140 K)		[127]
	2		c(2 × 2)	100a 83a							[129]
(111)	3		N	88, 75a 54		0.31					[127]
	2		N	86 79							[129]
Co (1000)	1			67					0.04 (300 K)	No interaction with CO up to 300 K)	[130]
Ni (100)	1			96					0.06 (300 K)		[131]
			D		94 i_s	0.17			0.25 (300 K)		[132]
		C	p(2 × 2) lc				74				[35]
			c(2 × 2)								[133]
(110)	2		(2 × 1) lc			0.58					[134]
			(1 × 2) hc								
			(1 × 2) hc		90 i_s	0.53			0.25 (300 K)		[132]
				137					0.8 (300 K)		[135]

	1	B	(1 × 2)	123 i_t	82 i_s	0.4				[136]
				86a						
				85 i_t						[137]
					116 i_s					[138]
				98						[139]
			(1 × 2)							[140]
(111)	2	O	p(2 × 2)	90		0.17	1.84		2 H atoms on	[141]
				79		0.10 hc			one O site	
		3 CN	p(2 × 2)							[142]
	1			95				0.1		[143]
								(300 K)		
			D	94		0.19		0.25		[132]
								(300 K)		
			(2 × 2)							[144]
										[145]
Mo (100)	3			113						[146]
				84						
				67						
						0.68				[147]
	2		N hc			0.2 lc				[148]
						0.4 hc				
(110)	2			142						[146]
				117						
						0.48				[147]
(111)						0.42				[147]
Ru (0001)	2			92						[149]
				58 a						
(0001)	1			109 a						[150]
Rh (111)	1		N	105 a					H can be displaced by CO, O_2	[152]
	2		N	78				0.65		[153]
				34 a				(175 K)		
	1			75						[154]

Table 3. (Continued)

Metal (crystallographic plane)	N_s	Type of site	Structure of the adsorbed layer	E_D/ kJ mol^{-1}	Q_{H_2}/ kJ mol^{-1}	$\Delta\Phi_{max}$/ eV	E_{vib}/ meV	d_{H-M}/ Å	$s(T)$	Note	Ref.
Pd (110)				88							[151]
			(1 × 2)	82 a	102 i_s	0.36					[155]
(111)			(1 × 1)		87 i_s	0.18					[155]
9(111) × (111)					99 i_s	0.23					[155]
W (100)	2	A		120 a 100 a							[156]
	2			150 104						Both states are displaced by N_2 at $T \geqq 300$ K	[157]
	2									Both states are displaced by CO at $T > 300$ K	[158]
	2		N							Only one state displaced by CO	[159]
	2			135 110						H-D exchange from both states	[160]
		C lc C + Ahc	c(2 × 2) lc								[161]
	2								0.51 (300 K)		[162]
	2		(2 × 2) lc (1 × 1) hc								[163]
	2			134 105		0.8				No displacement by CO at $T < 284$ K	[164]
	2			134 109							[146]
					146 i_s	0.42					[165]

								0.6 (300 K)		[166]
								0.13 (78, 300 K)		[167]
					0.85				Displacement by CO at $T \leqq$ $\leqq$ 300 K	[168]
	2	B		130, 115 a 107, 103 a				0.6 (170– –400 K)		[169]
		B	c(2 × 2)							[170]
	2	A lc B hc				155 lc 130 hc				[44]
	2					159 lc 132 hc				[43]
					0.94					[171]
	2	A lc B hc				155 lc 130 hc				[41]
	2	C lc B hc								[172]
		B	c(2 × 2)		0.9					[173]
	2									[174]
					0.73					[147]
		B	c(2 × 2) lc			155 lc 130 hc	2.15 lc 2.05 hc			[175]
		A lc B hc	c(2 × 2) lc p(1 × 1) hc			159 lc 130 hc				[35]
			(1 × 1)							[176]
(110)	2			136 113					H-D exchange from both states	[160]
			(2 × 1)							[177]
	2		(2 × 1)							[163]
					—0.48					[171]
	2	A + 5 CN at all coverages				157 A 95 5 CN				[41]

Table 3. (Continued)

Metal (crystallographic plane)	N_s	Type of site	Structure of the adsorbed layer	E_D/ kJ mol^{-1}	Q_{H_2}/ kJ mol^{-1}	$\Delta\Phi_{max}$/ eV	E_{vib}/ meV	d_{H-M}/ Å	$s(T)$	Note	Ref.
	2			138 113							[146]
W (110)		B hc							0.07 (300 K)		[162] [166]
	2										[174]
						0.83 lc 0.27 hc					[147]
				272–343							[179]
(111)	4			153 91 127 59							[160]
	4			130 79 104 50							[180]
									0.26 (300 K)		[166]
						0.28					[171]
	1	A					160				[41]
					153 i_s	0.21					[165]
(211)	3		N	134 a 113 a 92 a		0.72				Later authors agreed 2 states (ref. [182])	[181]
	2			146 a 92 a							[182]
					167 i_s	0.37					[165]
						0.65					[147]
	2			146 67		0.61			0.57^+ (110 K) 0.05^+ (110 K)	^+s — values for the two states	[183]

Ir	(100)					0.1					[184]
						—0.8					[185]
	(110)	2	A + C	N	96	0.3			1 (130 K)		[186]
		1		N	92 a						[189]
		2			92 a, 57 a	—0.05					[187]
	(111)	1		N	92 a						[190]
						—0.6					[185]
	6(111)×(100)	1		N	92 a						[190]
Pt	(100)	5			94, 51 58, 38				0.15 (78 K)		[191]
		4–5			96 a, 54 a						[192]
		2			100, 67	—0.37					[193]
					63–67						[194]
				(2×2)							[195]
	(110)	2			79 a, 58 a						[192]
		2		55, 41					0.2 (78 K)		[191]
						—0.38 0 hc					[196]
	(111)	2		N	39, 25	0.002 —0.23 hc			0.1 (150 K)	H-D exchange proceeds	[197]
		3			73, 74 a 52 a 33 a				0.1 (78 K)	Surface with preadsorbed oxygen s = 0.2 (78 K)	[198]
		2			86 a, 75 a						[199]
										H-D exchange does not proceed	[7]
		1			73 a						[192]
						—0.3					[196]
		1			71	—0.56					[193]
			C				153 lc 68 hc	1.76			[40]

Table 3. (Continued)

Metal (crystallo-graphic plane)	N_s	Type of site	Structure of the ad-sorbed layer	E_D/ kJ mol^{-1}	Q_{H_2}/ kJ mol^{-1}	$\Delta\Phi_{max}$/ eV	E_{vib}/ meV	d_{H-M}/ Å	$s(T)$	Note	Ref.
Pt (111)				130							[201]
				125—134							[202]
9(111) × (111)										H-D exchange proceeds	[7]
				50		0.02 —0.35 hc			0.34 (120 K)	H-D exchange: activ. energy as on (111), the rate 10 × higher	[203]
6(111) × (111)				$94\ a^{++}$						$^{++}$value for steps	[199]
				$75\ a^{+++}$						$^{+++}$value for terraces	
6(111) × (100)						—0.3					[204]
	2			$94\ a^{++}$ $75\ a^{+++}$							

N_s — the number of adsorption states; E_D — the activation energy of desorption; a — approximate values of E_D, estimated from the peak temperatures (T_m), using the empirical equation: $E_D = 0.23T_m$ kJ mol^{-1} (this equation is based on the data from refs. [127, 149, 155, 192, 198, 205]; the standard deviation of the empirical constant is 0.03); Q_{H_2} — the initial heat of chemisorption; i_s — isosteric Q_{H_2}; i_t — isothermal Q_{H_2}; $\Delta\Phi_{max}$ — maximum value of the work function change, due to the hydrogen chemisorption ($\Delta\Phi = \Phi_{ads} - \Phi_{clean}$); E_{vib} — vibration energy of a particular surface bond; d_{H-m} — hydrogen-metal distance; s — the intitial value of the sticking probability coefficient, estimated at the temperature (T K); N — no extra spots were observed on LEED pattern; D — disordered layer; lc — low coverage value; hc — high coverage value; A, B, C, O — the types of the adsorption sites: top, bridge, center, over a hole; nCN — n coordinated adsorption site.

The phenomenon of "multiple state adsorption" is alternatively interpreted as being due to: i) various types of adsorbed species (H, H_2 etc., *e.g.* [160, 214 b]); ii) one type of adparticle, chemisorbed on various adsorption sites (*e.g.* [86, 168, 214, 215]); iii) mutual interaction of adparticles (*e.g.* [168, 215–217]). Obviously, several of these possibilities or even all of them might occur simultaneously. Some experimental evidence for i) can be extracted from work function measurements (*e.g.* [218, 219]). This evidence, however, is far from being unambiguous [9].

LEED results when various structures of adlayers have been observed, together with experiments on stepped surfaces [86] can be considered as experimental evidence in favour of the interpretation ii) above. Assumption iii) is usually verified by the agreement between experiment and theory in the field of desorption kinetics (*e.g.* [216, 217, 220]). Unfortunately, the conclusions drawn from the kinetic measurements are again not unequivocal (*e.g.* first order desorption kinetics does not prove definitely non-dissociative adsorption, since it can be interpreted also in terms of atomic adsorption on the basis of breakthrough or porthole mechanisms [213]).

The experimental evidence for the occurence of weak adsorption of hydrogen stems mostly from thermal desorption (TD) spectrometry [79, 259] which, unfortunately, often lacks chemical identification of the desorbed species. The importance of the latter factor can be exemplified by the results of TD studies where both total and partial pressures have been measured. For example it has been shown that the low temperature desorption peaks (the weakly bound species) can correspond not to the desorption of hydrogen but to the desorption of CO [221, 222], CO_2 [223], or CH_4 [221]. Sometimes the desorbed species do not originate from the sample surface under study [224]. The amount of the weakly hydrogen (even after low temperature adsorption) represents usually a small fraction of the total coverage (1–10%) [146, 168, 215, 221, 223, 225, 226].

The arguments for the occurence of these species at higher reaction temperatures based on Langmuir's description of the adsorption process [207] (which, of course, cannot be accurately applied to heterogeneous surfaces with a possible interaction between the adparticles) are not very convincing. Higher hydrogen pressure certainly can influence the coverage of weakly bound species, but only in the case when other reactants are weakly bound too, assuming all compete for the same adsorption sites. Thus the interpretation of the correlation between the hydrogenation activity of transition metals and the amount of weakly bound hydrogen [207] remains unclear. No definition of weakly bound hydrogen has been generally accepted. The published desorption temperatures and binding energies corresponding to weakly bound species even on one metal differ considerably from each other (*e.g.* Fe, 90–220 K [209, 227]; Pt, 120–250 K [208, 210, 228]); Pt, 5, 8, 24 kJ mol^{-1} [219, 228, 229]). Nevertheless, authors often discuss and compare experimental results as if they concerned one well defined adsorbed species only. The weakly bound state is often treated as an intermediate (precursor) state to strong chemisorption [217, 220].

The physico-chemical nature of the weakly bound hydrogen state is not yet fully understood [212–215]. Some authors speak about physically adsorbed hydrogen [221], others simply ascribe the low temperature peak(s) of the TD curves to chemisorbed molecular species [207, 208, 210], or speculate about theoretical reasons for molecular chemisorption of hydrogen on transition metal surfaces [218, 230–232]. Another group of authors advocates dissociative adsorption of weakly bound hydrogen [213, 233]. In some systems no hydrogen-deuterium exchange with weakly bound species has been observed at low temperatures, thus proving their molecular character [116, 160, 208, 215, 226]. At higher temperatures ($T \geqq 300$ K), however, no adsorbed hydrogen molecules on well defined surfaces could be detected [41, 43, 44, 175, 176, 199].

One is tempted to conclude that weakly adsorbed hydrogen is not of universal importance in catalytic reactions involving hydrogen. A possible example of its importance is in the selective hydrogenation of weakly bound reactants [209].

One phenomenon has not been included into the preceding considerations, namely the surface migration of chemisorbed species [72, 279, 307], because this phenomenon is probably less important for hydrogen activation itself. On the other hand, the general problem of transportation of the activated particle from the trapping site towards the center of activation for a chemical reaction [295] (these two sites being not necessarily identical) might influence the overall kinetics of the interactions on non ideal metallic surfaces (stepped surfaces, bimetallic or multimetallic systems etc.) and particularly on combined metal-non metal systems (supported metal catalysts) (see also section 3.). In Table 4 the chemisorption activity of all metals with respect to hydrogen is characterizid qualitatively. It can be seen from this table that besides the *d*- and *f*-metals also some alkali earth metals are active in hydrogen chemisorption. This has been exploited for gettering in vacuum technology. Since, however, the mentioned metals do not exhibit a catalytic activity, neither experimental, nor theoretical effort has been devoted to them from the catalytic point of view. These metals exhibit exceptional and non-selective chemisorption activity, leading mostly to the formation of bulk compounds. Moreover, all of them have low melting points. Both of these factors exclude the exploitation of these metals in practical catalysis, except the possible participation as promoters. However, the qualitative difference between these metals and the other "free electron" metals — alkali metals and Al

Table 4. The activity of metals in hydrogen chemisorption [5, 32, 206]

Li													
Na	Mg^{-}											Al^{-}	
K^{-}	Ca^{+}	Sc^{+}	Ti^{+}	V^{+}	Cr^{+}	$Mn^{\mp}$	Fe^{+}	Co^{+}	Ni^{+}	Cu^{-}	Zn^{-}	Ga	
Rb	Sr^{+}	Y^{+}	Zr^{+}	Nb^{+}	Mo^{+}	Tc	Ru^{+}	Rh^{+}	Pd^{+}	Ag^{-}	Cd^{-}	In^{-}	Sn^{-}
Cs	Ba^{+}	La^{+}	Hf^{+}	Ta^{+}	W^{+}	Re^{+}	Os^{+}	Ir^{+}	Pt^{+}	Au^{-}	Hg^{-}	Tl	Pb^{-}

$^{+}$ indicates chemisorption; $^{-}$ no chemisorption; both signs simultaneously indicate a discrepancy in the quoted papers; no sign — means that no information is available
(all this information has been obtained on evaporated metal films)

which are inactive for the chemisorption of molecular hydrogen — is not yet fully understood.

In the following considerations, attention will be concentrated onto the catalytically most interesting group of *d*-metals. Their exceptional position among other metals is probably due to the directionality and the high density of the *d*-states near the Fermi level (frontier orbitals), so that a sufficient number of orbitals of desired symmetries are available as frontier orbitals, enabling the stabilization of some intermediate products along the reaction path [103].

2. *Localized Character of Chemisorption — the Role of d-Electrons*

Experimental evidence for the localized character of the chemisorption bond, and the information about the symmetry of the corresponding surface orbitals can be found in the spatial anisotropy of photoemission [51, 163, 176, 234], in the angular distribution of electron stimulated desorption [156] and in the anisotropy of field ionization [105, 235, 236]. The last evidence represents an illustrative example of information in terms of individual atomic events without any averaging procedure. The high probability of field ionization (the high brightness of the imaged surface metal atom) is presumed to be a consequence of the large overlap of the *s*-orbital of the imaging gas atom with the surface orbital emerging from the surface; these orbitals being approximated by the directions towards the nearest missing neighbours of the original metal lattice [105, 125, 235]. This model has successfully explained not only the regional brightness of the FIM image around the (100) plane of the fcc metals and the anisotropy both around the (111) plane of the fcc metals and around the (0001) plane of the hcp metals [105, 125, 235], but also the anisotropy of the reaction zone around the (111) plane of platinum in the low temperature oxidation of hydrogen [237]. This last anisotropy is identical with that of the FIM image of the same region. Thus the directional orbitals can be obviously identified with the trapping centers for the active gases, *e.g.* hydrogen [9, 125, 236].
Additionally, this experiment [237] suggests that the trapping centers are probably not permanently blocked by chemisorbed species (in this particular case by preadsorbed oxygen), in agreement with some theoretical conclusions [103]. However, the latter effect probably does not occur with the same intensity on all transition metals (see later, the experiment with atomic hydrogen on nickel).

The role of *d*-orbitals as trapping centers for hydrogen molecules can also be evidenced by the mere comparison of the chemisorption activity of the *d*-metals and of the noble or alkali metals, having frontier orbitals of *s*-type only (see Table 4). The latter metals are able to chemisorb strongly hydrogen atoms (with approximately the same binding energy as do the transition metals), if they are supplied directly from the gas phase [9, 31, 235]. They are, however, not able to chemisorb hydrogen molecules. A similar conclusion can be arrived at from the experiment with a polycrystalline nickel surface saturated by gaseous molecular hydrogen. This surface is able to bind additional hydrogen, if supplied in the atomic form from the gas phase [235].

Interesting evidence of the role of *d*-orbitals as trapping centers has been also provided by the experiments with stepped surfaces [238].

Obviously, the most direct information about the participation of the *d*-electrons in surface processes could be obtained from the photoemission spectroscopy. However, the theoretical identification of contributions of individual energy bands to the experimental spectral curve is still far from being unambiguously solved (see Section 2.C.3 and other examples in refs. [239, 240]). Consequently, there is still not full agreement among various authors in the interpretation of the experimental photoemission curves. Most authors infer the participation of the *d*-electrons from the decrease of the peak height near the Fermi level energy (*e.g.* refs. [241, 242]), and they conclude that the contribution of the *d*-electrons to the "equilibrium" chemisorption bond increases in the sequence Ni < Pd < Pt. On the other hand, the change of peak height might also result from a matrix element effect [67]. This problem has been studied also experimentally by estimating the symmetry corresponding to particular bands from the angular distribution of the photoemitted electrons. It has been concluded that on Ni (111) plane the *d*-electrons do not participate in the hydrogen-metal bond [243], whereas on W (100) plane they do [176, 244].

The decisive role of frontier orbitals in chemical reactions is undisputable. Unfortunately, the theory of the surface electronic structure of the transition

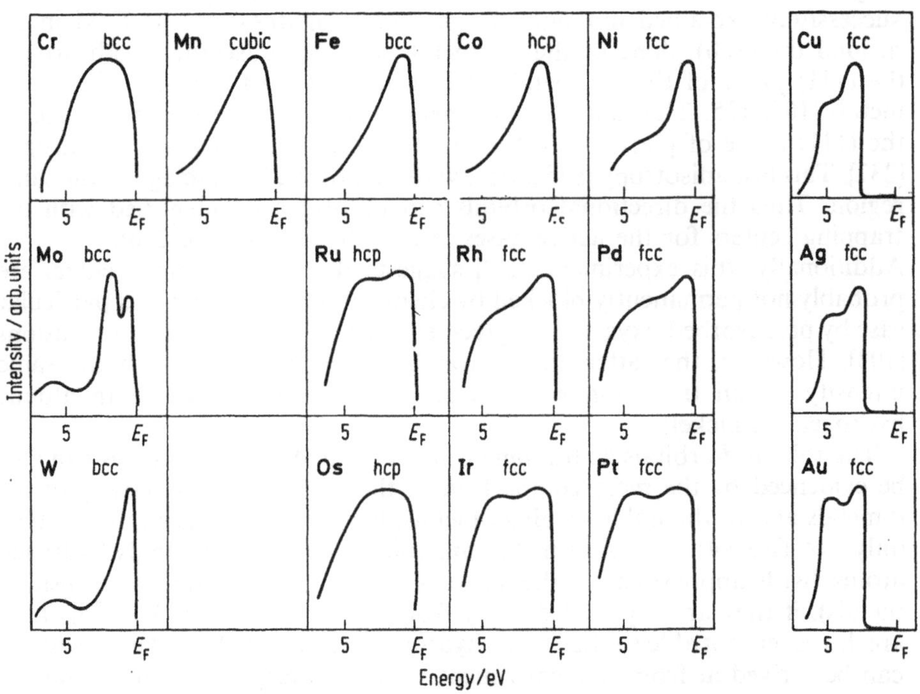

Figure 3. The schematic photoelectron spectra of transition metal surfaces. (After [239, 245 to 258])

metals provides us with divergent results strongly dependent on the models and computational methods [78, 239, 240] losing thereby the universal character of theoretical results. Experimentally accessible information on this subject can be found to some extent in photoelectron spectra, but they are related in a complicated way to the real density of states. The exploitation of these spectra implies several serious problems: the reference level problem (which is important for the interpretation of spectra, involving the comparison of theory and experiment); the problem of spatial anisotropy of photoemission and the problem of separating the surface and bulk features in the experimental curves. Having in mind all these problems, the qualitative character of the photoelectron spectra has been taken from refs. [239, 245 to 258], and depicted in Figure 3 for a series of transition metals. Strictly speaking, one should compare only the spectra of those metals which have the same crystallographic structure. However, since we are interested in the electronic structure of the metal surface in relation to its chemical reactivity (disregarding for the sake of simplicity the physical basis of a particular electronic structure), one can use Figure 3 for qualitative considerations on the catalytic activation of gas molecules.

3. The Delocalized Electrons

There exists also ample experimental evidence for the participation of delocalized electrons in surface chemical interactions. It is well known that these processes influence the electrical resistance, galvanomagnetic effects etc. of thin metallic films (see *e.g.* refs. [235, 259]). Moreover, one can speculate about the difference in the selectivity between heterogeneous and homogeneous metal catalysts. This difference might originate from the high density of delocalized electrons in metal surfaces, in contrast to their absence in coordination complexes. The effect of delocalized electron density onto the reactants is probably less selective than the perturbation of the reactants due to the localized interaction.

E. Discussion of Theoretical and Experimental Results

The energy of the interaction of gas molecules with metal surfaces is a quantity of primary interest both in theoretical and experimental studies. In spite of a great effort and a large quantity of published material, a direct comparison of the results is limited to several examples only, where identical systems have been studied (see Tables 2 and 3).

Figure 4 compares the calculated values of heat of chemisorption of hydrogen on (100) planes of tungsten and nickel with relevant experimental values. As it was already mentioned, the quantitative agreement is not very encouraging. The effect of the number of atoms in a metal cluster representing a particular crystallographic plane is shown in Figure 5 for (100) and (111) planes of nickel. The theoretical results are randomly scattered, the same degree of reliability being achieved with both small and large clusters. Qualitative trends could be followed in a limited number of cases only. The

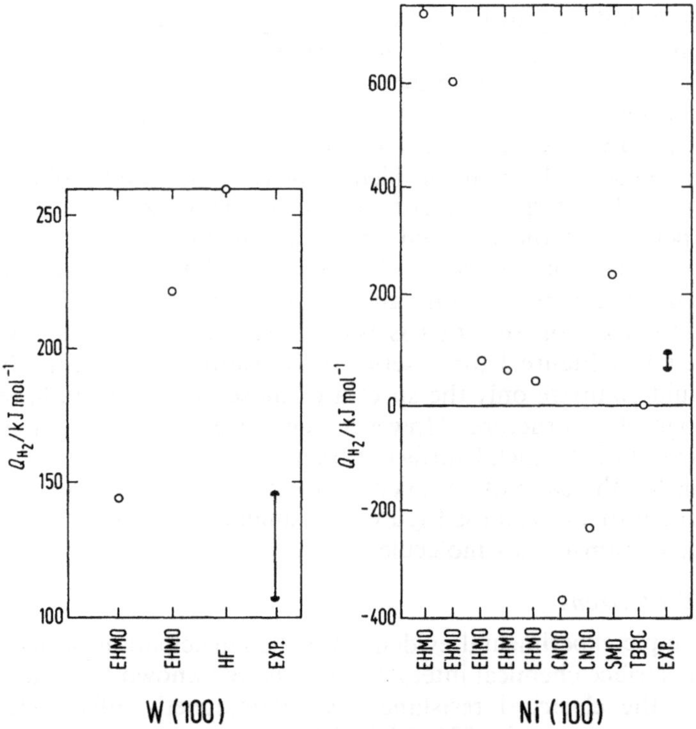

Figure 4. The comparison of theoretical and experimental values of hydrogen heat of adsorption on (100) planes of tungsten and nickel (see Tables 2 and 3); I— range of experimental values

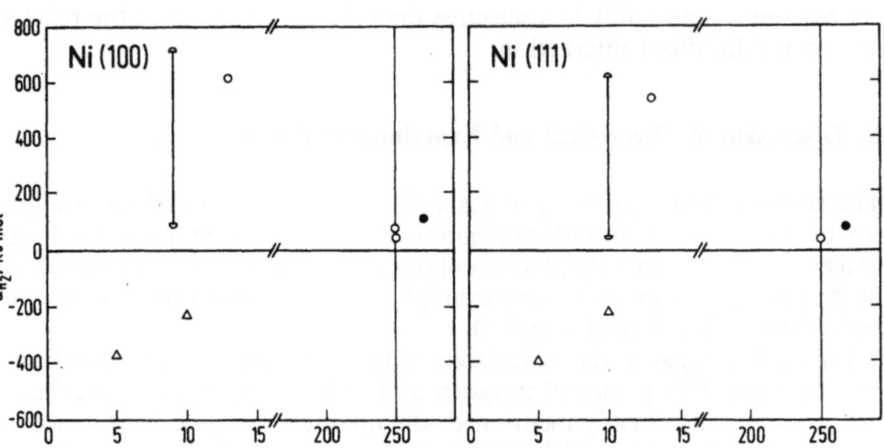

Figure 5. Theoretical values of heats of adsorption of hydrogen on Ni clusters, simulating (100) plane (**a**), and (111) plane (**b**), as a function of the number of Ni atoms in a cluster (see Tables 2 and 3); ○ — EHMO values, △ — CNDO values, ● — average experimental values

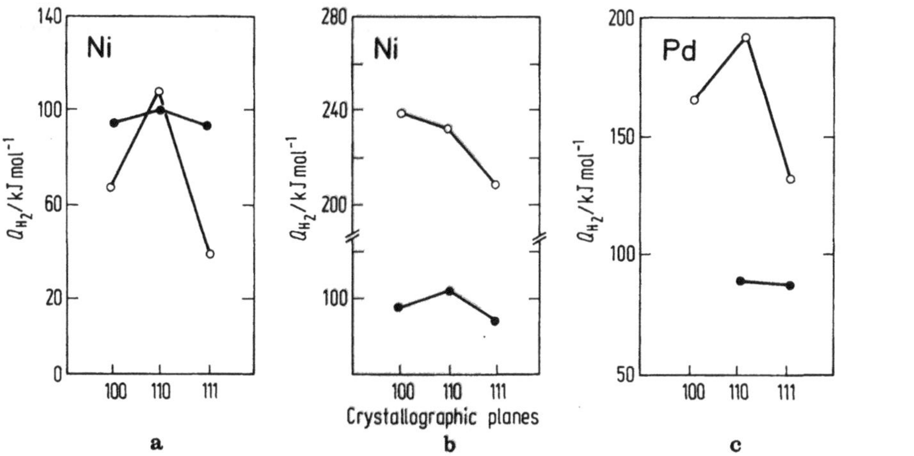

Figure 6. Crystallographic anisotropy of the hydrogen chemisorption heats. Ni — EMHO **(a)**, Ni — SMD **(b)**, Pd — SMD **(c)** (see Tables 2 and 3); ○ — theory, ● — experiment

crystallographic anisotropy of the heat of chemisorption exhibits similar trends for both experimental and theoretical curves, the theoretical values being shifted mostly by a factor of 2 upwards (Figure 6). The general trend in the horizontal rows of the Periodic System is the decreasing heat of adsorption from left to right, experimentally observed for hydrogen [86, 129, 260] and also for other gases [86, 260]. This trend is reproduced in a more or less satisfactory shape in the "surface molecule" approach, while the calculations based on the jellium model give qualitative disagreement with experiment (Figure 7). This can be expected because of the localized character

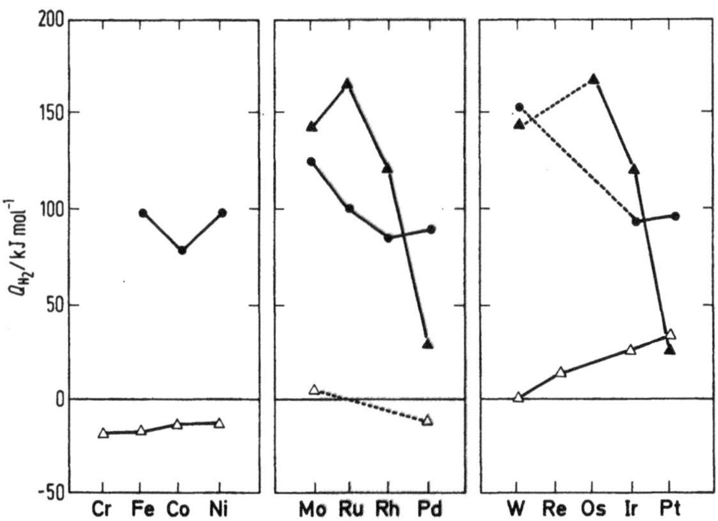

Figure 7. General trends of hydrogen chemisorption heats in horizontal rows of the Periodic System △ — theory ref. [56], ▲ — theory ref. [69], ● — experiment (average values from all planes for the most strongly bound species (see Table 3)

of chemisorption and the delocalized character of jellium model interactions. Complete disagreement between the experimental and theoretical trends has been found in the case of (100) and (110) planes of metals in the vertical columns of the Periodic System (Figure 8). However, both theoretical and experimental data are not sufficiently complete in order to draw final conclusions.

In contrast to the heat of chemisorption, the binding energy E_{H-M} (estimated from the experimental chemisorption heats by means of equation (1) exhibits crystallographic anisotropy to a much lesser extent, and the interval of E_{H-M} values for all the transition metals is also narrower than the interval of adsorption heats. This is merely a reflection of the procedure for obtaining the binding energy values by means of equation (1), since the hydrogen dissociation energy D_{H-H} is three to four times larger than the experimental heats of chemisorption: consequently, the variation of the binding energies on different crystal planes or different metals is relatively much less pronounced.

The possibilities of comparing the theoretical and experimental results for other quantities are even more limited. The calculated values of the vibration energy agree well with the electron energy loss spectra for the Ni (100) plane (corresponding to a four-coodinated adsorption site for hydrogen atom). The agreement is worse for tungsten and platinum (see Tables 2 and 3) where the jellium model has been used. However, one general unambiguous conclusion has been attained in these studies: no molecular hydrogen bound to the metal surface could be detected [40, 41, 43, 44, 175, 176]. A direct comparison of the experimental values of work function changes with the theoretically calculated charges on the adsorbed atoms is impossible (without speculative assumptions), because of the above mentioned uncertainty of the position of the chemisorbed hydrogen with respect to the surface plane (see e.g. ref. [9]).

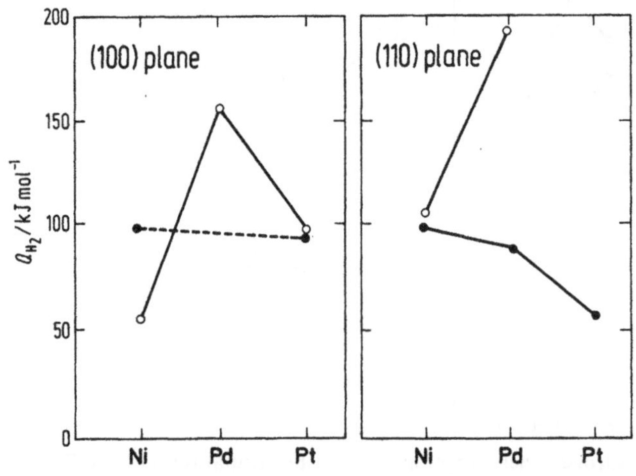

Figure 8. General trends of hydrogen chemisorption heats in vertical columns of the Periodic System (see Tables 2 and 3); ● — experimental values, ○ — theoretical values

The quantitative data collected in Tables 2 and 3 can be exploited for a discussion of the equillibrium state of an adsorbed hydrogen layer on a particular metal, and also to some extent for an assessment concerning the dynamics of surface processes. If the activation of dihydrogen is considered in terms of an energy balance, the value of the binding energies of the reaction intermediates in, for instance, the $H_2 + O_2$ reaction (H- and HO-) can be used to consider the probability of a particular reaction route [62]. For this purpose, the binding energies of the reaction intermediates may be estimated by, for instance, the interacting bonds method [62]. (The mechanism of the $H_2 + O_2$ reaction has also recently been studied on Ni (110) by a formal kinetic approach [261]).

Besides dihydrogen activation at metal surfaces, one can also exploit the dissolution of hydrogen in some metals as a specific form of activation, where the source of hydrogen need not always be a hydrogen molecule only (selective membrane catalysts [262]). The dissolved hydrogen can form various hydride phases, the catalytic activity and selectivity of which might differ from each other [263].

1. Localized-Delocalized-Electrons-Interplay Model

Having in mind the results and considerations outlined above, the following dynamic model of the surface chemical interaction of a hydrogen molecule with a transition metal surface (*i.e.* the activation of dihydrogen) can be constructed [9, 125]. Analogous considerations can be extended to other molecules too.

The characteristic features of transition metal surfaces from the viewpoint of the chemical interactions which have to be represented by a model are: i) the occurence of partly occupied *d*-orbitals localized at the surface atoms; ii) the occurence of a high density of delocalized electrons. A transition metal surface has the frontier orbitals energy, (*i.e.* the work function), equal

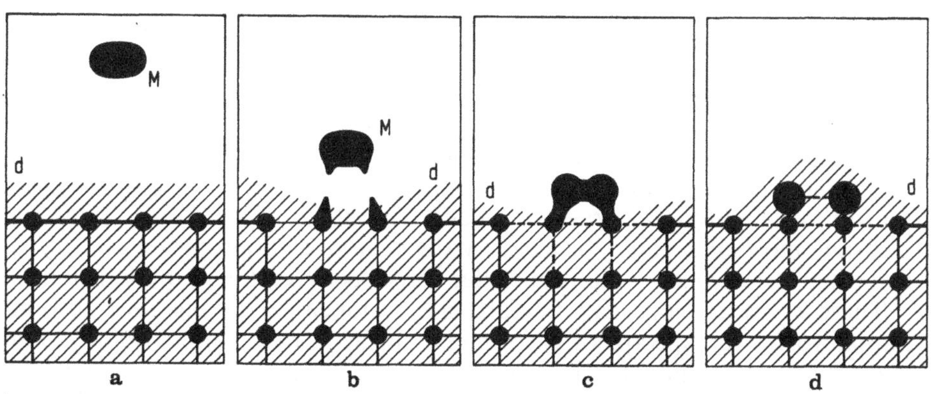

Figure 9. The Localized-Delocalized-Electron-Interplay model for the interaction of a gas molecule with a transition metal surface. (M — approaching gas molecule, d — the contour of the delocalized electron density, black circles — the ion cores, the prime lines — partly localized *d*-electrons (the thickness of the lines simulates the bond strength)

to 4–6 eV. The frontier orbital energy of the hydrogen molecule, *i.e.* its first ionization potential, is equal to 15 eV.

When a hydrogen molecule approaches a transition metal surface (Figure 9) then a surface hole forms in the delocalized electron density distribution due to the exchange and correlation repulsive forces, enabling a more intimate contact of molecule frontier orbitals with the localized orbitals of the metal. The localized bonds are then formed and consequently the electron structure of the whole surface complex is changed. This stage might be considered as a transition (precursor) state. As a consequence, the delocalized electron density relaxes into a new density distribution around the complex, embedding the trapped molecule or part of it, and thereby weakening both the intramolecular and metal-adparticle bonds (embedding of the "surface molecule") (Figure 9). The molecule has to stay on the surface long enough to enable: i) the relaxation of the delocalized electrons into the new density distribution; ii) the interaction of the "immersed" and thus weakened bond(s) of the reactant molecule with the neighbouring surface complexes or with another approaching molecule; iii) the complete dissociation of the "immersed" bond(s) so that the fragments can migrate apart, react with other species, or recombine again, the latter perhaps giving an isomeric product. The first of the above listed processes is probably rather fast [49], but any one of the others might be rate determining.

The desired selectivity of a given catalyst can be attained if the particular bond(s) are: i) perturbed by the formation of the localized surface bond(s) in complete analogy to homogeneous catalysis; ii) "immersed" into the *appropiate* electron density region. Obviously, because of the simplicity and symmetry of the hydrogen molecule, the other reactants play a decisive role, particularly in catalytic selectivity.

Many catalytic effects can be understood in terms of this model: the crystallographic anisotropy of catalytic properties, resulting from the different orientation and/or number of localized orbitals and from the different density distribution of delocalized electrons on various crystallographic planes on which a "smoothing" effect might cause partial "uncovering" of localized electrons at the edges of terraces; the role of constituents in bimetallic and alloy catalysts which might act as trapping centers and regulators of the density of delocalized electrons; catalysis by electron donor-acceptor complexes [264] which might be due to the cooperation of the "reservoir" of the delocalized electrons and the proper *d*-electrons of the central ion; catalysis by graphite loaded with alkali metals [264], where the reactant molecules could be held in the region of appropriate density of delocalized electrons by intermolecular forces between the graphite layers.

These considerations can be further developed by taking into account the photoelectron spectra of clean metal surfaces, reflecting to some extent their density of states distribution. In Figure 3 the region near the Fermi level is depicted because of its important role in chemical reactivity. On the basis of this figure one can formulate some general trends in the Periodic System of metals: i) the metals on the left side of Figure 3 having a relatively narrow *d*-band (2–4 eV), exhibit a strong interaction with gas molecules (high adsorption heats [86, 129, 260] and see also Table 3, dissociative chemisorption of CO and N_2 at low temperatures, cracking of hydrocarbons), probably on account of strong localized interaction with atomic-like surface orbitals; ii)) the metals on the right side of Figure 3 having a broad *d*-band (5–8 eV), exhibit weaker interactions with gas molecules (lower adsorption heats [86, 129, 260] and see also Table 3, nondissociative adsorption of CO and hydrocarbons at low temperatures), probably on account of screening effects by increased density of delocalized electron states near the bottom of the *d*-band (*c.f.* the appearance of the second maximum in the curves for these metals, particularly in the lower right part of Figure 3). Recently some new results in the photoemission spectra of Pd [309] can be considered as further evidence for the usefulness of this model. It has been shown [309] that a palladium monolayer on the Nb (110) plane exhibits a commesurate crystallographic structure with Nb. This system posesses an electronic structure similar to the noble metals (the first maximum on the photoelectron distribution curve lies at about 1.5 eV below the Fermi level) and this well correlates with a low activity of this Pd layer in hydrogen chemisorption. On the other hand, when the thickness of the Pd overlayer exceeds monolayer coverage, a close-packed fcc (111) mesh forms and the electronic structure typical for a transition metal can be detected (high density of states near the Fermi level). These changes of the electronic structure of the Pd overlayer have a dramatic effect onto its interaction with hydrogen [309].

On the basis of the Localized-Delocalized-Electrons-Interplay model and of the preceding considerations, one can formulate a qualitative general rule for the choice or tailoring of a metallic catalyst for a given reaction: an optimum interaction of a gas molecule (an optimum from the viewpoint of the particular reaction) is achieved when a certain ratio between the density of the localized and delocalized electrons occurs in the region near the Fermi level. The absolute values of these densities are probably less important.

The desired ratio of the above mentioned densities can be attained by [236]: i) increasing the density of the delocalized electrons (*e.g.* alkali metal promotion of an iron catalyst [119, 265], transition metal-noble metal alloys [239, 256]; ii) blocking of the strongly interacting localized electrons (*e.g.* molybdenum, tungsten and nickel catalysts covered with carbides [120, 266] or sulphides [267, 268]); iii) lowering the density of the delocalized electrons (*e.g.* highly electronegative oxygen on platinum or rhodium surfaces probably lowers particularly the density of delocalized electrons, thereby relatively strengthening the localized interaction; this results in the fission of the C—C bonds in hydrocarbon molecules [269] on platinum, and in efficient CO dissociation on rhodium [200]).

All these considerations represent probably an oversimplified image of the surface interactions. The influence of the modifying additives is certainly more complex. For example, they may cause a reconstruction of the surface layer, resulting in a change of the active site geometry; they can alter the charge density distribution; in some cases the exceptional reactivity of the additives towards possible contaminants may protect the cleanliness of the active metal surface [270]. Consequently, the above listed rules have to be considered as one of the possible approximations only.

3. Nonmetals

A. General Considerations

Transition metals with their high chemisorption activity for hydrogen are good catalysts for a wide range of reactions involving hydrogen. On the other hand, nonmetals which posses low chemisorption activity for hydrogen (sometimes even not chemisorbing it at all) are seldom used as catalysts in these reactions (or in special cases only: ZnO, Cr_2O_3 etc.). The main interest in the interaction of nonmetals with hydrogen has been motivated by two factors (ignoring some exceptional selective hydrogenations): i) in the case of semiconductors, there has been the expectation that band theory, elaborated primarily for its electronic applications, will be equally successful when applied to catalytic phenomena [188, 271–276]; ii) in the case of insulators, there is their practical importance as carriers for supporting industrial catalysts [188, 274, 276] (being either inert or taking part in reactions involving hydrogen *via* the spillover effect). Unfortunately experiments on well defined surfaces of these materials (single crystallographic planes with a known atomic structure) are very scarce. Most of the hitherto published work in this field has been obtained on more or less industrial catalysts with unknown chemical composition and structure of the surface layers [188, 274, 276]. This has considerably complicated the interpretation of the results because surface properties (chemisorption and catalytic activity) of these materials are very sensitive to the presence of defects and to traces of contaminants resulting from various procedures of sample preparation. This may possibly explain some controversial results [271, 273]. Most of the fundamental papers quoted even in recent reviews on these topics [188, 274–276] suffer from these drawbacks, being often based on classical experimental methods only (*e.g.* volumetric measurements on powder catalysts [271, 277]), and thus the conclusions on the molecular mechanism of these interactions are often only speculative ones. Because of all these reasons no experimental data have been explicitly included into this section.

The activation of molecules by nonmetals represents an even more complicated phenomenon than catalysis by metals. This is due to the fact that most semiconductors and insulators consist at least of two elemental constituents and can thus offer a wider variety of possible interactions with gaseous reactants than can the "simple" metallic surfaces (the elemental semiconductors having negligible importance for the catalytic activation of dihydrogen [188, 278]). Hydrogen chemisorption usually proceeds heterolytically on these materials [281, 282, 286, 293]

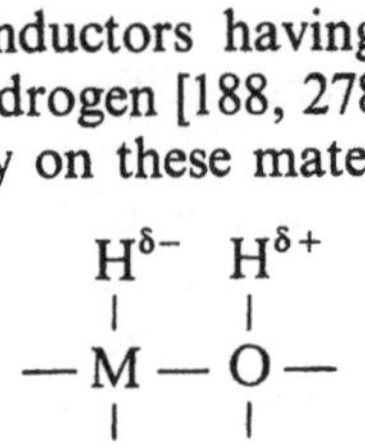

and thus additional problems with the reactivity of chemisorbed species arise.

Similarly as with the metals, the surface interactions on nonmetals can be divided into localized and cooperative processes [271, 275]. The localized interactions on nonmetals comprise not only the interactions with "dangling" bonds and/or unoccupied coordination sites (needed for the formation of covalent type bonds) but also strong interactions with some lattice atoms leading to the participation of these lattice atoms in the products of the catalytic reactions (*e.g.* lattice oxygen in redox reactions, protons in Brönsted acid-base sites [271, 275]. The cooperative phenomena (important in the case of semiconductors) are most often treated within the frame of mere charge transfer between the reactants and the catalyst [271, 275, 276].

B. Semiconductors

The results from theory and experiments in the field of physics of semiconductors led in the 1950's to studies of modifications of their bulk electronic structure (their semiconductive properties) in a predictable way. This approach, based on cooperative phenomena, has made it possible also to study the influence of these modifications on the catalytic activity of semiconductors. The electronic structure has been modified by [271]: i) incorporation of altervalent ions into the lattice; ii) pretreatment of the catalyst in oxidizing or reducing atmosphere; iii) preadsorption of donor or acceptor gases. If the catalyst is considered as a mere donor or acceptor of electrons, its activity would be expected to correlate with its semiconductive properties [188, 271–276]. Two basic theories have been elaborated in this direction, namely the boundary layer theory [271, 272, 274, 276] and, more generally, Wolkenstein's theory of chemisorption (taking into account) also weak adsorption by a one electron bond) [271, 273, 274, 276].

The boundary layer theory of chemisorption has to some extent successfully explained the influence of dopants on the chemisorbed amount (introducing the concept of depletive and cumulative chemisorption [271–274, 276]). However, the catalytic activity in hydrogen-deuterium exchange could not be simply correlated with the semiconductive properties of oxides [188, 279]. Obviously this approach is an oversimplification (neither the surface states nor the mobility of electrons, holes and ions are considered; the mutual interaction of defects is ignored, etc. [271]). As a further step, the details of the "active site" (the chemical composition and the geometric arrangement) should be considered, that is, one has to include also localized interactions. This, of course, represents already for binary compounds considerable complications both from the experimental and from the theoretical points of view. In spite of long and intensive effort, the problem of the nature of hydrogen chemisorption on materials like ZnO [279–283], MoO_3 [284], Cr_2O_3 [277], Co_3O_4 [286], MoS_2 [287, 288] remain largely unsolved. Interestingly — a rare exception in the research of catalysis by nonmetals — the study of hydrogen chemisorption on a well defined surface of ZnO (a UHV FEM study [289]) contradicts the other work on ZnO [279–283] in that no hydrogen adsorption could be detected in the temperature range 78–300 K unless hydrogen has been atomized in the gas phase.

Thus it appears that there is no firm experimental basis such as knowledge of the geometric and electronic structure of the adsorption site and of the adsorption complex, for the construction of reliable theoretical models of localized interactions on nonmetal surfaces (*e.g.* for the ligand field approach see ref. [271, 291, 292], or for Wolkenstein's weak chemisorption see refs. [271, 273].

The initial optimistic expectations concerning the general theory of catalytic phenomena based on the band theory of semiconductors have not been fulfilled. The assumption about the fundamental role of the theory of semiconductors in heterogeneous catalysis both by semiconductors and by metals (industrial metal catalysts being often covered by nonmetallic layers [273, 276]), has not been confirmed either (*e.g.* ref. [116]).

With metals by comparison, the current situation is rather different. In the late 1960's powerful theoretical methods in the solid state physics of metals became available, and this factor together with contemporary sophisticated experimental techniques caused a shift of prevailing interest towards a study of adsorption and catalytic properties of metallic single crystal planes, as has been discussed in the preceding sections.

C. Insulators

The catalytic properties of insulators are markedly different from those of conductors. They are mostly poor adsorbers of dihydrogen and consequently also poor hydrogenation catalysts [188]. This is paticularly true for insulating oxides such as MgO or Al_2O_3. In the case of oxides with Brönsted acidity, hydrogen from OH groups can participate in catalytic reactions [188]. Similarly as with semiconductors, only a limited number of studies has been performed on single crystallographic planes, and even in these cases the surface conditions were not well defined [188, 292]. Thus we have to deal here again with results obtained on polycrystalline or amorphous materials in a highly dispersed form [188, 293], and consequently we cannot expect unambiguous conclusions [293, 294]. Thus the nature of hydrogen chemisorption on insulators like MgO [188, 279], SiO_2 [188], TiO_2 [285] and Al_2O_3 [188, 284] is still not well understood. These materials, particularly refractory insulators, are widely used in the processes of heterogeneous activation of dihydrogen as supports and stabilizers for the active phases (metals). In these cases it is important to know whether the carrier is an inert constituent of this multicomponent catalyst, or if there is a transfer of hydrogen atoms from the "active" component (metal) onto the bare surface of the insulator. This is important both from the point of view of judging the role of the individual components of the catalyst in the reaction, and from the view point of the surface area estimation of the metallic phase by means of hydrogen chemisorption [294]. The first experimental evidence for the transfer of hydrogen atoms from the metal to the nonmetallic surface, based on the increased reducibility of oxides due to the presence of an adjacent metallic surface, was obtained already many years ago [235, 296, 297], and confirmed recently on a well defined surface [298]. The term "spillover effect" for this phenomenon has been introduced and the phenomenon itself analysed in ref. [299]. Further experimental evidence for the occurence of this spillover effect in chemisorption and catalysis has been collected in refs. [293, 294, 300–306]. However, as it is formulated in ref. [293] "the catalytic effects of spillover are not unequivocal, the evidence is persuasive but not conclusive that hydrogen spillage onto a poor adsorber partakes in or induces catalysis". This underlines the need for further studies of this effect on well defined surfaces with modern experimental techniques including strict control of possible promotors — water, grease — and elimination of contaminants [294].

4. Concluding Remarks

All the transition metals are able to activate dihydrogen. The selectivity of a given metal in a particular reaction is certainly influenced by the bond strength of hydrogen atoms to the metal surface. However, the trapping configuration, the symmetry and the binding energy of the second reactant is probably the determining factor. The same general conclusion of the decisive role of the other reactants can also be drawn from the results of catalytic research on nonmetals [308]. Thus the successful application of the approaches based in principle on bond energy considerations can be understood.

It is always advisable to look carefully at the applicability of any theoretical approach to a particular surface problem because of the nonuniversal character of these approaches. Closer cooperation between theory and experiment is needed because of: i) the inevitable role of theory in the interpretation of the experimental data; ii) the semiempirical character of many theoretical approaches; iii) the necessity of working with "identical" systems (to enable efficient comparison and discussion of the results).

The suggested formulation of the geometric and electronic factors in heterogeneous catalysis in terms of the Localized-Delocalized-Electrons-Interplay model has qualitatively explained the activation of hydrogen molecules by transition metal surfaces. Obviously, these two factors can no longer be treated separately because of their integral character. Together with the photoelectronic spectra of the clean metal surfaces, this model possesses some predictive power, too.

Acknowledgement

The author would like to express his gratitude to Dr. S. Černý and K. Kuchynka, whose many suggestions markedly improved the article.

References

1. The Physical Basis for Heterogeneous Catalysis. Drauglis, E., Jaffee, R. J. (eds.). New York: Plenum Press 1975, p. 567
2. Fisher, T. E.: Physics Today **1974**, No. 5, 24
3. Bonzel, H. P.: Surface Sci. **68**, 236 (1977)
4. Grimley, T. B. In: The Nature of the Surface Chemical Bond. (Rhodin, T. N.; Ertl, G., eds.). Amsterdam: North-Holland 1979, p. 3
5. Tamaru, K.: Dynamic Heterogeneous Catalysis. London: Academic Press 1978
6. Polizzoti, R. S.; Ehrlich, G.: J. Chem. Phys. **71**, 259 (1979)
7. Somorjai, G. A.: Surface Sci. **34**, 156 (1973)
   Bernasek, S. L.; Sickhaus, W. L.; Somorjai, G. A.: Phys. Rev. Lett. **30**, 1202 (1973)
8. Somorjai, G. A.: Surface Sci. **89**, 496 (1979)
9. Knor, Z. In: Surface and Defect Properties of Solids. (Roberts, M. W.; Thomas, J. M., eds.). **6**, 139 (1977). London: The Chemical Society
10. Baetzold, R. C.: J. Chem. Phys. **55**, 4355 (1971)
11. Anderson, A. B.; Hoffmann, R.: J. Chem. Phys. **61**, 4545 (1974)
12. Fassaert, D. J. M.; Verbeek, H.; van der Avoird, A.: Surface Sci. **29**, 501 (1972)
13. Blyholder, G.: J. Chem. Phys. **62**, 3193 (1975)
14. Lang, N. D.; Kohn, W.: Phys. Rev. **B 1**, 4555 (1970); **B 3**, 1215 (1971)
15. Bullett, D. W.: J. Phys. **C 8**, 2695 (1975)
16. Ying, S. C.; Smith, J. R.; Kohn, W.: Phys. Rev. **B 11**, 1483 (1975)
17. Huntington, H. B.; Turk, L. A.; White, W. W.: Surface Sci. **48**, 187 (1975)
18. Schrieffer, J. R.: J. Vacuum Sci. Technol. **13**, 335 (1976)
19. Johnson, K. H.; Smith, F. C.: Phys. Rev. **B 5**, 831 (1972)
20. Slater, J. C.; Johnson, K. H.: Phys. Rev. **B 5**, 844 (1972)
21. Johnson, K. H.; Messmer, R. P.: J. Vacuum Sci. Technol. **11**, 236 (1974)
22. Rösch, N.; Menzel, D.: Chem. Phys. **13**, 243 (1976)
23. Burwell, R. L. In: Survey of Progress in Chemistry **8**, 1 (1977). New York: Academic Press
24. Krebs, H. J.; Bonzel, H. P.: Surface Sci. **88**, 269 (1979)
25. Goodman, D. W.; Kelley, R. D.; Madey, T. E.; Yates, J. T. In: Hydrocarbon Synthesis from Carbon Monoxide and Hydrogen. Advances in Chemistry Series **178**, 1 (1979). (Kugler, E. L.; Steffgen, F. W., eds.). Washington: American Chemical Society
26. Kharson, M. S.; Kiperman, S. L.: React. Kinet. and Catal. Lett. **1**, 239 (1974)
27. Boreskov, G. K.: Kinet. Katal. **18**, 1111 (1977); **26**, 5 (1980)
28. Ponec, V.: Catal. Rev.-Sci. Eng. **18**, 151 (1978)
29. Dwyer, D.; Yoshida, K.; Somorjai, G. A.: In: ref. 25, p. 65
30. Jorgensen, W. L., Salem, L.: The Organic Chemist's Book of Orbitals. New York: Academic Press 1973
31. Ehrlich, G. In: Proc. 3rd Internatl. Congr. Catalysis. (Amsterdam 1964) (Sachtler, W. M. H.; Schuit, G. C.; Zwietering, P., eds.). Vol. I, 113 (1965). Amsterdam: North-Holland
32. Muscat, J. P.; Newns, D. M. In: Progress in Surface Science. (Davison, S. G., ed.). **9**, 1 (1978). Oxford: Pergamon Press

33a. Muetterties, E. L.; Rhodin, T. N.; Band, E.; Brucker, C. F.; Pretzer, W. R.: Chem. Revs. **79**, 91 (1979)

33b. Ehrlich, G.: J. Chem. Phys. **31**, 1111 (1959)

33c. Sachtler, W. M. H.; van Reijen, L. L.: Shokubai **4**, 147 (1962)

34. Sayers, C. M.: Surface Sci. **99**, 471 (1980)
35. Andersson, S. In: Topics in Surface Chemistry. (Kay, E.; Bagus, P. S., eds.). New York: Plenum Press 1978, p. 291
36. Huber, K. P.; Herzberg, G.: Molecular Spectra and Molecular Structure. Vol. 4. Constants of Diatomic Molecules. New York: van Nostrand-Reinhold Co. 1979
37. Bertolini, J. C.; Tardy, B.: Surface Sci. **102**, 131 (1981)
38. Jayasooriya, U. A.; Chester, M. A.; Howard, M. W.; Kettle, S. F. A.; Powell, D. B.; Sheppard, N.: Surface Sci. **93** 526 (1980)
39. Ho, W.; Di Nardo; J., Plummer, E. W.: J. Vacuum Sci. Technol. **17**, 134 (1980)

40. Baró, A. M.; Ibach, H.; Bruchmann, H. D.: Surface Sci. **88**, 384 (1979)
41. Backx, C.; Feuerbacher, B.; Fitton, B.; Willis, R. F.: Phys. Lett. **60** (A), 145 (1977)
42. Chabal, Y. J.; Sievers, A. J.: Phys. Rev. Lett. **44**, 944 (1980)
43. Adnot, A.; Carrette, J. D.: Phys. Rev. Lett. **39**, 209 (1977)
44. Froitzheim, H.; Ibach, H.; Lehwald, S.: Phys. Rev. Lett. **36**, 1549 (1976)
45. Backx, C.; Feuerbacher, B.; Fitton, B.; Willis, R. F.: Surface Sci. **63**, 193 (1977)
46. Yu, R. Y.; Helms, C. R.; Spicer, W. E.; Chye, P. W.: Phys. Rev. **B 15**, 1629 (1977)
47. Muscat, J. P.; Newns, D. M.: Surface Sci. **89**, 282 (1979)
48. Schrieffer, J. R.; Soven, P.: Physics Today **28**, 24 (1975)
49. Shirley, D. A.: J. Vacuum Sci. Technol. **12**, 280 (1975)
50. Bell, B.; Madhukar, A.: J. Vacuum Sci. Technol. **13**, 345 (1976)
51. Gadzuk, J. W.: Proc. 2nd Internatl. Conf. Solid Surfaces 1974. Japan J. Appl. Phys. **1974** Suppl. 2, Part 2, p. 851
52. Williams, A. R.; Lang, N. D.: Surface Sci. **68**, 138 (1977)
53. Itoh, H.: Japan J. Appl. Phys. **15**, 2311 (1976)
54. Tsang, Y. W.; Falicov, L. M.: J. Phys. **C 9**, 51 (1976)
55. Davidson, E. R.; Fain, S. C.: J. Vacuum Sci. Technol. **13**, 209 (1976)
56. Wang, S. W.; Weinberg, W. H.: Surface Sci. **77**, 14 (1978)
57. Zasucha, V. A.; Roev, L. M.: Teoret. Experim. Khim. **7**, 8 (1971)
58. Kölbel, H.; Tillmetz, K. D.: Ber. Bunsenges. Phys. Chem. **76**, 1156 (1972)
59. Anderson, A. B.: J. Am. Chem. Soc. **99**, 696 (1977)
60. Baetzold, R. C.: Surface Sci. **51**, 1 (1975)
61. Fassaert, D. J. M.; van der Avoird, A.: Surface Sci. **55**, 313 (1976)
62. Sobyanin, V. A.; Bulgakov, N. N.; Gorodetskii, V. V.: React. Kinet. Catal. Lett. **6**, 125 (1979)
63. Shopov, D., Andreev, A., Petrov, D.: J. Catal. **13**, 123 (1969)
64. Itoh, H.: ref. 51, p. 497
65. Upton, T. H.; Goddard, W. A.: Phys. Rev. Lett. **42**, 472 (1979)
66. Fassaert, D. J. M.; van der Avoird, A.: Surface Sci. **55**, 291 (1976)
67. Doyen, G.; Ertl, G.: J. Chem. Phys. **68**, 5417 (1978)
68. Bullett, D. W.; Cohen, M. L.: J. Phys. **C 10**, 2101 (1977)
69. Martin, A. J.: Surface Sci. **74**, 479 (1978)
70. Anders, L. W.; Hansen, R. S.; Bartell, L. S.: J. Chem. Phys. **59**, 5277 (1973)
71. Bullett, D. W.; Cohen, M. L.: J. Phys. **C 10**, 2083 (1977)
72. Kyo, S. K.; Gomer, R.: Phys. Rev. **B 10**, 4161 (1974)
73. Kobayashi, N.; Yoshida, S.; Kato, H.; Fukui, K.; Tarama, K.: Surface Sci. **79**, 189 (1979)
74. Kumamoto, D.; van Himbergen, J. E.; Silbey, R.: Chem. Phys. Lett. **68**, 189 (1979)
75. Lang, N. D.; Williams, A. R.: Phys. Rev. Lett. **34**, 531 (1975)
76. Gunnarson, O.; Hjelmberg, H.; Lundqvist, B. I.: Phys. Rev. Lett. **37**, 292 (1976)
77. Messmer, R. P. In: ref. 4, p. 53
78. Lundqvist, B. I.; Hjelmberg, H.; Gunnarson, O. In: Photoemission and the Electronic Properties of Surfaces. (Feuerbacher, B.; Fitton, B.; Willis, R. F., eds.). Chichester: J. Wiley 1978, p. 227
79. Tompkins, F. C.: Chemisorption of Gases on Metals. London: Academic Press 1978
80. Grimley, T. B. In: Electronic Structure and Reactivity of Metal Surfaces. (Derouane, E. G.; Lucas, A. A., eds.). New York: Plenum Press 1976, pp. 35, 113
81. Gadzuk, J. W. In: Surface Physics of Materials. (Blakely, J. M., ed.). Vol. 2. New York: Academic Press 1975, p. 339
82. Davenport, J. W.; Einstein, T. L.; Schrieffer, J. R.; Soven, P. In: ref. 1, p. 295
83. Lyo, S. K.; Gomer, R. In: Interactions on Metal Surfaces. (Gomer, R., ed.). Berlin, Heidelberg, New York: Springer 1975, p. 41
84. Gomer, R. In: Critical Reviews in Solid State Sciences. (Schuele, D. E.; Hoffman, R. W., eds.). Cleveland: CRC Press, Inc. 1974, p. 247
85. Salem, L.; Leforestier, C.: Surface Sci. **82**, 390 (1979)
86. Ertl, G. In: ref. 4, p. 315
87. Smith, J. R.: Phys. Rev. Lett. **25**, 1023 (1970)

88. Boudreaux, D. S.; Juretschke, H. J. In: Structure and Properties of Metal Surfaces. Honda Memorial Series on Material Science. No. 1. Tokyo: Maruzen Co. 1973, p. 94
89. Bennett, A. J. In: ref. 84, p. 261
90. Ref. 1, p. 578
91. Lee, T. H.; Rabelais, J. W.: Surface Sci. **75**, 29 (1978)
92. Baetzold, R. C.: J. Catal. **29**, 129 (1973)
93. Russo, C.; Kaplow, R.: J. Vacuum Sci. Technol. **15**, 479 (1978)
94. Brivio, G. P.; Grimley, T. B.: Surface Sci. **89**, 226 (1979)
95. Norskov, J. K.; Lundqvist, B. I.: Surface Sci. **89**, 251 (1979)
96. Doyen, G.: Surface Sci. **89**, 238 (1979)
97. Müller, H.; Brenig, W.: Z. Phys. **B 34**, 165 (1979)
98. Ehrlich, G.: Brit. J. Appl. Phys. **15**, 349 (1964)
99. Kobayashi, H.; Yoshida, S.; Fukui, K.; Tarama, K.: Chem. Phys. Lett. **53**, 457 (1978)
100. Muscat, J. P.; Newns, D. M.: Phys. Rev. Lett. **43**, 2025 (1979)
101. Ellis, D. E.; Adachi, H.; Averill, F. W.: Surface Sci. **58**, 497 (1976)
102. Charlot, M. F.; Kahn, O.: Surface Sci. **81**, 90 (1979)
103. Melius, C. F.: Chem. Phys. Lett. **39**, 287 (1976)
Melius, C. F.; Moskowitz, J. W.; Mortola, A. P.; Baillie, M. B.; Ratner, M. A.: Surface Sci. **59**, 279 (1976)
104. Desjonguères, J. M.; Cyrot-Lackmann, F.: J. Chem. Phys. **64**, 3707 (1976); J. Phys. **F 5**, 1368 (1975); Surface Sci. **53**, 429 (1975)
105. Knor, Z.; Müller, E. W.: Surface Sci. **10**, 21 (1968)
106. Wood, J. H.: Phys. Rev. **117**, 714 (1960)
107. Jennings, P. J.; Painter, G. S.; Jones, R. O.: Surface Sci. **60**, 255 (1976)
108. Tapilin, V. M.: Kinet. Katal. **12**, 1426 (1971); **13**, 154 (1972)
109. Mahanty, J.; March, N. H.: J. Phys. **C 9**, 2905 (1976)
110. Flores, F.; March, N. H.; Moore, I. D.: Surface Sci. **69**, 133 (1977)
111. see ref. 74
112. Brown, J. S.; Brown, R. C.; March, N. H.: Phys. Lett. **47 A**, 489 (1974)
113. Deb, B. M.: Rev. Modern Phys. **45**, 22 (1973)
114a. Szymerska, I.; Lipski, M.: J. Catal. **41**, 197 (1976)
114b. Carley, A. F.; Rassias, S.; Roberts, M. W.: J. Chem. Research (S) **1979**, 208
115a. Bernasek, S. L.; Staudt, G. E.: J. Catal. **45**, 372 (1976);
115b. Issett, L. C.; Blakely, J. M.: Surface Sci. **58**, 397 (1976);
115c. Sickafus, E. N.: Surface Sci. **19**, 181 (1970)
116a. McCarroll, J. J.: Surface Sci. **53**, 297 (1975);
116b. Edmonds, T.; McCarroll, J. J. In: ref. 35, p. 261
117. Schwarz, J. A.: Surface Sci. **87**, 525 (1979)
118. Brodén, G.; Bonzel, H. P.: Surface Sci. **84**, 106 (1979)
119. Ertl, G.; Weiss, M.; Lee, S. B.: Chem. Phys. Lett. **60**, 391 (1979)
120. Stefan, P. M.; Helms, C. R.; Perino, S. C.; Spicer, W. E.: J. Vacuum Sci. Technol. **16**, 577 (1979)
121. Somorjai, G. A.: Pure and Appl. Chem. **50**, 963 (1978); Catal. Rev.-Sci. Eng. **18**, 173 (1978)
122. Carley, A. F.; Joyner, R. W.; Roberts, M. W.: Chem. Phys. Lett. **27**, 580 (1974)
123. Hagstrum, H. D.: Surface Sci. **54**, 197 (1976)
124. Forbes, R. G. In: Proc. 7th Internatl. Vacuum Congr. and 3rd Internatl. Conf. on Solid Surfaces (Vienna 1977). (Dobrozemsky, R.; Rüdenauer, F.; Viehböck, F. P.; Breth, A., eds.). p. 433
125. Knor, Z.: Surface Sci. **70**, 286 (1978)
126. Knor, Z.: Collect. Czech. Chem. Commun. **44**, 3434 (1979)
127. Bozso, F.; Ertl, G.; Grunze, M.; Weiss, M.: Appl. Surface Sci. **1**, 103 (1977)
128. Benziger, J.; Madix, R. J.: Surface Sci. **94**, 201 (1980)
129. Yoshida, K.: Japan. J. Appl. Phys. **19**, 1873 (1980)
130. Bridge, E. M.; Comrie, C. M.; Lambert, R. M.: J. Catal. **58**, 28 (1979)
131. Lapujoulade, J.; Neil, K. S.: Surface Sci. **35**, 288 (1973)
132. Christmann, K.; Schober, O.; Ertl, G.; Neumann, M.: J. Chem. Phys. **60**, 4528 (1974)

133. Andersson, S. see in ref. Castner, D. G.; Somorjai, G. A.: Chem. Revs. **79**, 233 (1979)
134. Taylor, T. N.; Estrup, P. J.: J. Vacuum Sci. Technol. **11**, 244 (1974)
135. Madix, R. J.: Catal. Rev.-Sci. Eng. **15**, 293 (1977)
136. Ertl, G.; Küppers, D.: Ber. Bunsenges. Phys. Chem. **75**, 1017 (1971)
137. Lapujoulade, J.; Neil, K. S.: see ref. 260
138. Germer, L. H.; McRae, A. U.: see ref. 260
139. McCarty, J.; Falconer, J.; Madix, R. J.: see ref. 260
140. Küppers, J.: loc. cit. 133
141. Christmann, K.; Behm, R. J.; Ertl, G.; van Hove, M. A.; Weinberg, W. H.: J. Chem. Phys. **70**, 4168 (1979)
142. Casalone, G.; Cattania, M. G.; Simonetta, M.; Tescari, M.: Chem. Phys. Lett. **61**, 36 (1979)
143. Lapujoulade, J.; Neil, K. S.: J. Chem. Phys. **57**, 3535 (1972)
144. Bertolini, J. C.; Dalmai-Imelik, G.: loc. cit. 133
145. van Hove, M. A.; Ertl, G.; Weinberg, W. H.; Christmann, K.; Behm, H. J.: loc. cit. 133
146. Mahnig, M.; Schmidt, L. D.: Z. Phys. Chem. (Frankfurt am Main) **80**, 71 (1972)
147. Chrzanowski, E.: Acta Phys. Polonica **A 44**, 711 (1973)
148. Dooley, G. J.; Haas, T. W.: J. Chem. Phys. **52**, 993 (1970)
149. Danielson, L. R.; Dresser, M. J.; Donaldson, E. E.; Dickinson, J. T.: Surface Sci. **71**, 599 (1978)
150. See ref. 117
151. Madix, R. J.; Ertl, G.; Christmann, K.: Chem. Phys. Lett. **62**, 38 (1979)
152. Castner, D. G.; Sexton, B. A.; Somorjai, G. A.: Surface Sci. **71**, 519 (1978)
153. Yates, J. T.; Thiel, P. A.; Weinberg, W. H.: Surface Sci. **84**, 427 (1979)
154. Williams, E. D.; Thiel, P. A.; Weinberg, W. H.; Yates, J. T.: J. Chem. Phys. **72**, 3496 (1980)
155. Conrad, H.; Ertl, G.; Latta, E. E.: Surface Sci. **41**, 120 (1974)
156. Madey, T. E.; Czyzewski, J. Z.; Yates, J. T.: Surface Sci. **49**, 465 (1975)
157. Yates, J. T.; Madey, T. E.: J. Vacuum Sci. Technol. **8**, 63 (1971)
158. Yates, J. T.; Madey, T. E.: J. Chem. Phys. **54**, 4969 (1971)
159. Leggett, M. R.; Armstrong, R. A.: Surface Sci. **24**, 404 (1971)
160. Tamm, P. W.; Schmidt, L. D.: J. Chem. Phys. **54**, 4775 (1971)
161. Yonehara, K.; Schmidt, L. D.: Surface Sci. **25**, 238 (1971)
162. Madey, T. E.: Surface Sci. **36**, 281 (1973)
163. Feuerbacher, B.: Surface Sci. **47**, 115 (1975)
164. Madey, T. E.; Yates, J. T. In: Structure et Propriètes des Surfaces des Solides. Nr. 187. Paris: Ed. C.N.R.S. 1970, p. 155
165. Domke, M.; Jähnig, G.: Drechsler, M.: Surface Sci. **42**, 389 (1974)
166. King, D. A. In: ref. 124, p. 769
167. Tamm, P. W.; Schmidt, L. D.: J. Chem. Phys. **51**, 5352 (1969), **52**, 1150 (1970)
168. Vorburger, T.; Sandstrom, D. R.; Waclawski, B. J.: Surface Sci. **60**, 211 (1976)
169. King, D. A.; Thomas, G.: Surface Sci. **92**, 201 (1980)
170. Jaeger, R.; Menzel, D.: Surface Sci. **100**, 561 (1980)
171. Barford, B. D.; Rye, R. R.: J. Chem. Phys. **60**, 1046 (1974)
172. Jaeger, R.; Menzel, D.: Surface Sci. **63**, 232 (1977)
173. Estrup, P. J.; Anderson, J.: J. Chem. Phys. **45**, 2257 (1966)
174. Plummer, E. W.; Bell, A. E.: J. Vacuum Sci. Technol. **9**, 583 (1972)
175. Willis, R. F.: Surface Sci. **89**, 457 (1979)
176. Feuerbacher, B.; Willis, R. F.: Phys. Rev. Lett. **36**, 1339 (1976)
177. Matysik, K. J.: Surface Sci. **29**, 324 (1972)
178. Blanchet, G. B.; Estrup, P. J.; Stiles, P. J.: Phys. Rev. Lett. **44**, 171 (1980)
179. Gomer, R.: loc. cit. 260
180. Madey, T. E.: Surface Sci. **29**, 571 (1972)
181. Adams, D. L.; Germer, L. H.; May, J. W.: Surface Sci. **22**, 45 (1970)
182. Rye, R. R.; Barford, B. D.: Surface Sci. **27**, 667 (1971)
183. Rye, R. R.; Barford, B. D.; Cartier, P. G.: J. Chem. Phys. **59**, 1693 (1973)
184. Rhodin, T. N.; Brodén, G.: Surface Sci. **60**, 466 (1976)

185. Derochette, J. M.; Marien, J.: phys. stat. sol. (a) **39**, 281 (1977)
186. Ibbotson, D. E.; Wittig, T. S.; Weinberg, W. H.: J. Chem. Phys. **72**, 4885 (1980)
187. Ibbotson, D. E.; Wittig, T. S.; Weinberg, W. H.: Surface Sci. **97**, 297 (1980)
188. Dowden, D. A. In: Surface Science. Vol. 8. Vienna: International Atomic Energy Agency 1975, p. 215
189. Nieuwenhuys, B. E.; Somorjai, G. A.: Surface Sci. **72**, 8 (1978)
190. Nieuwenhuys, B. E.; Hagen, D. I.; Rovida, G.; Somorjai, G. A.: Surface Sci. **59**, 155 (1976)
191. McCabe, R. W.; Schmidt, L. D. In: ref. 124, p. 1201
192. Lu, K. E.; Rye, R. R.: Surface Sci. **45**, 677 (1974)
193. Nieuwenhuys, B. E.: Surface Sci. **59**, 430 (1976)
194. Netzer, F. P.; Kneringer, G.: loc. cit. 260
195. Morgan, A. E.; Somorjai, G. A.: loc. cit. 133
196. Marien, J.: Bull. Soc. Roy. Sci. Liege **45**, 103 (1976)
197. Christmann, K.; Ertl, G.; Pignet, T.: Surface Sci. **54**, 365 (1976)
198. McCabe, R. W.; Schmidt, L. D.: Surface Sci. **65**, 189 (1977)
199. Collins, D. M.; Spicer, W. E.: Surface Sci. **69**, 85 (1977)
200. Castner, D. G.; Blackadar, R. L.; Somorjai, G. A.: J. Catal. **66**, 257 (1980)
201. Weinberg, W. H.; Lampton, V. A.: loc. cit. 260
202. Weinberg, W. H.; Monroe, D. R.; Lampton, V. A.; Merrill, R. P.: loc. cit. 260
203. Christmann, K.; Ertl, G.: Surface Sci. **60**, 365 (1976)
204. Collins, D. M.; Spicer, W. E.: Surface Sci. **69**, 114 (1977)
205. Goodman, D. W.; Madey, T. E.; Ono, M.; Yates, J. T.: J. Catal. **50**, 279 (1977)
206. Hayward, D. O. In: Chemisorption and Reactions on Metallic Films. (Anderson, J. R., ed.). Vol. 1. New York: Academic Press 1971, p. 225
207. Aben, A. B.; van der Eijk, H.; Oelderijk, J. M. In: Proc. 5th Internatl. Congr. Catal. (Palm Beach 1972) (Hightower, J. W., ed.). Amsterdam: North-Holland 1973, p. 717
208. Tsuchiya, S.; Nakamura, M.: J. Catal. **50**, 1 (1977)
209. Popova, N. M.; Babenkova, L. B.; Savel'eva, G. A.: Adsorption and Interaction of Simple Gases with the VIII Group Metals (Russ.). Alma-Ata: Publ. House "Nauka" 1979
210. Tsuchiya, S.; Amenomiya, Y.; Cvetanovič, R. J.: J. Catal. **19**, 245 (1970)
211. Emmett, P. H.; Takezawa, N.: J. Res. Inst. Catalysis, Hokkaido Univ. **26**, 37 (1978)
212. Lynch, J. F.; Flanagan, T. B.: J. Phys. Chem. **77**, 2628 (1973)
213. Konvalinka, J. A.; Scholten, J. J.: Catal. **48**, 374 (1977)
214a. Christmann, K.: Bull. Soc. Chim. Belg. **88**, 519 (1979)
214b. Burch, R. In: ref. 9 **8**, 1 (1980)
215. Madey, T. E.; Yates, J. T.: Surface Sci. **63**, 203 (1977)
216. King, D. A.: Surface Sci. **47**, 384 (1975)
217. Cassuto, A.; King, D. A.: Surface Sci. **102**, 388 (1981)
218. Mignolet, J. C. P.: Bull. Soc. Chim. Belg. **67**, 358 (1958)
219a. Dus, R.; Tompkins, F. C.: J. Chem. Soc. I **71**, 930 (1975)
219b. Dus, R.: J. Catal. **42**, 334 (1976)
220. King, D. A.: Surface Sci. **64**, 43 (1977)
221. Benndorf, C.; Thieme, F.: Z. Phys. Chem. (Frankfurt am Main) **87**, 40 (1973)
222. Hickmott, T. W.; Ehrlich, G.: J. Phys. Chem. Solids **5**, 47 (1958)
223. Völter, J.; Procop, H.: Z. Phys. Chem. (Leipzig) **249**, 344 (1972)
224. Norton, P. R.; Creber, D. K.; Davies, J. A.: J. Vacuum Sci. Technol. **17**, 149 (1980)
225. Goodman, D. W.; Yates, J. T.; Madey, T. E.: Surface Sci. **93**, L 135 (1980)
226. Han, H. R.; Schmidt, L. D.: J. Phys. Chem. **75**, 227 (1971)
227. Wedler, G.; Borgmann, D.: Ber. Bunsenges. Phys. Chem. **78**, 67 (1974)
228. Stephan, J. J.; Ponec, V.; Sachtler, W. M. H.: J. Catal. **37**, 81 (1975)
229. Candy, J. P.; Fouilloux, P.; Renouprez, A. J.: J. C. S. Faraday I **76**, 616 (1980)
230. Dowden, D. A.: J. Chim. Phys. **51**, 780 (1954)
231. Dowden, D. A. In: Chemisorption. (Garner, W. E., ed.). London: Butterworth 1957, p. 3
232. Bond, G. C.: Catalysis by Metals. New York: Academic Press 1962

233. Gundry, P. M.; Tompkins, F. C.: Trans. Faraday Soc. **52**, 1609 (1956)
234. Waclawski, J.; Vorburger, T. V.; Stein, R. J.: J. Vacuum Sci. Technol. **12**, 301 (1975)
235. Knor, Z.: Advances in Catalysis **22**, 51 (1972)
236. Knor, Z.: Kinet. Katal. **21**, 17 (1980)
237. Gorodetskii, V. V.; Sobyanin, V. A.; Bulgakov, N. N.; Knor, Z.: Surface Sci. **82**, 120 (1979)
238. Lehwald, S.; Ibach, H.: Surface Sci. **89**, 425 (1979)
239. Hüfner, S. In: Photoemission in Solids. (Ley, L.; Cardona, M., eds.). Berlin: Springer-Verlag 1979, p. 173
240. Tanabe, T.; Adachi, H.; Imoto, S.: Japan J. Appl. Phys. **16**, 861 (1977)
241. Demuth, J. E.: Surface Sci. **65**, 369 (1977)
242. Conrad, H.; Ertl, G.; Küppers, J.; Latta, E. E.: Surface Sci. **58**, 578 (1976)
243. Himpsel, F. J.; Knapp, J. A.; Eastman, D. E.: Phys. Rev. **B 19**, 2872 (1979)
244. Anderson, J.; Lapeyre, G. J.; Smith, R. J.: Phys. Rev. **B 17**, 2436 (1978)
245. Feuerbacher, B.; Christensen, N. E.: Phys. Rev. **B 10**, 2373 (1974)
246. Eastman, D. E.: Solid State Commun. **7**, 1697 (1969); J. Chem. Phys. **40**, 1387 (1969)
247. Baer, Y.; Hedén, P. F.; Hedmon, J.; Klasson, M.; Nordling, C.; Siegbahn, K.: Physica Scripta **1**, 55 (1970)
248. Pierce, D. T.; Spicer, W. E.: Phys. Rev. **B 5**, 2125 (1972)
249. Smith, N. V.; Wertheim, G. K.; Hüfner, S.; Traum, M. M.: Phys. Rev. **B 10**, 3197 (1974)
250. Hüfner, S.; Wertheim, G. K.; Wernick, J. H.: Phys. Rev. **B 8**, 4511 (1973)
251. Cinti, R. C.; Al Khour, E.; Chakraverty, E. K.: Phys. Rev. **B 14**, 3286 (1976)
252. Heimann, P.; Neddermeyer, H.: J. Phys. **F 6**, L 257 (1976)
253. Pessa, M.; Heimann, P.; Neddermeyer, H.: Phys. Rev. **B 14**, 3488 (1976)
254. Ley, L.; Dabousi, O. B.; Kowalczyk, S. P.; McFeely, F. R.; Shirley, D. A.: Phys. Rev. **B 16**, 5372 (1978)
255. Lin, S. F.; Pierce, D.; Spicer, W. E.: Phys. Rev. **B4**, 326 (1971)
256. Nemoshkalenko, V. V.; Aleshin, V. G.: Electron Spectroscopy of Crystals. New York: Plenum Press 1979
257. Mehta, H.; Fadley, C. S.: Phys. Rev. **B 20**, 2280 (1979)
258. Dahlbäck, N.; Nilsson, P. O.; Pessa, M.: Phys. Rev. **B 19**, 5961 (1979)
259. Ponec, V.; Knor, Z.; Černý, S.: Adsorption on Solids. London: Butterworth 1974
260. Toyoshima, I.; Somorjai, G. A.: Catal. Rev.-Sci. Eng. **19**, 105 (1979)
261a. Savtchenko, V. I.; Boreskov, G. K.; Dadayan, K. A.: Kinet. Katal. **20**, 741 (1979);
261b. Dadayan, K. A.; Boreskov, G. K.; Savtchenko, V. I.; Sadovskaya, E. M.; Jablonski, I. S.: Kinet. Katal. **20**, 795 (1979); Izv. Akad. Nauk SSSR, Ser. Fiz. **43**, 1794 (1979)
262a. Gryaznov, V. M.; Smirnov, V. S.: Usp. Khim. **43**, 1716 (1974); Gryaznov, V. M.: Dokl. Akad. Nauk SSSR **189**, 794 (1969); Kinet. Katal. **12**, 640 (1970)
263a. Palczewska, W.: Advances in Catalysis **24**, 235 (1975);
263b. Borodzinski, A.; Dus, R.; Frak, R.; Janko, A.; Palczewska, W.: 6th Internatl. Congr. Catalysis, London 1976, Preprint A 7
264. Tamaru, K.: Advances in Catalysis **20**, 327 (1969); Amer. Scientist **60**, 474 (1972)
265. Urabe, K.; Oh-Ya, A.; Ozaki, A.: J. Catal. **54**, 436 (1978)
266. Kojima, I.; Mitazaki, E.; Inove, Y.; Yasumori, I.: J. Catal. **59**, 472 (1979)
267a. Nguyen, T. T. A.; Cinti, R. C.: Surface Sci. **68**, 566 (1977)
267b. Nguyen, T. T. A.; Cinti, R. C.; Capiomont, Y.: Surface Sci. **87**, 613 (1979)
268. Capehart, T. W.; Rhodin, T. N.: Surface Sci. **83**, 367 (1979)
269. Kadlecová, M.; Kadlec, V.; Knor, Z.: Collect. Czech. Chem. Commun. **36**, 1205 (1971)
270. Schlapbach, L.; Seiler, A.; Stucki, F.: Mat. Res. Bull. **14**, 785 (1979)
271. Clark, A.: The Theory of Adsorption and Catalysis. New York: Academic Press 1970
272. Hauffe, K.: Advances in Catalysis **7**, 213 (1955)
273. Wolkenstein, F. F.: The Electron Theory of Catalysis on Semiconductors. Moscow: Gos. Izd. Fiziko-Matem. Lit. 1960 (in Russian); Paris: Masson 1961 (in French); New York: Pergamon Press 1963 (in English); Berlin: VEB Deutscher Verlag 1964 (in German)
274. Krylov, O. V.; Catalysis by Nonmetals Leningrad: Izd. Khimiya 1967 (in Russian); New York: Academic Press 1970 (in English)

275. Morrison, S. R.: Chemtech. September **1977**, 570
276. Contact Catalysis. (Szabó, Z. G.; Kalló, D., eds.). Budapest: Akadémiai Kiadó 1976
277. Wittgen, P. P. M. M.; Groeneveld, C.; Janssens, J. H. G. J.; Wetzels, M. L. J. A.; Schuit, G. C. A.: J. Catal. **59**, 168 (1979)
278. Dowden, D. A. In: Colloquio Sobre Quimica Fisica de Procesos en Superficies Solidas (Madrid 1964). Madrid: Consejo Superior de Investigationes Cientificas 1965, p. 177
279. Hayward, D. O.; Trapnell, B. M. W.: Chemisorption. London: Butterworth 1964
280. Barański, A.; Gałuszka, J.: J. Catal. **44**, 259 (1976)
281. Conner, W. C.; Kokes, R. J.: J. Catal. **36**, 199 (1975)
282. Boccuzzi, F.; Borello, E.; Zecchina, A.; Bossi, A.; Camia, M.: J. Catal. **51**, 150 (1978)
283. John, C. S. In: Catalysis. (Kemball, C., Dowden, D. A., eds.). **3**, 171 (1980). London: Chemical Society
284. Millan, W. S.; Crespin, M.; Cirillo, A. C.; Abdo, S.; Hall, W. K.: J. Catal. **60**, 404 (1979)
285. Knotek, M. L.: Surface Sci. **91**, L 17 (1980); **101**, 339 (1980)
286. Siegel, S.: J. Catal. **30**, 139 (1973)
287. Wright, Ch. J.; Sampson, Ch.; Fraser, D.; Moyes, R. B.; Wells, P. B.; Riekel, Ch.: J. C. S. Faraday I **76**, 1585 (1980)
288. Okuhara, T.; Tanaka, K. I.; Miyahara, K.: J. Catal. **48**, 229 (1977)
289. Marien, J.: phys. stat. sol. (a) **38**, 513 (1976)
290. Dowden, D. A.; Wells, D. In: Proc. 2nd Internatl. Congr. Catal. (Paris 1960). Paris: Technip 1961, p. 1499
291. Dowden, D. A.: J. Res. Inst. Catal., Hokkaido Univ. **14**, 1 (1966)
292. Lawrenko, W. A.; Tikusch, W. L.; Senkow, W. S.; Krawez, W. A.; Nasarenko, K. W.; Werestschak, W. M.: Z. Phys. Chem. (Leipzig) **259**, 129 (1978)
293. Dowden, D. A. In: ref. 283, p. 136
294. Sermon, P. A.; Bond, G. C.: Catalysis Revs. **8**, 211 (1973)
295. Bodenstein, M.: Z. Phys. Chem. **B 2**, 345 (1939)
296a. Ponec, V.; Knor, Z.; Černý, S. In: Proc. 3rd Internatl. Congr. Catalysis. (Amsterdam 1964) (Sachtler, W. M. H.; Schuit, G. C. A.; Zwietering, P., eds.). Amsterdam: North-Holland 1965, p. 353
296b. Ponec, V.; Knor, Z.; Černý, S.: J. Catal. **4**, 485 (1965)
297. Khobiar, S.: J. Phys. Chem. **68**, 411 (1964)
298. de Bokx, P. K.; Labohm, F.; Gijzeman, O. L. J.; Bootsma, G. A.; Geus, J. W.: Appl. Surface Sci. **5**, 321 (1980)
299a. Boudart, M.; Vannice, M. A.; Benson, J. E.: Z. Phys. Chem. (Frankfurt am Main) **64**, 171 (1969)
299b. Boudart, M.: Advances in Catalysis **20**, 153 (1969)
300. Dalmai-Imelik, G.; Leclerq, C.; Massardier, J.; Maubert-Franco, A.; Zalhout, A. In: Proc. 2nd Internatl. Conf. on Solid Surfaces (1974). Japan J. Appl. Phys. Suppl. 2, Pt. 2, 489 (1974)
301. Levy, R. B.; Boudart, M.: J. Catal. **32**, 304 (1974)
302. Gadgil, K.; Gonzales, R. D.: J. Catal. **40**, 190 (1975)
303. van Meerbeek, A.; Jelli, A.; Fripiat, J. J.: J. Catal. **46**, 320 (1977)
304. Fleisch, T.; Abermann, R.: J. Catal. **50**, 268 (1977)
305. Kramer, R.; Andrade, M.: J. Catal. **58**, 287 (1979)
306. Bianchi, D.; Lacroix, M.; Pajonk, G.; Teichner, S. J.: J. Catal. **59**, 467 (1979)
307. King, D. A.: J. Vacuum Sci. Technol. **17**, 241 (1980)
308. Tanaka, K. I.; Okuhara, T.: J. Catal. **65**, 1 (1980)
309. El-Batanouny, M.; Strongin, M.; Williams, G. P.; Colbert, J.: Phys. Rev. Lett. **46**, 269 (1981)

Subject Index

Author Index Volume 1–3

Catalysis · Science and Technology

Editors: J. R. Anderson, M. Boudart

Volume 1

1981. 107 figures. X, 309 pages
ISBN 3-540-10353-8
Distribution rights for all socialist countries:
Akademie-Verlag, Berlin

Contents/Information:

H. Heinemann: **History of Industrial Catalysis**
The first chapter reviews industrial catalytic developments, which have been commercialized during the last fourty years. Emphasis is put on heterogeneous catalytic processes, largely in the petroleum, petrochemical and automotive industries, where the largest scale applications have occurred. Homogeneous catalytic processes are briefly treated and polymerization catalysis is mentioned. The author concentrates on major inventions and novel process chemistry and engineering (79 references).

J. C. R. Turner: **An Introduction to the Theory of Catalytic Reactors**
The second chapter introduces to the catalytic chemist those aspects of chemical reaction engineering involved in any industrial application of a catalytic chemical reaction (19 references).

A. Ozaki, K. Aika: **Catalytic Activation of Dinitrogen**
The third chapter is a comprehensive and critical review of studies on the catalytic activation of dinitrogen, including chemisorption and coordination of dinitrogen, kinetics and mechanism of ammonia synthesis, chemical and instrumental characterization of active catalysts, and homogeneous activation of dinitrogen including metal complexes (353 references).

M. E. Dry: **The Fischer-Tropsch Synthesis**
The fourth chapter concentrates mainly on the development of the Fischer-Tropsch process from the late 1950's to 1979. During this period the Sasol plant was the only Fischer-Tropsch process in operation and hence a large part of this review deals with the information generated at Sasol. The various types of reactors are compared and discussed (198 references).

J. H. Sinfelt: **Catalytic Reforminf of Hydrocarbons**
The fifth chapter discusses the catalytic reforming of hydrocarbons from the point of view of the individual types of chemical reactions involved in the process and the nature of the catalysts employed. Some consideration is also given to technological aspects of catalytic reforming (103 references).

Volume 2

1981. 145 figures. X, 282 pages
ISBN 3-540-10593-X
Distribution rights for all socialist countries:
Akademie-Verlag, Berlin

Contents/Information:

G.-M. Schwab: **History of Concepts in Catalysis**
The concept of catalysis can be attributed to J. Berzelius (1838), whose formulation was based on the manifold observations made in the 17th and 18th centuries. This article traces the development of this and related theories along with the scientific research and empirical material from which they are drawn.

J. Haber: **Crystallography of Catalyst Types**
Structural properties of metals and their substitutional and interstitial alloys, transition metal oxides as well as alumina, silica, aluminosilicates and phosphates are discussed. Implications of point and extended defects for catalysis are emphasized and the problem of the structure and composition of the surface as compared to the bulk is considered.

G. Froment, L. Hosten: **Catalytic Kinetics: Modelling**
The text reviews the methodology of kinetic analysis for simple as well as complex reactions. Attention is focused on the differential and integral methods of kinetic modelling. The statistical testing of the model and the parameter estimates required by the stochastic character of experimental data is described in detail and illustrated by several practical examples. Sequential experimental design procedures for discrimination between rival models and for obtaining parameter estimates with the greatest attainable precision are developed and applied to real cases.

A. J. Lecloux: **Texture of Catalysts**
Useful guidelines and methods for a systematic investigation and a coherent description of catalyst texture are proposed in this contribution. Such a description requires the specification of a very large number of parameters and implies the use of "models" involving assumptions and simplifications. The general approach for determining the porous texture of solids is based on techniques, whose results are cross analyzed in such a way that a self-consistent picture of the porous texture of solids is obtained.

K. Tanabe: **Solid Acid and Base Catalysts**
This chapter deals with the types of solid acids and bases, the acidic and basic properties, and the structure of acidic and basic sites. The chemical principles of the determination of acid-base properties and the mechanism for the generation of acidity and basicity are also described. How acidid and basic properties are controlled chemically is discussed in connection with the preparation method of solid acids and bases.

New Syntheses with Carbon Monoxide

Editor: **J. Falbe**

1980. 118 figures, 127 tables. XIV, 465 pages
(Reactivity and Structure, Volume 11)
ISBN 3-540-09674-4

Contents: *B. Cornils:* Hydroformylation. Oxo Synthesis, Roelen Reaction. – *H. Bahrmann, B. Cornils:* Homologation of Alcohols. – *A. Mullen:* Carbonylations Catalyzed by Metal Carbonyls/Reppe Reactions. – *C. D. Frohning:* Hydrogenation of the Carbon Monoxide. – *H. Bahrmann:* Koch Reactions. – *A. Mullen:* Ring Closure Reactions with Carbon Monoxide.

V. N. Kondratiev, E. E. Nikitin

Gas-Phase Reactions

Kinetics and Mechanisms

1981. 1 portrait, 64 figures, 15 tables. XIV, 241 pages
ISBN 3-540-09956-5

The science of contemporary gas kinetics owes much to the pioneering efforts of *V. N. Kondratiev.* In this book, he and his co-author *E. E. Nikitin* describe the kinetics and mechanisms of gas reactions in terms of current knowledge of elementary processes of energy transfer, uni-, bi- and trimolecular reactions.

Their consideration of formal chemical kinetics is followed by a discussion of the mechanisms of elastic collisions, and of unimolecular, combination and bimolecular reactions. In addition, they have devoted several chapters to the kinetics of the more complicated photochemical reactions, reactions in discharge and radiation- chemical reactions, the general theory of chain reactions, and processes in flames. Particular attention is paid to non-equilibrium reactions, which occur as a result of the Maxwell-Boltzmann distribution principle.

This comprehensive and critical presentation of gase phase kinetics will prove an excellent source of information for chemists and physicists in research and industry as well as for advanced students in chemistry and chemical physics (540 refernces).

Bei Fragen zur Produktsicherheit wenden Sie sich bitte an:
If you have any questions regarding product safety,
please contact:

Walter de Gruyter GmbH
Genthiner Straße 13
10785 Berlin
productsafety@degruyterbrill.com